MYCOTOXINS, ENDEMIC NEPHROPATHY AND URINARY TRACT TUMOURS

INTERNATIONAL AGENCY FOR RESEARCH ON CANCER

The International Agency for Research on Cancer (IARC) was established in 1965 by the World Health Assembly, as an independently financed organization within the framework of the World Health Organization. The headquarters of the Agency are at Lyon, France.

The Agency conducts a programme of research concentrating particularly on the epidemiology of cancer and the study of potential carcinogens in the human environment. Its field studies are supplemented by biological and chemical research carried out in the Agency's laboratories in Lyon and, through collaborative research agreements, in national research institutions in many countries. The Agency also conducts a programme for the education and training of personnel for cancer research.

The publications of the Agency are intended to contribute to the dissemination of authoritative information on different aspects of cancer research. A complete list is printed at the back of this book.

This volume is the outcome of an international symposium held in Lyon on 6–8 June 1991, organized by the IARC.

WORLD HEALTH ORGANIZATION
INTERNATIONAL AGENCY FOR RESEARCH ON CANCER

MYCOTOXINS, ENDEMIC NEPHROPATHY AND URINARY TRACT TUMOURS

Edited by

M. Castegnaro, R. Pleština, G. Dirheimer, I.N. Chernozemsky & H. Bartsch

IARC Scientific Publications No. 115

International Agency for Research on Cancer
Lyon, 1991

Published by the International Agency for Research on Cancer,
150 cours Albert Thomas, 69372 Lyon Cedex 08, France

Distributed by Oxford University Press, Walton Street,
Oxford OX2 6DP, United Kingdom

Distributed in the USA by Oxford University Press, New York

ISBN 92 832 2115 X

ISSN 0300-5085

Printed in the United Kingdom

CONTENTS

Risk assessment for human exposure to ochratoxin A

FOREWORD

The first written descriptions of a hitherto unrecognized kidney disease, now referred to as Balkan endemic nephropathy, appeared in 1953–56 in Bulgaria, Yugoslavia and Romania. The disease was later associated with a high frequency of carcinoma of the renal pelvis, ureter and urinary bladder. It often affects several members of the same family, but studies of migrant populations and of twins have not established a genetic basis for the disease but rather point to environmental determinants.

A number of hypotheses have been investigated with respect to the possible involvement of various environmental factors, including heavy metals, moulds, radiation, organic matter and silica in drinking-water and viruses. None of these suspected exposures has gained satisfactory epidemiological support. The possibility that mycotoxins may be linked to this disease, however, is supported more and more strongly by evidence from studies in the laboratory and in human populations, although causality has not been established. A number of mycotoxins have been found to be nephrotoxic; ochratoxin A has been shown also to produce kidney and liver tumours in mice and rats. Ochratoxin A has been found not only in foods consumed by families in which cases of Balkan endemic nephropathy have been reported but also as a contaminant in food and feed in other parts of Europe, in North America and, recently, in some African countries.

Because of interest at the IARC since the early 1980s in research on the etiology of Balkan endemic nephropathy, and in view of the large number of national and international studies that have been carried out in an effort to elucidate the etiology of this disease and of the tumours associated with it, it was considered timely to organize an international meeting in Lyon in June 1991. The meeting was attended by scientists from the Balkan countries and from many other countries. They discussed the present state of research and suggested future approaches for determining the causes of Balkan endemic nephropathy and the associated urinary tract tumours, and especially for evaluating the role of nephrotoxic mycotoxins in causation of the disease.

The IARC acknowledges the financial support provided by the International Programme on Chemical Safety, Geneva, the US National Institute for Environmental Health Sciences, Research Triangle Park, the Commission of the European Communities, Brussels, and the German Ministry for Youth, the Family, Women and Health, Bonn.

We should also like to thank the session chairmen—B. Armstrong, P. Bach, H. Bartsch, E. Creppy, G. Dirheimer, H. van Egmond, J. Estève, G. Manolov, K. Netter, R. Pleština, A. Rahimtula and R. Wolf—who also accepted to review the manuscripts collected in this volume.

We are pleased to dedicate this volume to the late Palle Krogh in Denmark, whose pioneering work laid important foundations for our current knowledge on nephrotoxic mycotoxins, and especially ochratoxin A (see next page).

L. Tomatis, Director
M. Castegnaro
H. Bartsch

DEDICATION

This volume is dedicated to the memory of
Palle Krogh, DVM, PhD
in recognition of his impressive contribution to the field of research into nephrotoxic mycotoxins.

Palle Krogh (1935–90) grew up in the Danish–German border district, where he was able to pick up early in life both the Scandinavian and the Central European cultures. This experience provided the basis for his later life as a cosmopolite and an international scientist.

In addition to the Doctorate of Veterinary Medicine, he took a PhD degree at the Royal Veterinary and Agricultural University (Copenhagen) in 1978, and he was honoured as Doctor *honoris causa* by several universities abroad.

Palle Krogh achieved international renown for his research on ochratoxin A—particularly so when, in 1974, he called attention to certain striking similarities between the changes in renal structure and function seen in a disease known as Balkan endemic nephropathy and those seen in ochratoxin A-induced porcine nephropathy, suggesting a common causal relation. He furthermore noted epidemiological similarities, particularly with respect to endemicity.

Palle Krogh was always dedicated to the highest scientific standards.

LIST OF PARTICIPANTS

Dr B. Armstrong
International Agency for Research on Cancer
150 cours Albert Thomas
69372 Lyon Cedex 08
France

Dr P.H. Bach
School of Science
Polytechnic of East London
Romford Road
London E15 4LZ
United Kingdom

Dr H. Bartsch
International Agency for Research on Cancer
150 cours Albert Thomas
69372 Lyon Cedex 08
France

Dr A. Breitholtz Emanuelsson
Department of Biochemistry
Royal Institute of Technology
100 44 Stockholm
Sweden

Dr C. Brera
Istituto Superiore de Sanità
Laboratorio Alimenti
viale Regina Elena 299
00161 Rome
Italy

Dr M. Castegnaro
International Agency for Research on Cancer
150 cours Albert Thomas
69372 Lyon Cedex 08
France

Dr J.R.P. Cabral
International Agency for Research on Cancer
150 cours Albert Thomas
69372 Lyon Cedex 08
France

Dr S. Čeović
Department of Epidemiology
Medical Centre
55000 Slavonski Brod
Yugoslavia

Dr E.E. Creppy
Laboratoire de Toxicologie
Université Bordeaux II
3 ter, Place de la Victoire
33076 Bordeaux
France

Dr R. Dietrich
Hygiene und Technologie der Milch
Schellingstrasse 10
8000 Munich 40
Germany

Dr G. Dirheimer
Institut de Biologie Moléculaire et Cellulaire
Centre National de la Recherche Scientifique
15 rue René Descartes
67084 Strasbourg
France

Dr D.L. Dorossiev
National Centre of Cardiac Rehabilitation
1320 Bankja
Bulgaria

Dr H.P. van Egmond
Laboratory for Residue Analysis
National Institute of Public Health and Environmental Protection
PO Box 1
3720 BA Bilthoven
The Netherlands

Dr J. Estève
International Agency for Research on Cancer
150 cours Albert Thomas
69372 Lyon Cedex 08
France

Dr H.K. Frank
Hans Sachs Strasse 3
7505 Ettlingen
Germany

Dr A.A. Frohlich
Department of Animal Science
University of Manitoba
Winnipeg, Manitoba
Canada R3T 2NZ

Dr R. Fuchs
Department of Toxicology
Institute for Medical Research and Occupational Health
University of Zagreb
Ksaverska cesta 2
PO Box 201
41000 Zagreb
Yugoslavia

Dr P. Galtier
Laboratoire de Pharmacologie-Toxicologie
Institut National de Recherche Agronomique
BP 3
180 chemin de Tournefeuille
31931 Toulouse Cedex
France

Dr P. Goliński
Department of Chemistry
Agricultural University of Poznań
60.625 Poznań
Poland

Dr J.B. Greig
Medical Research Council Toxicology Unit
Medical Research Council Laboratories
Woodmansterne Road
Carshalton
Surrey SM5 4EF
United Kingdom

Dr R. Hadidane
c/o Dr E.E. Creppy
Laboratoire de Toxicologie
Université Bordeaux
33076 Bordeaux
France

Dr B. Hald
Department of Veterinary Microbiology
Royal Veterinary and Agricultural University
13 Bülowsvej
1870 Frederiksberg C
Denmark

Miss C. Harrison
Department of Tropical Paediatrics
School of Tropical Medicine
Pembroke Place
Liverpool L3 5QA
United Kingdom

Dr A. Hennig
Institute of Microbiology, Toxicology and Histology
Federal Institute for Meat Research
E.C. Baumann Strasse 20
8650 Kulmbach
Germany

Dr J.L. Hermann
International Programme on Chemical Safety
World Health Organization
1211 Geneva 27
Switzerland

Dr E. Hietanen
Department of Clinical Physiology
Turku University Hospital
20520 Turku
Finland

Dr J.E. Huff
National Institute of Environmental Health Sciences
PO Box 12233
Research Triangle Park
NC 27709
USA

Dr K. Hult
Department of Biochemistry
Royal Institute of Technology
100 44 Stockholm
Sweden

Dr A. Kane
Laboratoire des Mycotoxines
Institut de Technologie alimentaire
BP 2765
Dakar
Senegal

Dr T. Kuiper-Goodman
Toxicological Evaluation Division
Bureau of Chemical Safety
Food Directorate
Health Protection Branch
Health and Welfare Canada
Tunney's Pasture
Ottawa
Canada K1A 0L2

Dr J. Lacey
Agricultural and Food Research Council
Institute of Arable Crops Research
Rothamsted Experimental Station
Harpenden
Herts AL5 2JQ
United Kingdom

Dr M. Lang
International Agency for Research on Cancer
150 cours Albert Thomas
69372 Lyon Cedex 08
France

Dr C. Malaveille
International Agency for Research on Cancer
150 cours Albert Thomas
69372 Lyon Cedex 08
France

Dr G. Manolov
National Oncological Centre
Medical Academy
6 Plovdivsko pole Street
Darvenitza
1156 Sofia
Bulgaria

Dr P.G. Mantle
Department of Biochemistry
Imperial College of Science, Technology and Medicine
London SW7 2AY
United Kingdom

Dr S.M. Maxwell
Department of Tropical Paediatrics
School of Tropical Medicine
Pembroke Place
Liverpool L3 5QA
United Kingdom

Dr M. Miraglia
Istituto Superiore de Sanità
Laboratorio Alimenti
viale Regina Elena 299
00161 Rome
Italy

Dr K.J. Netter
Institut für Pharmakologie
Philips Universität Marburg
Karl von Frisch Strasse
3550 Marburg
Germany

Dr I.G. Nikolov
National Oncological Centre
Medical Academy
6 Plovdivsko pole Street
1156 Sofia
Bulgaria

Dr J.H. Olsen
Danish Cancer Registry
Danish Cancer Society
Rosenvængets Hovedvej 35
Box 839
2100 Copenhagen
Denmark

Dr L. Parvanova
National Oncological Centre
Medical Academy
6 Plovdivsko pole Street
1156 Sofia
Bulgaria

Dr S. Pepeljnjak
Faculty of Pharmacy and Biochemistry
Institute of Microbiology
A. Kovacica 1
41000 Zagreb
Yugoslavia

Dr T. Petkova-Bocharova
National Oncological Centre
Medical Academy
6 Plovdivsko pole Street
1156 Sofia
Bulgaria

Dr A. Pfohl-Leszkovicz
Institut de Biologie Moléculaire et Cellulaire
Centre National de la Recherche Scientifique
15 rue René Descartes
67084 Strasbourg
France

Dr R. Pleština
International Programme on Chemical Safety
World Health Organization
1211 Geneva 27
Switzerland

Dr S. Pleština
Nephrology Department
University Hospital of Zagreb
Rebro
41000 Zagreb
Yugoslavia

Dr Z. Radovanović
Institute of Epidemiology
Faculty of Medicine
Visegradska 26
11000 Belgrade
Yugoslavia

Dr A. D. Rahimtula
Biochemistry Department
Memorial University
St John's
Newfoundland
Canada A1B 3X9

Dr V. Roman
Epidemiological Department
Institute of Oncology
SOS Fundeni 252, S2
72425 Bucharest
Romania

Dr J. Schlatter
Federal Office of Public Health
Division of Food Science
c/o Institute of Toxicology
Schorenstrasse 16
8603 Schwerzenbach
Switzerland

Dr B. Šoštarić
Institute for Medical Research and Occupational Health
University of Zagreb
Ksaverska cesta 2
PO Box 291
41000 Zagreb
Yugoslavia

Dr F.C. Stormer
Department of Toxicology
National Institute of Public Health
Postuttak Oslo 1
Oslo
Norway

Dr I. Thorup
Institute of Toxicology
National Food Agency
Morkhoj Bygade 19
2860 Soborg
Denmark

Dr L. Tomatis
International Agency for Research on Cancer
150 cours Albert Thomas
69372 Lyon Cedex 08
France

Dr Y. Ueno
Department of Toxicology and Microbial Chemistry
Faculty of Pharmaceutical Sciences
Science University of Tokyo
Ichigaya
Tokyo 162
Japan

Dr C.R. Wolf
Molecular Pharmacology Group
Imperial Cancer Research Fund
Hugh Robson Building
George Square
Edinburgh EH8 9XD
United Kingdom

EPIDEMIOLOGY AND ETIOPATHOLOGY OF BALKAN ENDEMIC NEPHROPATHY AND URINARY TRACT TUMOURS

Mycotoxins, Endemic Nephropathy and Urinary Tract Tumours
Ed. M. Castegnaro, R. Pleština, G. Dirheimer, I.N. Chernozemsky & H. Bartsch
Lyon, International Agency for Research on Cancer, pp. 3-4

BALKAN ENDEMIC NEPHROPATHY AND THE ASSOCIATED TUMOURS OF THE URINARY SYSTEM: A SUMMARY OF EPIDEMIOLOGICAL FEATURES IN BULGARIA

I.N. Chernozemsky

National Oncological Centre, Sofia, Bulgaria

Balkan endemic nephropathy, a fatal kidney disease, is a serious cause for public concern and has provoked research on its origin for several decades. It has been observed up to the present in a few regions of Yugoslavia (Serbia, Croatia and Bosnia), Bulgaria (Vratza District) and Romania (Banat region), where it persists, affecting thousands of people. Its initial clinical manifestation is relatively nonspecific; progressive renal failure dominates subsequently, accompanied by anaemia, without hypertension. Morphologically, the tubular structures in the kidneys are the most heavily affected, and a gradual replacement of the renal parenchyma leads to extraordinary shrinkage of the kidneys. Epidemiological studies have shown that middle-aged women are more often affected than men, and the disease is almost never seen in children or adolescents. The distribution of Balkan endemic nephropathy is like a mosaic in terms of regions, subregions, villages and families or households.

A very high incidence of urothelial neoplasms is seen in affected regions, in some cases representing the leading cancer site. Typically, they are pelvic or ureteral, multiple, bilateral tumours with slow progression. Their epidemiological features are strikingly similar to those of Balkan endemic nephropathy.

Age-standardized population data on incidence and mortality are available for only certain periods for the Vratza District. During the late 1970s and early 1980s, 41 villages were found to be affected by Balkan endemic nephropathy, 15 of them heavily. In this hyperendemic area, the age-adjusted incidence rate for Balkan endemic nephropathy per 100 000 people was 506 for females and 315 for males. The age-adjusted incidence rate of urothelial neoplasms was 105 per 100 000 for females (i.e., nearly half of all neoplasms) and 89 for males. The relative risk for patients with Balkan endemic nephropathy of developing urinary tract tumours was 90 times higher than that for the populations of nonendemic villages. In the Vratza District, there has been a trend for increased spread of these diseases, resulting in a gradual involvement of new households and villages.

Various hypotheses for the etiology of this disease pattern have been put forward; however, solid proof is lacking for any of them. Radiation and heavy metal poisoning have been almost completely abandoned as possible causes, but infection by slow viruses or other microorganisms still cannot be ruled out. Both diseases seem, however, to be determined by environmental factors. Chromosomal damage, but no specific cytogenetic marker, have been associated with these diseases. Current research implicates the involvement of mycotoxins, and particularly ochratoxin A, and these results support a previous suggestion that Balkan endemic nephropathy has a mycotoxin-related etiology.

Mycotoxins, Endemic Nephropathy and Urinary Tract Tumours
Ed. M. Castegnaro, R. Pleština, G. Dirheimer, I.N. Chernozemsky & H. Bartsch
Lyon, International Agency for Research on Cancer, pp. 5-10

EPIDEMIOLOGICAL ASPECTS OF BALKAN ENDEMIC NEPHROPATHY IN A TYPICAL FOCUS IN YUGOSLAVIA

S. Čeović[1], R. Pleština[2], M. Miletić-Medved[1], A. Stavljenić[3], J. Mitar[1] & M. Vukelić[1]

[1]*Medical Centre, Slavonski Brod, Yugoslavia;* [2]*International Programme on Chemical Safety, World Health Organization, Geneva, Switzerland; and* [3]*Medical Faculty, University of Zagreb, Zagreb, Yugoslavia*

Summary

Balkan endemic nephropathy is a noninflammatory bilateral kidney lesion that affects rural populations in several circumscribed areas of the Balkans. Its etiology is still not understood, but recently it has been associated with exposure to nephrotoxic mycotoxins. It has been known to be present since the mid-1950s in 14 villages in an endemic area of Croatia, where approximately 10 000 people are at risk. Its prevalence fluctuates between 0.4 and 8.3%, showing a slight decline in recent years, but it has not disappeared from any of the endemic villages. The occurrence of the disease in several ethnic groups contradicts the hypothesis of a primary hereditary basis for Balkan endemic nephropathy. Recently, evidence has been found of an extremely high incidence of urinary tract tumours in the endemic area, and particularly of urothelial tumours of the pelvis and ureter. There may therefore be a common causative agent for these two rare diseases.

Introduction

Balkan endemic nephropathy is a specific nosologic entity, known at present to exist only in three Balkan countries—Bulgaria, Romania and Yugoslavia. The disease is primarily chronic and noninflammatory, affecting both kidneys, and no specific treatment is available. It affects exclusively adults in rural populations living in villages located along large rivers.

Although the disease was recognized in the mid-1950s and has been studied extensively since, its etiology and many epidemiological features remain obscure. In this paper we hope to throw more light on some of the most characteristic epidemiological features of the disease, in order to provide better understanding of its nature and etiology; in particular, we hope to show the possibility of a causative association between the disease and the accompanying urothelial tumours and exposure to nephrotoxic mycotoxins.

Subjects and Methods

The endemic area in Croatia, Yugoslavia, near the city of Slavonski Brod, comprises 14 villages with a total population of approximately 10 000 inhabitants. The control area chosen includes villages in the county of Slavonski Brod where the disease has never been found and the city of Slavonski Brod itself, with approximately 100 000 inhabitants.

Data on both Balkan endemic nephropathy and urothelial tumours were collected either from hospital records (for the period 1951–90) or during regular screening campaigns undertaken in several hyper-endemic, hypoendemic and nonendemic villages.

Since 1975, in 12 out of 14 endemic villages containing 8239 inhabitants (as per the census of 1981), 70–80% of the population were examined medically, and blood and urine samples were collected and analysed for creatinine, haemoglobin and ochratoxin A in blood and for protein in urine. In one of the hyperendemic villages (Kaniža), such screening was carried out annually; in four others, it was done every two years, and in the remaining seven at longer intervals. During this period, over 30 000 medical examinations and sets of biochemical tests were performed. In the same period, 6500 medical examinations and biochemical tests were performed in 20 nonendemic villages at less frequent intervals. The 20–30% who were not examined medically consisted of children under three years of age and people who were absent from the village at the time of the campaign, most of whom were working outside the village or even outside the country.

As there are no clear-cut clinical criteria for recognizing the early stages of the disease, the results of biochemical tests, together with a positive family history, were utilized as the basis for calculating the prevalence and incidence of the disease (WHO, 1965). On the basis of these recommendations, all people who had repetitive proteinuria of unknown origin, haemoglobin levels lower than 120 g/l (for men) or 113 g/l (for women) and a serum creatinine level above 133 μmol/l, with no known cause for such impairment, were considered to have nephropathy. If only two of the above-mentioned criteria were fulfilled, the subject was suspected of having nephropathy, as recommended in the WHO protocol. If in subsequent screening campaigns another one or two of the parameters was present, the person was considered to have nephropathy.

The final diagnosis was made after a thorough medical examination at the hospital.

Results

The results of screening campaigns initiated in 1957 and intensified since 1975 have shown that the prevalence of Balkan endemic nephropathy ranges between 0.4 and 8.3% (Table 1), varying from village to village and from year to year, although in some hyperendemic villages the prevalence seems to be declining steadily. The disease affects women more than men, the ratio being approximately 3:2.

The incidence of newly recorded Balkan endemic nephropathy patients (Table 2) ranges between 1.0 and 2.4 per thousand, progressing until the 1980s, with a sharp decline during the last decade. There is a clear trend for older people to be diagnosed with the disease for the first time.

In overall observation periods of 34 years, specific mortality ranged from 2.0 to 2.7 per thousand in four hyperendemic villages to 0.2–0.9 per thousand in the seven hypoendemic villages. The mortality ratio between women and men was 3:2. The average age at death was 45.1 years for the period 1957–60, gradually increasing and reaching 63.4 years during the last four-year period. The phenomenon of late appearance of the disease is reflected in the finding of death at later ages, and the average time between diagnosis and death is steadily increasing (Table 3).

The finding that the disease occurs in similar proportions in different ethnic groups (Čeović *et al.*, 1985) contradicts the hypothesis that the disease has a primarily genetic etiology. Out of 37 newly diagnosed cases of nephropathy between 1975 and 1990 in the hyperendemic village of Kaniža, 18 were in Croatian aborigines, 16 in Ukrainian immigrants and three in immigrants from other regions. Similar proportions are found in other endemic villages.

Hospital records over the last ten-year period indicate that urinary tract tumours are much more frequent in endemic than in nonendemic areas (Table 4). Pelvic and ureteral tumours were particularly prevalent (Table 5): these two otherwise rare tumours comprised 50% of all urinary tract tumours as compared to 12.5% in nonendemic areas. Patients from endemic areas with pelvic or ureteral tumours were generally younger (mean age, 61.5 years) than similar patients from nonendemic areas (mean age, 69.9 years). Tumours other than those of the urinary bladder were more frequent among women than men, a particularly striking difference being found for pelvic and ureteral tumours.

Table 1. Prevalence (%) of nephropathic patients in villages in an endemic area near Slavonski Brod

Village	No. of inhabitants	Year																
		1957	1975	1976	1977	1978	1979	1980	1981	1982	1983	1984	1985	1986	1987	1988	1989	1990
Banovci	372	–	–	2.0	–	1.2	–	1.4	–	3.2	–	3.9	–	2.9	2.2	–	1.0	–
Bebrina	577	1.9	2.4	–	2.6	–	1.8	–	1.9	–	1.9	–	–	1.3	2.6	–	2.2	–
Brodski Varoš	1570	–	0.8	–	0.6	–	0.5	–	0.8	–	1.2	–	–	–	–	–	–	–
Dubočac	287	–	–	–	–	–	–	–	–	–	–	–	2.2	–	–	–	–	–
Kaniža	755	–	4.4	1.9	4.1	3.3	2.2	2.0	2.2	1.5	1.6	2.8	1.9	1.7	2.5	1.0	1.4	1.0
Lužani	1233	–	–	–	–	–	–	–	–	–	–	–	1.3	–	–	–	–	1.0
Pričac	139	4.0	–	–	8.3	–	–	–	–	–	2.5	–	–	–	–	1.8	–	–
Slavonski Kobaš	1 461	2.3	–	–	2.4	–	–	–	–	–	–	1.7	–	–	–	0.8	–	–
Stupnički Kuti	473	–	–	–	–	–	–	–	–	–	–	–	–	1.9	–	–	–	–
Šumeće	552	–	–	1.1	–	1.9	–	0.7	–	0.7	–	0.7	–	0.7	0.6	–	–	–
Zbjeg	453	–	–	–	–	–	–	–	–	–	–	–	–	–	0.8	–	–	–
Živike	366	–	–	–	–	–	–	–	–	–	0.4	–	–	–	–	–	–	–

Table 2. Incidence of newly diagnosed nephropathy patients from hospital records

Period	Population at risk	No. of new cases	per 1000
1957–66	11 235	193	1.72
1967–76	10 717	252	2.35
1977–86	10 094	179	1.77
1987–90	10 094	42	1.04

Table 3. Average time between diagnosis and death

Period	No.	Mean (years)		
		Diagnosis (M_1)	Death (M_2)	M_2–M_1
1957–60	37	44.5	45.1	0.6
1967–70	71	49.7	51.6	1.9
1977–80	81	61.1	64.8	3.7
1987–90	32	57.8	63.4	5.6

Table 4. Numbers of urinary tract tumours in inhabitants of endemic (population, 10 346) and nonendemic (population, 98 713) areas

Site	1981–85		1986–90	
	Endemic	Nonendemic	Endemic	Nonendemic
Kidney	3	4	7	8
Pelvis	2	1	14	1
Ureter	5	3	10	2
Bladder	7	16	14	21
Total	17	24	45	32

Table 5. Distribution (%) of different urinary tract tumours in endemic and nonendemic areas

Area	Kidney	Pelvis	Ureter	Bladder
Endemic	16.1	25.8	24.2	33.9
Nonendemic	21.4	3.6	8.9	66.1

Discussion

The main problem encountered in past and present estimates of the prevalence and incidence of Balkan endemic nephropathy is the lack of reliable clinical and biochemical tests, which are needed for making an early diagnosis of kidney impairment. Scientists working in different endemic areas frequently use different criteria for such diagnoses, and it is therefore difficult to compare their results. The recommendations of an international group of scientists convened by WHO (1965) are seldom fully respected. The study described in this paper was done by the same team using the same methods throughout. Therefore, the results are fully comparable.

As the etiology of the disease is still unknown and there are no clear-cut clinical criteria for recognizing the early stages of the disease, most epidemiological studies lack firm support. There is no doubt, however, that nephropathy is still present in villages in the endemic region, its incidence being steady in some and declining in the others. No evidence exists, however, that the disease has spread beyond the known endemic region. An apparent increase in the time between diagnosis and death can be attributed to improved medical care of the patients, including haemodialysis.

An unusually high incidence of urinary tract tumours in the endemic area has been recognized for a long time (Petković *et al.*, 1971; Petković, 1975). There is sufficient evidence that otherwise rare urothelial tumours of the pelvis and ureter are nearly 50 times more frequent in endemic than in nonendemic areas, and the number of these tumours appears to be increasing while that of nephropathy patients is decreasing. This phenomenon deserves special attention, and further studies should be carried out to verify the many indications that these two rare diseases have a common causative agent.

Similar types of nephropathy do not appear to occur in areas outside the Balkan region. In the absence of specific, well-validated diagnostic parameters, however, the existence of the disease outside the presently known endemic regions cannot be ruled out, and studies along this line should be encouraged.

References

Čeović, S., Hrabar, A. & Radonić, M. (1985) An etiological approach to Balkan endemic nepropathy based on the investigation of two genetically different populations. *Nephron*, **40**, 175–179

Petković, S.D. (1975) Epidemiology and treatment of renal and ureteral cancer. *J. Urol.*, **114**, 858–865

Petković, S., Mutavdžić, M., Petronić, V. & Marković, V. (1971) Les tumours du bassinet et de l'uretere. Recherches cliniques et etiologiques. *J. Urol. Nephrol.*, **77**, 429–439

WHO (1965) Memoranda. The endemic nephropathy of South-eastern Europe. *Bull. World Health Organ.*, **32**, 431–448

Mycotoxins, Endemic Nephropathy and Urinary Tract Tumours
Ed. M. Castegnaro, R. Pleština, G. Dirheimer, I.N. Chernozemsky & H. Bartsch
Lyon, International Agency for Research on Cancer, pp. 11-20

EPIDEMIOLOGICAL CHARACTERISTICS OF BALKAN ENDEMIC NEPHROPATHY IN EASTERN REGIONS OF YUGOSLAVIA

Z. Radovanović

Institute of Epidemiology, Faculty of Medicine, Belgrade, Yugoslavia

Summary

Foci of Balkan endemic nephropathy have been found scattered along several rivers in many parts of Serbia, as well as in north-eastern Bosnia. There is no indication that the medical geography of the disease changes over time. As for the intensity of the endemic process, both the poor quality of the routinely collected data and an inconsistent methodological approach obscure the real epidemiological situation. The incidence of deaths, the most reliable measure of outcome, has been stable over the last few decades, slightly exceeding 3 per 1000 person-years of observation in the most heavily affected endemic villages. The point prevalence rate of hyper-ß-2-microglobulinuria, an indicator of tubular damage, also remained unchanged over that period, at a value as high as 20–25%. Over time, however, the course of the disease became more protracted and its onset moved towards older ages. These two facts may indicate a less intense contact with the agent(s) and, consequently, suggest a decreased burden of Balkan endemic nephropathy in the near future. The natural history of Balkan endemic nephropathy is still not well understood, however. The disease seems to have had an endemo-epidemic pattern in the past, and the possibility of another epidemic wave in the future cannot be completely ruled out.

Introduction

The purpose of this paper was to review critically the basic descriptive epidemiological features of Balkan endemic nephropathy (BEN), which are relevant for generating etiological hypotheses. Emphasis was placed on the situation in eastern parts of Yugoslavia, but experience in the other endemic regions was also taken into account.

Medical Geography

The Republic of Serbia, situated in the eastern parts of Yugoslavia, is the region most heavily affected by BEN. Its 91 severely affected settlements are scattered over 13.3% (24 of 180) Serbian municipalities. If one takes into account 285 other villages classified as probably endemic, 6.2% of all settlements in Serbia may be affected by BEN (Radovanović, 1979, 1985). North-western Serbian endemic foci extend to the central Yugoslav republic of Bosnia, where six municipalities (with several dozen villages) have been affected (Fajgelj *et al.*, 1976). To the west, Bosnian foci approach the single Croatian municipality affected, Slavonski Brod, with 14 endemic villages (Čeović, 1982).

All of the foci of BEN are located along tributaries of the Danube River. In Serbia and Bosnia, most of the endemic villages are in flooded areas (Danilović, 1973) and/or areas with a high level of underground water and shallow wells (Gaon *et al.*, 1976). The frequency of BEN may differ within a single endemic village according to the level of flooding of its hamlets (Milojčić *et al.*, 1962).

A vast majority (88%) of all Serbian foci are situated at an altitude of less than 200 m (Radovanović, 1979), and the landscape of endemic sites at higher altitude is basically the same as that in lowland areas. The affected villages usually lie in alluvial plains, and only seldom on river or lake terraces a few meters above the alluvial plains (Radovanović & Perić, 1979). In Bosnia, the disease is unknown at altitudes higher than 130 m (Gaon, 1973). All Croatian foci are at an altitude of about 100 m, and the Romanian endemic villages are situated at the bottom of river valleys, at an altitude of 200–300 m (Biberi Moroeanu, 1967a). The situation seems to be different only in Bulgaria, where BEN occurs in a well drained hilly region (Puchlev, 1967), in a range from 45 to 600 m, but with no correlation between the altitude and the incidence of the disease (Gabev, 1974).

Neither in eastern Yugoslavia, nor in any other part of the Balkans, is there any evidence of the spread of endemic foci. Data on shrinkage of affected areas, however, are also lacking.

Frequency of Balkan Endemic Nephropathy

Mortality rates from BEN have been estimated at 1–3/1000 per year (WHO, 1965). This estimate, a quarter of a century old, still holds true for most endemic settlements; occasional discrepant results (Zaharia *et al.*, 1965a; Hrabar *et al.*, 1977) are due to methodological shortcomings. In Petka, the best studied and one of the most affected endemic villages in Serbia, the risk of death from BEN has not changed over time, slightly exceeding 3/1000 per year (Udicki *et al.*, 1970; Radovanović *et al.*, 1991a).

Morbidity data are rather more controversial. The differences seen occasionally are due to misunderstanding of basic epidemiological concepts, as exemplified by 'incidence' rates of 16% (Verbev *et al.*, 1965) and 30% (Stojimirović, 1974). Often, however, the insidious onset of the disease precludes a reliable assessment of incidence rates. We therefore used incidence of deaths as the measure of outcome, in order to assess accrual rates. A follow-up study revealed that the incidence density of BEN was 3.3/1000 person-years of observation (Radovanović *et al.*, 1991a).

Prevalence data differ widely, depending upon the diagnostic criteria used. In a seriously affected Serbian village, we found the point prevalence rate for hypercreatininaemia to be 10.7% in a group of 900 inhabitants examined (Radovanović *et al.*, 1985). This rate corresponds well to an earlier estimate (WHO, 1965) that various degrees of kidney failure, as assessed by an increased level of blood nitrogen, affect 5–13% of the population in endemic villages. Problems arise when loose criteria (e.g., proteinuria, anaemia) are used to ascertain cases of BEN. Such approaches have led to the conclusion that more than one-third (Milojčić, 1960; Verbev *et al.*, 1965) or even more than one-half of all inhabitants are victims of BEN (Ignjatović, 1962; Danilović & Stojimirović, 1967a).

Apparently, proteinuria is not identified in cases of BEN, but an excess of positive findings, as assessed by comparison of the prevalence rates of (total or tubular) proteinuria in endemic and nonendemic regions, can be attributed to the causative agent(s) of BEN. Hyper-ß-2-microglobulinuria, an indicator of tubular damage, may affect less than 1% of the general population, but the point prevalence rate of this condition in endemic villages is 5–10% and can be as high as 25% (Radovanović *et al.*, 1985). This figure may indicate the extent of exposure to nephrotoxic agents in a BEN-affected environment.

Chronological Features of Balkan Endemic Nephropathy

Attempts have been made to trace BEN back to the beginning of this century (Bulić, 1967a; Georgiev & Dimitrov, 1965) or to the period between the two wars (Danilović, 1967; Gaon, 1967a). No single piece of firm evidence, however, supports this idea. Life expectancy in rural areas at that time was generally rather low; malaria was very prevalent in lowland regions; and tuberculosis affected several members of single households. The usual source of information on deaths at that time—records kept by local priests—does not provide any diagnostic clue, and causes of death, if recorded at all, were vague (e.g., 'exhaustion', 'water disease').

Hard facts point to the mid-1950s as the period when the disease was recognized almost simultaneously not only in the three countries concerned, but in each major affected region of those countries. Epidemiological investigations revealed that it was not a new disease, and cases could be traced back to the past (Milojčić, 1960) or, more precisely, to the early 1940s (Danilović, 1958). Apparently, the natural history of BEN has a cyclic pattern and what happened in the mid-1950s was an amplification of an already existing endemic process.

Many authors have addressed the issue of trends in BEN since the time when it was originally described. The results have been surprisingly inconsistent, and sometimes the same authors provided conflicting evidence on a single focus of BEN (e.g., Gaon & Telebak, 1979; Gaon *et al.*, 1979). An extensive search of the relevant literature (Radovanović, 1985) revealed that the absence of dramatic change over time is the only inference that can be drawn from these studies.

As already pointed out, mortality rates from BEN in the most thoroughly studied Serbian focus have remained the same (Udicki *et al.*, 1970; Radovanović *et al.*, 1991a), as have the rates of total and tubular proteinuria, at least since the early 1970s (Jevremović *et al.*, 1991). Some changes in the natural history of BEN definitely did take place, however. Thus, a 'family outbreak' which led to the original description of BEN in Serbia (Danilović *et al.*, 1957) resulted in three deaths (the mother and two adolescent daughters) and two cases of severe kidney disease (the father and a son) within less than three months. Nowadays, the disease has a more chronic course (due only partially to the availability of haemodialysis and transplantation facilities), and kidney failure due to BEN is seldom seen in people under the age of 40. These two facts—protracted clinical course of BEN and a shift in the age distribution to older ages—may indicate less intense exposure to the agent(s). These changes seem to have been greater in more prosperous villages, but there is no indication whatsoever if better housing, improved water supply, more hygienic sewage disposal and other factors related to living conditions and life style are responsible. The present trend may lead to the ultimate disappearance of the disease. However, keeping in mind the epidemic waves in the early 1940s and mid-1950s, another rise in incidence may (at least theoretically) still be ahead.

Demographic Data

In endemic areas, a peculiar pattern of proteinuria can be found even in children of pre-school age (Hall *et al.*, 1967). In the past, however, clinical forms of BEN usually did not occur before the third decade of life (WHO, 1965), and it is still rare to diagnose BEN-related uraemia in people under 40. Occasional reports of BEN-affected children (Biberi Moroeanu, 1967b; Gaon *et al.*, 1974; Zaharia *et al.*, 1965b) have been based on different (usually only laboratory or pathohistological) criteria.

A consensus has been reached among scientists (WHO, 1965) that in Romania females are more affected, while in Yugoslavia and Bulgaria there is no difference between the two sexes. Thus, Romanian authors reported more cases of BEN among females (Biberi Moroeanu, 1967a; Bruckner *et al.*, 1979), while Bulgarian and Yugoslav authors usually declared that BEN was equally frequent in males and females (Danilović, 1974a; Puchlev, 1974), although the results of field studies consistently pointed to higher prevalence among females (Tanchev *et al.*, 1965; Danilović *et al.*, 1974; Gabev, 1974; Strahinjić & Stefanović, 1980; Velimirović, 1980). The difference between the two sexes may be due largely to differences in diagnostic criteria, which are highly dependent on proteinuria and anaemia, both of which are more likely to occur in females. This point of view is supported by reports that males and females have an equal chance of dying from BEN (Milojčić *et al.*, 1965; Udicki *et al.*, 1970; Gabev, 1974; Hrabar *et al.*, 1977), although this was not found by others (Gaon, 1967b; Gaon *et al.*, 1974). The frequency of tubular damage, as assessed by hyper-ß-2-microglobulinuria, is no different in males and females (Radovanović *et al.*, 1985).

BEN is a disease of peasants. People with other occupations, e.g., railwaymen (Bruckner *et al.*, 1965) and teachers (Čeović, 1971), can be affected if they are exposed long enough to an endemic environment. There is no convincing evidence that people never involved in agricultural activities develop BEN. Rare reports on BEN victims who were born and spent their entire lives outside endemic villages come from authors who advocate genetic determination of the disease (Bulić, 1967b) or a peculiar definition (Trnavčević *et al.*, 1981).

The risk for developing BEN is not associated with any ethnic or confessional group in a consistent manner. Gypsies appear to be relatively little affected (Radovanović, 1985), but no numerical data or significant differences have been provided.

Other Epidemiological Features of Balkan Endemic Nepropathy

BEN is a familial disease, affecting blood relatives as well as in-laws (Veljković *et al.*, 1967). Immigrants from nonendemic areas develop BEN (Danilović, 1974b; Naumović *et al.*, 1974), and emigrants may become affected many years after leaving endemic foci (Danilović & Stojimirović, 1967b; Gaon, 1967c). Crucial for both immigrants and emigrants is long-term exposure to the BEN environment (Gaon, 1967c; Danilović, 1974b; Puchlev, 1974).

One of the most peculiar features of BEN is its association with tumours of urinary organs in the same patients (Petrinska-Venkovska, 1960) or in the same population (Petković *et al.*, 1968). In some of the Serbian endemic foci, the risk for developing upper urothelial tumours was more than 100 times higher than that in nonendemic areas (Petković *et al.*, 1971). Familial clustering of such tumours, affecting several siblings, has been reported (Radovanović *et al.*, 1984a, 1990), and the incidence density of upper urothelial tumours reaches 1.4 per 1000 person-years of observation (Radovanović *et al.*, 1991b). The risk of nonurinary cancer death also appears to be significantly higher in BEN-affected settlements (Radovanović *et al.*, 1984b).

In endemic villages, families belonging to different socioeconomic strata are equally affected (Gaon, 1967b), and, as shown originally in Bulgaria (Puchlev *et al.*, 1960), there is no difference related to standard of living, customs, hygienic conditions or habits between endemic and neighbouring nonendemic villages. Diet and consumption of alcoholic beverages do not differ in BEN-affected and unaffected settlements (Danilović & Stojimirović, 1967a). Well-water samples from BEN foci used to be contaminated to 80%, but the same percentage of positive results was also obtained in comparable control areas (Mušicki *et al.*, 1971; Jevtić & Tasić, 1977). Epizoological investigations have given controversial, inconsistent results (Radovanović, 1985).

References

Biberi Moroeanu, S. (1967a) Epidemiological observations on the endemic nephropathy in Rumania. In: Wolstenholme, G.E.W. & Knight, J., eds, *The Balkan Nephropathy* (CIBA Foundation Study Group No. 30), London, Churchill, pp. 4–13

Biberi Moroeanu, S. (1967b) Discussion. In: Wolstenholme, G.E.W. & Knight, J., eds, *The Balkan Nephropathy* (CIBA Foundation Study Group No. 30), London, Churchill, p. 103

Bruckner, I., Zosin, C., Lazarescu, R., Paraschiv, D., Manescu, N., Serban, M. & Titeica, M. (1965) A clinical study of nephropathy of an endemic character in the People's Republic of Romania. In: *International Symposium on Endemic Nephropathy*, Sofia, Bulgarian Academy of Sciences Press, pp. 25–35

Bruckner, I., Nichifor, E. & Rusu, G. (1979) Endemic nephropathy in Romania. In: Strahinjić, S. & Stefanović, V., eds, *Proceedings of the 4th Symposium on Endemic (Balkan) Nephropathy*, Niš, University of Niš, pp. 11–14

Bulić, F. (1967a) Discussion. In: Wolstenholme, G.E.W. & Knight, J., eds, *The Balkan Nephropathy* (CIBA Foundation Study Group No. 30), London, Churchill, p. 15

Bulić, F. (1967b) The possible role of genetic factors in the aetiology of the Balkan nephropathy. In: Wolstenholme, G.E.W. & Knight, J., eds, *The Balkan Nephropathy* (CIBA Foundation Study Group No. 30), London, Churchill, pp. 17–22

Čeović, S. (1971) *A Contribution to the Epidemiology of Endemic Nephropathy in Brodska Posavina*, MSc Thesis, University of Zagreb (in Serbo-Croat)

Čeović, S. (1982) Public health characteristics and importance of Balkan endemic nephropathy. *Med. Vjes.*, **14**, 20–24 (in Serbo-Croat)

Danilović, V., Djurisić, M., Mokranjac, M., Stojimirović, B., Zivojinović, J. & Stojaković, P. (1957) Familial kidney affections caused by a chronic lead-intoxication in the village of Sopić. *Srpski Arhiv*, **85**, 1115–1125 (in Serbo-Croat)

Danilović, V. (1958) Chronic nephritis due to ingestion of lead-contaminated flour. *Br. Med. J.*, **i**, 27–28

Danilović, V. (1967) Discussion. In: Wolstenholme, G.E.W. & Knight, J., eds, *The Balkan Nephropathy* (CIBA Foundation Study Group No. 30), London, Churchill, p.15

Danilović, V. (1973) Importance and results of the previous studies on endemic nephropathy. In: *Proceedings of the Symposium on Endemic Nephropathy*, Belgrade, Serbian Academy of Sciences and Arts, pp. 47–63 (in Serbo-Croat)

Danilović, V. (1974a) Endemic nephropathy. In: *Proceedings of the 2nd Seminar in Nephrology* (Documenta 1977), Belgrade, Galenika, pp. 407–413 (in Serbo-Croat)

Danilović, V. (1974b) Endemic nephropathy in Yugoslavia. In: *WHO International Meeting of Investigators on Endemic Nephropathy, Beograd and Lazarevac, November 1974* (NCD/WP/74.4) (unpublished)

Danilović, V. & Stojimirović, B. (1967a) A critical reapprisal of the previous studies on endemic nephropathy and our understanding of the future investigations of this problem. In: *I Symposium on Endemic Nephropathy*, Niš, University of Niš Press, pp. 27–30 (in Serbo-Croat)

Danilović, V. & Stojimirović, B. (1967b) Endemic nephropathy in Kolubara, Serbia. In: Wolstenholme, G.E.W. & Knight, J., eds, *The Balkan Nephropathy* (CIBA Foundation Study Group No. 30), London, Churchill, pp. 44–50

Danilović, V., Naumović, T. & Velimirović, D. (1974) Endemic nephropathy in the Lazarevac community. In: Puchlev A., Dinev, I., Milev, B. & Doichinov, D., eds, *Proceedings of the II International Symposium on Endemic Nephropathy*, Sofia, Bulgarian Academy of Sciences, pp. 281–283

Fajgelj, A., Filipović, A. & Popović, N. (1976) A programme of action aimed at the solution of the problem of Balkan endemic nephropathy in SR Bosnia and Hercegovina: results of the first year of its realisation. *Acta Med. Saliniana*, **5**, 57–65 (in Serbo-Croat)

Gabev E. (1974) Some epidemiologic characteristics of endemic nephropathy. In: Puchlev A., Dinev, I., Milev, B. & Doichinov, D., eds, *Proceedings of the II International Symposium on Endemic Nephropathy*, Sofia, Bulgarian Academy of Sciences, pp. 271–275

Gaon, J. (1967a) Discussion. In: Wolstenholme, G.E.W. & Knight, J., eds, *The Balkan Nephropathy* (CIBA Foundation Study Group No. 30), London, Churchill, p. 15

Gaon, J. (1967b) Endemic nephropathy in Bosnia. In: Wolstenholme, G.E.W. & Knight, J., eds, *The Balkan Nephropathy* (CIBA Foundation Study Group No. 30), London, Churchill, pp. 51–71

Gaon, J. (1967c) Discussion. In: Wolstenholme, G.E.W. & Knight, J., eds, *The Balkan Nephropathy* (CIBA Foundation Study Group No. 30), London, Churchill, p. 26

Gaon, J. (1973) Longitudinal studies of endemic nephropathy in two villages of Bosnia and Hercegovina and some possibilities of protection of the population. In: *Proceedings of the Symposium on Endemic Nephropathy*, Belgrade, Serbian Academy of Sciences and Arts, pp. 47–63 (in Serbo-Croat)

Gaon, J. & Telebak, B. (1979) Possibilities of application of some preventive measures for control of endemic nephropathy. In: Strahinjić, S. & Stefanović, V., eds, *Proceedings of the 4th Symposium on Endemic (Balkan) Nephropathy*, Niš, University of Niš, p. 270

Gaon, J., Alibegović, S. & Pokrajčić, B. (1974) Trend of endemic nephropathy occurrence over several years in two Bosnian villages. In: *WHO Meeting of Investigators on Endemic Nephropathy, Belgrade and Lazarevac, November 1974* (NCD/WP/74.6) (unpublished)

Gaon, J., Aganović, I., Djurić, E. & Zlomusica, J. (1976) Water as a possible carrier of endemic Balkan nephropathy. *Folia Med.*, **11**, 97–112

Gaon, J., Dedić, I., Aganović, I., Mandić, M. & Telebak, B. (1979) Epidemiological investigation of endemic nephropathy in SR Bosnia and Hercegovina from 1957 until now. In: *Proceedings of the 2nd Symposium on Endemic Nephropathy*, Belgrade, Serbian Academy of Sciences and Arts, pp. 9–25 (in Serbo-Croat)

Georgiev, G. & Dimitrov, T. (1965) A study on endemic nephritis in the village of Bistretz, Vratza district. In: *International Symposium on Endemic Nephropathy*, Sofia, Bulgarian Academy of Sciences Press, pp. 245–249

Hall, P.W., Gaon, J., Griggs, R.C., Piscator, M., Popović, N., Vasiljević, M. & Zimonjić, B. (1967) The use of electrophoretic analysis of urinary protein excretion to identify early involvement in endemic (Balkan) nephropathy. In: Wolstenholme, G.E.W. & Knight, J., eds, *The Balkan Nephropathy* (CIBA Foundation Study Group No. 30), London, Churchill, pp. 72–83

Hrabar, A., Šuljaga, K., Borčić, B., Aleraj, B., Čeović, S. & Čvoriščeć, D. (1977) BEN morbidity and mortality in the village of Kaniza. In: *Proceedings of the III Symposium on Endemic Nephropathy, Niš, 1975* (Documenta 1977), Belgrade, Galenika, pp. 119–121 (in Serbo-Croat)

Ignjatović, B. (1962) A contribution to the aetiological studies of endemic nephritis. *Nar. Zdravlje*, **18**, 329–334 (in Serbo-Croat)

Jevremović, I., Janković, S., Radovanović, Z., Danilović, V., Velimirović, D., Naumović, T., Vacca, C., Stamenković, M., Bukvić, D., Stojanović, V., Trbojević, S. & Hall, P.W. (1991) Beta-2-microglobulinuria in a population exposed to Balkan endemic nephropathy. II. Inferences from repeated cross-sectional studies. *Kidney Int.*, **40** Suppl. 34) (in press)

Jevtić, Z. & Tasić, M. (1977) Studies of water supply in endemic nephropathy foci. In: *Proceedings of the III Symposium on Endemic Nephropathy, Niš, 1975* (Documenta 1977), Belgrade, Galenika, pp. 102–108 (in Serbo-Croat)

Milojčić, B. (1960) Epidemic of chronic nephritis of unknown aetiology. *Br. Med. J.*, **i**, 244–245

Milojčić, B., Udicki, S., Krajinović, S &, Obradović, M. (1962) Results of epidemiological observations on the development of the chronic nephritis epidemic in the village of Sopić (Crna Bara hamlet). *Higijena*, **14**, 124–129 (in Serbo-Croat)

Milojčić, B., Udicki, S. & Obradović, M. (1965) Evolution of endemic nephritis on the territory of Sopić village (Crna Bara hamlet) from 1957 to 1965. *Glasnik ZZZZ*, **7–8**, 18–22 (in Serbo-Croat)

Mušicki, B., Vasić, V. & Ćirić, A. (1971) Results of bacteriological investigations of drinking water in endemic and non-endemic settlements in the South Morava draining area. In: *Proceedings of the 2nd International Symposium on Endemic Nephropathy*, Niš, University of Niš, pp. 11–12 (in Serbo-Croat)

Naumović, T., Velimirović, D. & Danilović, V. (1974) Endemic nephropathy in immigrants living in the foci of disease in Lazarevac community. In: Puchlev, A., Dinev, I., Milev, B. & Doichinov, D., eds, *II International Symposium on Endemic Nephropathy*, Sofia, Bulgarian Academy of Sciences Press, pp. 319–321

Petković, S., Mutavdzić, M., Petronić, V. & Marković, V. (1968) Geographical distribution of urothelial tumours in Yugoslavia. *Urologia*, **35**, 425–433

Petković, S., Mutavdzić, M., Petronić, V. & Marković, V. (1971) Tumours of the renal pelvis and ureter. Clinical and etiological research. *J. Urol. Nephrol.*, **77**, 429–439 (in French)

Petrinska-Venkovska, S. (1960) Morphologic studies on endemic nephropathy. In: Puchlev, A., ed., *Endemic Nephritis in Bulgaria*, Sofia, Medicine and Gymnastics, pp. 72–90 (in Bulgarian)

Puchlev A. (1967) Endemic nephropathy in Bulgaria. In: Wolstenholme, G.E.W. & Knight, J., eds, *The Balkan Nephropathy* (CIBA Foundation Study Group No. 30), London, Churchill, pp. 28–39

Puchlev A. (1974) Problem of endemic nephropathy—report of a temporary adviser. In: *WHO International Meeting of Investigators on Endemic Nephropathy, Beograd and Lazarevac, November 1974* (NCD/WP/74.8) (unpublished)

Puchlev, A., Astrug, A., Popov, N. & Docev, D. (1960) Clinical studies of endemic nephritis. In: Puchlev, A., ed., *Endemic Nephritis in Bulgaria*, Sofia, Medicine and Gymnastics, pp. 7–71 (in Bulgarian)

Radovanović, Z. (1979) Topographical distribution of the Balkan endemic nephropathy in Serbia (Yugoslavia). *Trop. Geogr. Med.*, **31**, 185–189

Radovanović, Z. (1985) Epidemiology of endemic nephropathy. In: *Proceedings of the Third Symposium on Endemic Nephropathy*, Belgrade, Serbian Academy of Sciences and Arts, pp. 3–27 (in Serbo-Croat)

Radovanović, Z. & Perić, J. (1979) Hydrogeological characteristics of endemic nephropathy foci. *Public Health (Lond.)*, **93**, 76–81

Radovanović, Z., Naumović, T. & Velimirović, D. (1984a) Clustering of the upper urothelial tumours in a family. *Oncology*, **41**, 396–398

Radovanović, Z., Gledović, Z. & Janković, S. (1984b) Balkan nephropathy and malignant tumors. *Neoplasma*, **31**, 225–229

Radovanović, Z., Djordjević, G., Raičević, R., Velimirović, D., Velimirović, A., Janković, S. & Miljković, V. (1985) Endemic nephropathy in a defined locality—implications of a cross-sectional epidemiological study. *Med. Istraz.*, **18**, 75–80 (in Serbo-Croat)

Radovanović, Z., Velimirović, D. & Naumović, T. (1990) Upper urothelial tumours and the Balkan nephropathy—inference from the study of a family pedigree. *Eur. J. Cancer*, **26**, 391–392

Radovanović, Z., Danilović, V., Velimirović, D., Naumović, T., Jevremović, I., Janković, S., Vacca, C. & Hall, P.W. (1991a) Beta2-microglobulinuria as a predictor of death in a population exposed to Balkan endemic nephropathy. *Kidney Int.*, **40** (Suppl. 34) (in press)

Radovanović, Z., Janković, S. & Jevremović, I. (1991b) Incidence of tumours of urinary organs in a focus of Balkan endemic nephropathy. *Kidney Int.* (in press)

Stojimirović, B. (1974) Recent data on the incidence and distribution of endemic nephropathy along the Kolubara river. In: Puchlev, A., A., Dinev, I., Milev, B. & Doichinov, D., eds, *II International Symposium on Endemic Nephropathy*, Sofia, Bulgarian Academy of Sciences Press, pp. 297–299

Strahinjić, S. & Stefanović, V. (1980) Endemic nephropathy around the South Morava river. *Bull. Acad. Serb. Sci. Arts*, **LXIX**, Cl. Sci. Med., **12**, 105–110

Tanchev, Y., Naidenov, D., Dimitrov, T. & Nikolov, B. (1965) Health centre control of patients suffering from endemic nephropathy in the Vratza district. In: *International Symposium on Endemic Nephropathy*, Sofia, Bulgarian Academy of Sciences Press, pp. 142–148

Trnavčević, S., Suša, S., Janković, D., Dumović, B., Skatarić, V., Savić, D. & Opalić, S. (1981) Occurrence of endemic glomerulonephritis in urban populations. In: *Proceedings of the 1st Congress of Yugoslav Nephrologists, Belgrade*, pp. 67–71 (in Serbo-Croat)

Udicki, S., Obradović, M. & Krajinović, S. (1970) Mortality in an area affected by endemic nephropathy. *Glasnik ZZZZ*, **2**, 119–123 (in Serbo-Croat)

Velimirović, D. (1980) Field investigation of the spread and frequency of endemic nephropathy in the region of Kolubara. *Bull. Acad. Serb Sci. Arts*, **LXIX**, Cl. Sci. Med., **12**, 79–85

Veljković, A., Milovanović, B. & Živanović, D. (1967) Epidemiological investigation of families with cases of endemic nephropathy. In: *Proceedings of the 1st Symposium on Endemic Nephropathy*, Niš, University of Niš Press, pp. 97–101

Verbev, P., Gubev, E., Ivanov, N., Donchev, D., Tanchev, Y., Stoyanova, M. & Teofilova, S. (1965) Epidemiological study of endemic nephropathy in Bulgaria. In: *International Symposium on Endemic Nephropathy*, Sofia, Bulgarian Academy of Sciences Press, pp. 115–121

WHO (1965) The 'endemic nephropathy' of south-eastern Europe. *Bull. World Health Organ.*, **32**, 431–448

Zaharia, C.A, Birzu, I., Popescu, G. & Torjescu, V. (1965a) Endemic renal disease (endemic atrophy) in Romania. In: *International Symposium on Endemic Nephropathy*, Sofia, Bulgarian Academy of Sciences Press, pp. 57–72

Zaharia, C.A. *et al.* (1965b) Endemic nephropathy (renal atropy). In: *International Symposium on Endemic Nephropathy*, Sofia, Bulgarian Academy of Sciences Press, pp. 73–76 (in German)

Mycotoxins, Endemic Nephropathy and Urinary Tract Tumours
Ed. M. Castegnaro, R. Pleština, G. Dirheimer, I.N. Chernozemsky & H. Bartsch
Lyon, International Agency for Research on Cancer, pp. 21-28

THE FIRST CLINICAL DESCRIPTION OF BALKAN ENDEMIC NEPHROPATHY (1956) AND ITS VALIDITY 35 YEARS LATER

Y. Tanchev[1] & D. Dorossiev[2,3]

[1]Formerly: Christo Botev District Hospital, Clinic for Nephrology and Haemodialysis, Vratza; and [2]National Centre for Cardiac Rehabilitation, Banja, Bulgaria

Summary

A high prevalence of renal disease in Vratza, a district in north-west Bulgaria, was studied in 1950–54 by Tanchev at the district hospital. A particular unknown renal condition was described at local meetings in 1953 and was referred to as 'endemic Vratza nephritis' in 1955. The first clinical description of this new nosological entity, published by Tanchev and colleagues in 1956, was based on 664 patients hospitalized for renal disease. Of 296 with chronic nephritis, 17 died in hospital and 103 died a few days later at home, all with uraemia, to give a total of 40.5%. Peasants formed the majority of the patients (85.7%), and 4–43 came from only 16 villages and 1–3 from 36 villages; none came from the remaining 21 villages in the district. Clusters of patients were thus noted in villages, families and even houses. The patients had the following common characteristics: from endemic areas; other renal ailments in the family; copper-yellow skin and orange palms and soles; normochromic anaemia; absence of acute onset, considerable albuminuria, hypertension and oedema; no compensatory polyuria; azotaemia progressing

[3]To whom correspondence should be addressed

insidiously to fatal uraemia; 83.5% died within one year of the appearance of symptoms. After similar ailments were described in Yugoslavia in 1957 and Romania in 1961, the condition became known as Balkan endemic nephropathy. The etiology of this disease remains unknown, and no treatment is available, although haemodialysis and kidney transplants have prolonged patients' survival.

Introduction

The attention of nephrologists was attracted in the 1950s by scientific publications from Bulgaria (Tanchev *et al.*, 1956), Yugoslavia (Danilović *et al.*, 1957a,b) and Romania (Fortza & Negoescu, 1961) describing a kidney disease found in geographically limited areas of these three Balkan countries. This relatively new chronic ailment, with unmarked onset, deviated in some clinical characteristics from other known kidney diseases. The progressive, untreatable course and fatal outcome in uraemia shortly after manifestation of symptoms made this puzzling disease one of the most important problems of contemporary renal pathology. In 1964, the disease was recognized as a new nosological entity and was referred to thereafter as Balkan endemic nephropathy (BEN). An association between BEN and urinary tract tumours was also recognized early on (Petrinska-Venkovska, 1960): the problem of BEN thus turned out to be not only nephrological but also oncological.

This paper has the limited purpose of looking back at the first clinical description of BEN (Tanchev *et al.*, 1956) and evaluating its validity after the severe scrutiny of the 35 years that have since passed. Eventual changes in the clinical features of BEN, in their interpretation and in the conclusions that can be drawn in the light of new scientific achievements related to this enigmatic phenomenon may benefit from this retrospective examination.

Background

Physicians working in villages in Vratza District in north-west Bulgaria shared the impression more than half a century ago that kidney diseases were more frequent in some settlements than in others. Some staff physicians at the Medical Department of the District Hospital supported their opinion, as a growing number of patients, hospitalized for chronic nephritis or pyelonephritis, displayed a constellation of signs that were nonspecific for classical diagnostic criteria. A working diagnosis of 'atypical nephritis' was accepted to describe this syndrome, which constituted a new diagnostic and therapeutic problem for the Department.

A systematic assessment of the problem was initiated by Y. Tanchev in 1950, in three directions: (i) a study of patients hospitalized at the

Medical Department for renal conditions; (ii) on-the-spot screening for atypical nephritis by teams of specialists sent out to the most heavily affected settlements; and (iii) an epidemiological study to establish the prevalence of atypical nephritis in the District and to identify potential etiological factors that might explain local differences.

It should be kept in mind that the whole project was to be implemented only five years after the end of the Second World War and 10 years of disrupted inflow of international medical information. Even contacts with neighbouring Yugoslavia were cut off, for political reasons. Moreover, the facilities available for comprehensive laboratory and pathomorphological examinations were relatively modest. An advantage of the District Hospital was, however, its situation in the affected area, with an experienced staff who had had direct contact with the patients and were thus motivated to help them. Although the three aspects of the study were carried out simultaneously, the data described below are related mainly to the clinical characteristics of 'atypical nephritis', known later as 'endemic Vratza nephritis', 'endemic nephritis' and 'endemic nephropathy' (Tanchev, 1985).

The Clinical Picture of Endemic Nephropathy

The five-year study (1950–54) comprised 664 patients who had been hospitalized for various renal conditions: 296 with chronic nephritis (44.6%), 179 with nephrolithiasis (27.0%), 71 with acute or chronic pyelonephritis (10.5%), 32 with acute nephritis (4.9%) and 86 with other renal diseases. Most of the patients with chronic nephritis were farmers (84%) of each sex (42% men); 85% were over 30 years of age (51% were aged 41–60); and all except 20 patients were from 74 settlements in Vratza District. The interval between the appearance of symptoms (malaise, weakness, fatigue, dry mouth) and hospitalization was relatively short—less than one year for 71% of patients—and varied from less than six months (47%) to more than three years (16.5%). None of the patients reported any acute manifestation of their illness. The fact that not all 296 patients with chronic nephritis would satisfy the criteria for BEN established later (Puchlev *et al.*, 1960) should be kept in mind when interpreting the results described below.

On admission to hospital, the patients had a typical appearance, which consisted of paleness, skin that was often copper-yellowish and orange palms and soles. Xanthodermia was also found later in apparently healthy individuals in endemic areas. Normo- or slightly hypochromic anaemia, with an average haemoglobin concentration below 96 g/l and an erythrocyte count of less than 3.5×10^{12}/l, was present in 71% of patients; frequently, much lower values were found and the anaemia did not respond to treatment. Blood pressure was generally not elevated—less than 21.5 kPa (160 mm Hg) systolic in 75% of patients and less than 12 kPa (90 mm Hg) diastolic in 69% of patients.

Virtually all patients had slight proteinuria, below 1 g/l and often less than 0.02 g/l or only traces. Urine sediment was scanty and haematuria rare, and casts were found in 9% of patients. Diuresis did not exceed 2000 ml/24 h in 92% of patients and was less than 500 ml/24 h in only 12 patients with uraemia. The specific gravity of the urine was less than 1010 in 40.8% of patients and frequently reached 1001–1002 spontaneously. Values in the concentration and dilution test (Volhard test) varied between 1017 and 1004–1003.

Blood urea levels greater than 8.1 mmol/l were found in 76.4% and greater than 20.2 mmol/l in 48% of patients. The ocular fundi, examined in 89 patients with hyperazotaemia, were pale. Retinitis albuminurica was seen in 65.3% but was inconsistent with blood pressure and blood urea levels, in contrast to classical forms of chronic nephritis. Fluoroscopy of the lungs and heart revealed no significant change; there was no sign of heart failure or oedema. The typical picture of severe uraemia appeared only during the terminal stage of the disease.

Seventeen of the 296 patients (5.7%) died of uraemia in hospital, and 110 were discharged in bad condition. Many of these patients had entered hospital in advanced stages of renal failure and were aware of their prognosis on discharge. Their desire to die in dignity at home was honoured, but this prevented post-mortem examination. If one adds to the hospital deaths the 103 that occurred within a few days of discharge, 'hospital mortality' amounted to 40.5%: 'atypical chronic nephritis' was proving to be a frightening killer.

Particularly informative was the time between initial symptoms and death for the 103 patients who died at home. The interval was less than six months for 62 patients (60.2%), up to one year for 24 (23.3%) and two years for five patients (4.9%). Overall, therefore, 83.5% were dead within one year and only 11.6% survived more than two years. Early preclinical diagnosis of BEN became an imperative research priority.

The first clinical description of BEN was an attempt to signal the existence of a deadly, probably endemic disease of unknown etiology which did not respond to conventional treatment. A constellation of features that emerged from the study seemed important for identifying patients with BEN: provenance from an endemic area; kidney disease in other family members; unmarked onset and insidious course; typical colouring of the skin, palms and soles; pronounced normochromic anaemia; absence of significant albuminuria, hypertension or oedema; and azotaemia progressing rapidly to fatal uraemia. These provisional criteria were later confirmed (Puchlev *et al.*, 1960), and they were used for diagnosis and screening in different combinations (Bruckner *et al.*, 1965; Puchlev *et al.*, 1965b). Their validity subsequently became generally accepted in BEN research and clinical nephrology (Tanchev *et al.*, 1965; Puchlev *et al.*, 1974).

Multiprofile medical screening in Beli Izvor to detect cases of BEN on the basis of the above criteria (Tanchev & Evstatiev, 1958) made it possible to draw a map of BEN foci in Vratza District (Figure 1). Of the 279 patients in the study, 239 (85.7%) were peasants residing in this District. The geographic limitations of their places of origin was impressive: 4–43 patients came from 16 settlements, 1–3 from 36 and none from the remaining 21. Some differences were striking. BEN morbidity was, for example, 1 per 1000 in the town of Vratza; 36 per 1000 in Bistretz, only 2 km from Vratza; and 37 per 1000 in Beli Izvor, 11 km from Vratza.

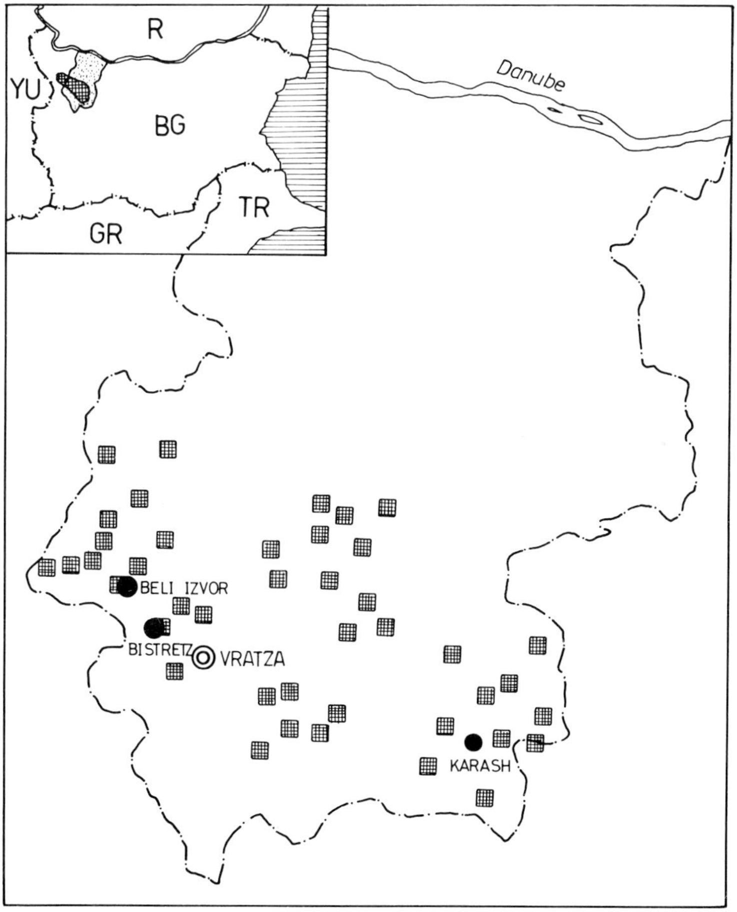

Figure 1. Vratza District in north-west Bulgaria

The inset shows the situation of the District (hatched area) within Bulgaria (BG, Bulgaria; YU, Yugoslavia; R, Romania; TR, Turkey; GR, Greece). On the main map, endemic foci for Balkan nephropathy are marked with squares; villages with the highest prevalences are marked with filled circles.

It was concluded from the study that this '*sui generis*' renal disease represented a new endemic phenomenon of seemingly epidemic and familial character and unknown etiology. Two hypotheses were proposed for its etiology: (i) chronic poisoning originating from water, soil or food (e.g., lead and other metals); or (ii) a viral infection with a late manifestation.

Early Medical and Social Effects of the Study

Preliminary findings from the first investigation were reported at a hospital conference by Tanchev in 1953 and two years later at two conferences in Varshetz and Vratza, which were attended by health officials and physicians from most other districts of the country. It was only after the first publication appeared (Tanchev *et al.*, 1956), however, that systematic measures were taken to control this deadly disease at regional and national levels. The Ministry of Health appointed a special committee for research on BEN in 1957.

The first microbiological experiments on animals to elucidate the etiology of BEN were initiated in Vratza by L. Ivanov in 1950. The results suggested tuberculosis, fungi or moniliasis as possible factors (Ivanov, 1956) but received no further support. A comprehensive monograph (Puchlev, 1960) and two international symposia on BEN in Bulgaria (Puchlev *et al.*, 1965a, 1974) summarized all of the available, albeit inconclusive, information on the etiology of BEN. They also stimulated large-scale hygienic measures, the provision of a water supply from nonendemic areas to the affected villages, screening and medical follow-up of populations at risk and particularly children and adolescents, and education to ensure a healthier life style. Furthermore, in 1959, the inhabitants of the village of Karash were resettled in nine nonendemic districts.

In an important communication, Petrinska-Venkovska (1960) reported 16 cases of urinary-tract tumours among 33 patients with BEN (48.4%). Tanchev *et al.* (1970) reported a similar finding among 16% of BEN patients, and other reports appeared subsequently. A further study examined the geographic correlation between BEN and renal tumours (Chernozemsky *et al.*, 1977).

A haemodialysis unit was installed in the Vratza District Hospital in 1972 to facilitate treatment of BEN patients. Haemodialysis and kidney transplants have prolonged the lives of these patients; however, this has resulted in the appearance of an increased frequency of urinary tract tumours (Tanchev, 1983). Studies on the three basic hypotheses for the etiology of BEN—exogenous poisoning, viral infection or genetic disease—are being continued to find common etiological links between BEN and oncogenesis.

A View on Further Research on Balkan Endemic Nephropathy

Sufficient information is still lacking with regard to the hypothesis that a certain factor, irrespective of its origin, affects simultaneously or consecutively the renal tubules and the production of erythropoietin and renin in the kidney. A role for erythropoietin suppression in BEN was suggested by Parashiv *et al.* (1965) and Kirov (1974) but has still not received due attention.

Structural damage or primary inhibition of function, for example at enzymatic levels entailing structural changes, were until recently inaccessible to investigation *in vivo*, particularly in human organs such as the kidney and its infrastructure (glomeruli, tubules, juxtaglomerular apparatus). The development of feasible, reliable methods for the early detection of preclinical stages of BEN, as for example through identification of specific microglobulins in the urine of suspects from endemic areas, is still not satisfactory.

The fascinating recent progress in medical techniques offers new perspectives in BEN research through unexploited fields of application, such as imaging by magnetic resonance and positron emission tomography. Access to such sophisticated, expensive tools will require international cooperation and coordination, an interdisciplinary approach and a standardized interpretation of results obtained from different points of view. These requirements might, hopefully, be met in the near future in the framework of an integrated, united Europe.

References

Bruckner, I., Zosin, C., Lazarescu, R., Paraschiv, D., Manescu, N., Serban, M. & Titeica, M. (1965) A clinical study on nephropathy of an endemic character in the Peoples' Republic of Romania. In: Puchlev, A., Popov, N., Astrug, A., Dotchov, D. & Dinev, I., eds, *International Symposium on Endemic Nephropathy*, Sofia, Bulgarian Academy of Science Press, pp. 23–25

Chernozemsky, I.N., Stoyanov, I.S., Petkova-Bocharova, T.K., Nicolov, I.G., Draganov, I.V., Stoichev, I.I., Tanchev, Y., Naidenov, D. & Kalcheva, N.D. (1977) Geographic correlation between the occurrence of endemic nephropathy and urinary tract tumours in Vratza District, Bulgaria. *Int. J. Cancer*, **19**, 1–11

Danilović, V., Djurišić, M., Mokranjac, M., Stojimirović, B., Živojinović, J. & Stojaković, P. (1957a) Porodična oboljenja bubrega u selu Šopić izvazvana hroničnom intoksikácijom olovom. *Srpski Arh. Tcelok. Lek.*, **85** (10), 1115–1125

Danilović, V., Djurišić, M., Mokranjac, M., Stojimirović, B., Živojinović, J. & Stojaković, P. (1957b) Chronic nephritis caused by poisoning with lead via the digestive tract (flour). *Presse Méd.*, **65** (90), 2039–2040 (in French)

Fortza, N. & Negoescu, M. (1961) Nefrita cronica azotemica endo-epidemica. *Stud. Cercet. Med.*, **1**, 217–221

Ivanov, L.M. (1956) Studies on the etiology of renal diseases in the District of Vratza. *Sovrem. Med.*, **7** (9), 30–34 (in Bulgarian)

Kirov, C. (1974) Studies on the pathogenesis of the anaemic syndrome in endemic nephropathy. In: Puchlev, A., Dinev, I., Milev, B. & Doichniov, D., eds, *Proceedings of the Second International Symposium on Endemic Nephropathy*, Sofia, Publishing House of the Bulgarian Academy of Sciences, pp. 152–155

Parashiv, D., Axinescu, M. & Cartianu, P. (1965) Haematology and myelogram before and after microtransfusion for victims of endemic nephropathy (Erghevitza type). In: Puchlev, A., Popov, N., Astrug, A., Dotchov, D. & Dinev, I., eds, *International Symposium on Endemic Nephropathy*, Sofia, Bulgarian Academy of Science Press, pp. 82–86 (in German)

Petrinska-Venkovska, S. (1960) Morphological studies on endemic nephritis. In: Puchlev, A., ed. *Endemic Nephritis in Bulgaria*, Sofia, Medizina i Fizkultura, pp. 72–90 (in Bulgarian)

Puchlev, A., ed. (1960) *Endemic Nephritis in Bulgaria*, Sofia, Medizina i Fizkultura (in Bulgarian)

Puchlev, A., Astrug, A., Popov, N. & Dotchev, D. (1960) Clinical studies on endemic nephritis. In: Puchlev, A., ed. (1960) *Endemic Nephritis in Bulgaria*, Sofia, Medizina i Fizkultura, pp. 7–71 (in Bulgarian)

Puchlev, A., Popov, N., Astrug, A., Dotchov, D. & Dinev, I., eds (1965a) *International Symposium on Endemic Nephropathy*, Sofia, Bulgarian Academy of Science Press

Puchlev, A., Popov, N., Astrug, A. & Dotchev, D. (1965b) Clinical studies on endemic nephritis in Bulgaria. In: Puchlev, A., Popov, N., Astrug, A., Dotchov, D. & Dinev, I., eds (1965a) *International Symposium on Endemic Nephropathy*, Sofia, Bulgarian Academy of Science Press, pp.

Puchlev, A., Dinev, I., Milev, B. & Doichniov, D., eds (1974) *Proceedings of the Second International Symposium on Endemic Nephropathy*, Sofia, Publishing House of the Bulgarian Academy of Sciences

Tanchev, Y. (1983) Possibility of treatment of Balkan endemic nephropathy through haemodialysis program. In: Strahinjić, S. & Stefanović, V., eds, *Current Research in Endemic (Balkan) Nephropathy, Proccedings of the Fifth Symposium on Endemic (Balkan) Nephropathy*, Niš, University of Niš, pp. 145–148

Tanchev, Y. (1985) Additional data on the history of the Balkan endemic nephropathy. *Vatr. Bol.*, **24** (1), 50–54 (in Bulgarian)

Tanchev, Y. & Evstatiev, Z. (1958) Study on the nephritides in the village of Bely Isvor, District of Vratza. *Sovrem. Med.*, **9** (1), 9–16 (in Bulgarian)

Tanchev, Y., Evstatiev, Z., Dorossiev, D., Pencheva, J. & Zvetkov, G. (1956) Studies on the nephritides in the District of Vratza. *Sovrem. Med.*, **7** (9), 14–29 (in Bulgarian)

Tanchev, Y., Naidenov, D., Dimitrov, T. & Nikolov, B. (1965) Health centre control of patients suffering from endemic nephropathy in Vratza District. In: Puchlev, A., Popov, N., Astrug, A., Dotchov, D. & Dinev, I., eds *International Symposium on Endemic Nephropathy*, Sofia, Bulgarian Academy of Science Press, pp. 142–148

Tanchev, Y., Naidenov, D., Dimitrov, T. & Karlova, E. (1970) Neoplastic diseases and endemic nephritis. *Vatr. Bol.*, **9**, 21–29 (in Bulgarian)

Mycotoxins, Endemic Nephropathy and Urinary Tract Tumours
Ed. M. Castegnaro, R. Pleština, G. Dirheimer, I.N. Chernozemsky & H. Bartsch
Lyon, International Agency for Research on Cancer, pp. 29-35

CHARACTERISTICS OF URINARY TRACT TUMOURS IN THE AREA OF BALKAN ENDEMIC NEPHROPATHY IN CROATIA

B. Šoštarić[1] & M. Vukelić[2]

[1]Department of Toxicology, Institute for Medical Research and Occupational Health, University of Zagreb, Zagreb; and
[2]Department of Pathology and Forensic Medicine, Slavonski Brod Medical Centre, Slavonski Brod, Yugoslavia

Summary

There is a general consensus that the incidence of urothelial tumours is much higher among the inhabitants of areas of Balkan endemic nephropathy than in populations in nonendemic areas. Data from different authors on tumour incidences in unrelated endemic areas in various countries vary widely, however. Lack of understanding of the possible etiological relationship between Balkan endemic nephropathy and the increased incidence of urothelial tumours justifies efforts to conduct retrospective studies using unique diagnostic criteria. In order to investigate the occurrence of urinary tract tumours in Slavonski Brod, a part of which has a high incidence of nephropathy, a retrospective study was carried out using data and tissue samples collected at the Department of Pathology and Forensic Medicine of the Slavonski Brod Medical Centre. During the last 16 years, 193 urinary tumours from the county were examined; the material available consisted of autopsy reports and surgical and biopsy specimens. Tumours were diagnosed histologically and coded according to guidelines of the WHO and US registries, as papillomas, papillary and solid carcinomas of the transitional-cell type, squamous-cell carcinomas and carcinomas *in situ*.

A 5.1 times higher incidence of urothelial tumours was seen among inhabitants of the nephropathic area of the county than in nonendemic areas. Tumours from cases in the endemic region also tended more often to be malignant and multiple; they were not infrequently bilateral.

Introduction

Certain villages within the county of Slavonski Brod were found several decades ago to have a high prevalence of nephropathy, and others were suspected. In an extensive series of consecutive, primarily clinico-epidemiological studies (Čeović *et al.*, 1979a,b, 1983), the entire area of the county was monitored for several years, and all hyperendemic foci were clearly recognized. Each affected village appeared to be a virtually separate focus of the disease. A sharply demarcated area within the county, bordered by the river Sava on the south and the Zagreb-Beograd highway on the north, encompasses all the affected villages; none of the villages outside of this area nor the city of Slavonski Brod are affected. The affected area is exclusively agricultural, where a traditional household-type economy involving 10 094 inhabitants is maintained. The rest of the villages in the county, with a comparable socioeconomic structure, including the moderately industrialized town of Slavonski Brod, are inhabited by 96 306 people.

The aim of this paper is to present the pathomorphological characteristics of the cases of urinary tract tumour found in the entire county during a 16-year period and to compare their incidence in an endemic and a nonendemic area of the same county

Materials and Methods

During the last 16 years, 193 cases of urinary tract tumour were diagnosed among permanent inhabitants of Slavonski Brod at the Department of Pathology and Forensic Medicine of the Slavonski Brod Medical Centre. The material studied included autopsy reports, both medical and legal, and surgical and biopsy specimens. All the cases were examined grossly and microscopically; histological slides were prepared and stained using standardized techniques (Lillie, 1954; Luna, 1979). The systematization, nomenclature and evaluation criteria used, such as grading and staging, were in accordance with widely accepted guidelines described in detail by Robbins and Cotran (1979) and in the literature cited therein.

Results

Data on tumour incidence in the two areas of the county are given in Table 1. The anatomical location of the diagnosed tumours, regardless of type, malignancy or staging, is given for endemic and nonendemic areas in Table 2. On gross examination, the papillary form was seen in

Table 1. Incidence of urothelial tumours in endemic and nonendemic areas for Balkan nephropathy in the county of Slavonski Brod, 1974–89

Area	No. of inhabitants	No. of tumours	% in relation to no. of inhabitants
Endemic	10 094	67	0.664
Nonendemic	96 306	126	0.131
Total	106 400	193	0.795

Proportion endemic:nonendemic 5.068

Table 2. Anatomical location of urothelial tumours for all recognized cases in endemic and nonendemic areas

Anatomical	Endemic area		Nonendemic area		Proportion
	No. of cases	%	No. of cases	%	
Renal pelvis	29	0.287	20	0.021	13.67
Ureter	9	0.089	13	0.013	6.847
Urinary bladder	23	0.228	86	0.089	2.561
Combination (renal pelvis and ureter)	6	0.059	7	0.007	8.429
Total	67	0.664	126	0.130	5.11

the majority of cases in all anatomical locations, regardless of whether they derived from nephropathic or non-nephropathic areas (Figure 1). Not infrequently, solid masses of a cauliflower-like appearance attached to the base were noted (Figure 2). Flat, plaque-like tumours rising just above the mucosal surface were seen occasionally. Histologically, all of the tumours from both areas were of epithelial origin. Benign papillomas were considerably less well represented, especially in the material from the endemic area. Among the malignant tumours, transitional-cell carcinomas (Figures 3 and 4) were by far the most frequent (95%). Squamous-cell carcinomas were seen in 5% of cases, while adenocarcinoma was not diagnosed. In a relatively large number of transitional-cell carcinomas, small foci of squamous metaplasia were noted, but there was no difference among the tumours from each area in this respect. In general, there was no difference in basic tumour type from the two areas, any differences being in the degree of malignancy and spread of the lesion (Table 3).

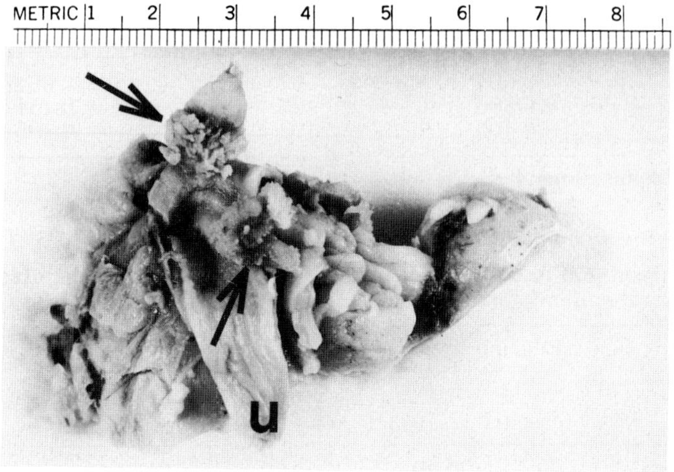

Figure 1. Kidney from a patient in an advanced stage of Balkan endemic nephropathy

The renal pelvis and ureter (u) are exposed. Note two tumour masses of papillary type growing up from mucosis of the pelvis (arrows).

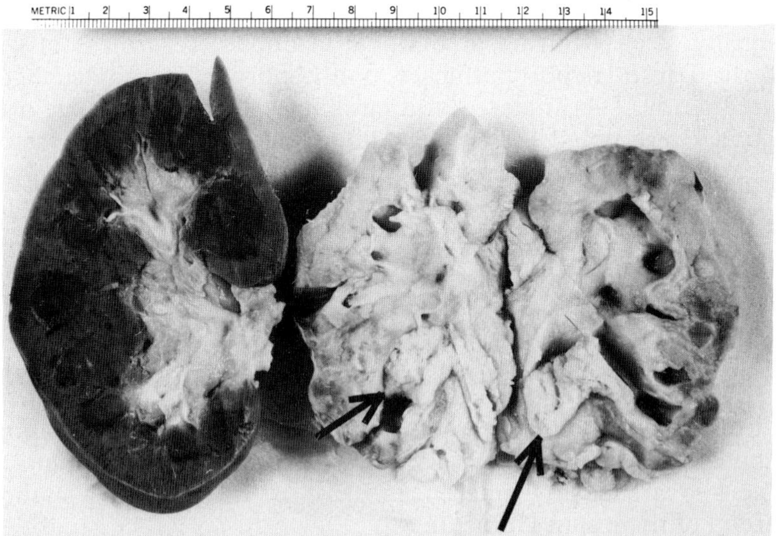

Figure 2. Exposed, cut surface of kidney of patient in an early stage of Balkan endemic nephropathy

Concomitant urothelial tumour of the renal pelvis (arrows) and associated pyelonephritis, right side; normal human kidney on the left

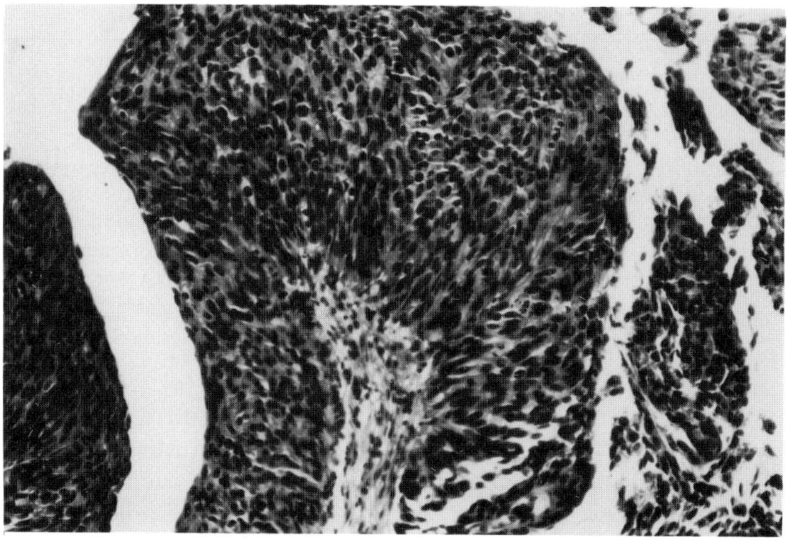

Figure 3. Photomicrograph of grade 2 papillary transitional-cell carcinoma

Note central fibrovascular tissue stroma and many layers of moderately disoriented epithelium.

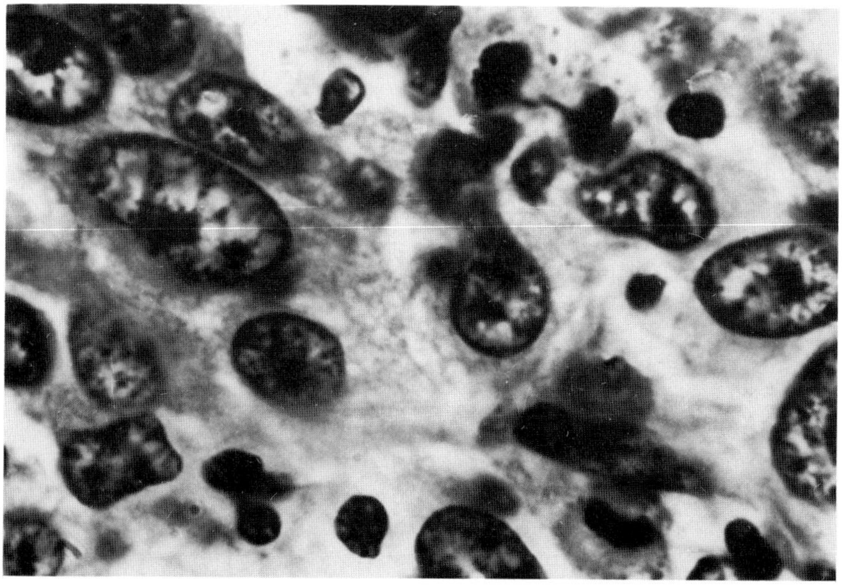

Figure 4. Higher magnification of one field from the slide shown on Figure 3

Moderate anaplasia and still recognizable urothelial type of epithelial cells are visible. Note the mitotic figure at the top.

Table 3. Differences in urothelial tumours between endemic and nonendemic areas for Balkan nephropathy

Endemic area	Nonendemic area
Higher incidence	Lower incidence
Younger age	Older age
Females more frequently affected	Males more frequently affected
Affects mostly renal pelvis and ureter	Affects mostly urinary bladder
Frequently multiple	Usually single lesions
Usually malignant	Usually nonmalignant

Discussion

A higher incidence of urothelial tumours was first noted in residents of an area of Balkan nephropathy in Bulgaria (Puhlev *et al.*, 1960) and was confirmed subsequently in virtually all other such areas (Čeović *et al.*, 1976; Petković *et al.*, 1977; Suša, 1979) and among residents of these areas who emigrated during their youth to nonendemic locations (Nikolić & Popović, 1983). In our study, the incidence of urothelial tumours, regardless of their anatomical location in the urinary passage or their malignancy, was roughly five times higher among inhabitants of the endemic area than among residents of nonendemic areas. Furthermore, their location is much higher, reaching an almost 14 times higher incidence in the endemic area. A progressive decrease in the incidence of tumours of the ureter and bladder was observed.

The correlation between the tumours and regions of Balkan endemic nephropathy is obvious, but their etiology is still obscure and is not apparent from our results. Our findings concur with the hypothesis that a common xenobiotic plays a role in the etiology of Balkan endemic nephropathy and tumours. The theory that tumours develop because of impaired kidney function and excretion of metabolites with an oncogenic effect on the urothelium, although plausible, is not supported by our results, because in many of the cases the kidneys were histologically unremarkable. Tumours associated with different stages of Balkan endemic nephropathy might have a previously pathogenetic origin, but this cannot be proved by means of pathomorphology. It is interesting that authors studying other nephropathic areas report about a 100 times higher incidence of upper urothelial tumours (Petković, 1985) and no increase in the incidence of tumours in the urinary bladder (Petković, 1979). Since the same authors have histological evidence favouring a viral etiology for the tumours in question (Petković *et al.*, 1985), a high selectivity of the possible viral agent for certain anatomical locations of the urothelium might be assumed.

References

Čeović, S., Radonić, M., Hrabar, A., Radošević,, Z., Bobinec, E., Pleština, R. & Habazin-Novak, V. (1979a) Endemic nephropathy in Brodska Posavina in a twenty year period. In: Strahinjić,, S. & Stefanović,, V., eds, *Endemic (Balkan) Nephropathy* (Proceedings of the 4th Symposium on the Endemic (Balkan) Nephropathy), Niš, University of Niš, pp. 223–229

Čeović, S., Hrabar, A. & Radonić, M. (1979b) Endemic nephropathy in different ethnic groups in the area of Slavonski Brod. In: Strahinjić, S. & Stefanović, V., eds, *Endemic (Balkan) Nephropathy* (Proceedings of the 4th Symposium on the Endemic (Balkan) Nephropathy), Niš, University of Niš, pp. 229–233

Čeović, S., Hrabar, A., Radonjić, M., Čeović, J., Mitar, I., Pleština, R., Bistrović, B. & Miletić-Medved, M. (1983) Distribution of Balkan endemic nephropathy in the region of Slavonski Brod, Croatia, Yugoslavia. In: Strahinjić, S. & Stefanović, V., eds, *Current Research in Endemic (Balkan) Nephropathy* (Proceedings of the 5th Symposium on Endemic (Balkan) Nephropathy), Niš, University of Niš, pp. 257–26

Čeović,, S., Grims, P., Hrabar, A. & Mitar, J. (1976) Praćenje pojave tumora mokraćnih organa u kraju endemske nefropatije i kontrolnom podrucju. In: Regionalni Medicinski Centar: "Dr Mustafa Mujbegović" Tuzla. *Acta Med. Saliniana*, **2**, 35–45

Lillie, R. D. (1954) *Histopathologic Technic and Practical Histochemistry*, New York, The Blakiston Company

Luna, L. G. (1979) *Manual of Histologic Staining Methods of the Armed Forces Institute of Pathology*, 3rd ed., New York, McGraw-Hill

Nikolić, J. & Popović, J. (1983) Tumours of the pyelum and ureter in emigrants from endemic nephropathy areas. In: Strahinjić, S. & Stefanović, V., eds, *Current Research in Endemic (Balkan) Nephropathy* (Proceedings of the 5th Symposium on Endemic (Balkan) Nephropathy), Niš, University of Niš, pp. 245–248

Petković, S. (1979) The pathology of renal pelvic and ureteral tumours in regions of endemic (Balkan) nephropathy. In: Strahinjić, S & Stefanović, V., eds, *Proceedings of the 4th Symposium on Endemic (Balkan) Nephropathy*, Niš, University of Niš, pp. 21–24

Petković, S. (1985) Tumori pielona i uretera u regionima endemske nefropatije. In: Petković, S., ed., *Proceedings, III Symposium on Endemic Nephropathy*, Beograd, Serbian Academy of Sciences and Arts, pp. 279–289

Petković, S., Mutavdić, V., Petronić, V., Marković, M., Isvaneski, S., Micić, J., Hadži-Dokić, J., Nikolić, J. & Argirović, D. (1977) Osobitosti karcinogenog dejstva na urotelijum u regionima endemske nefropatije. In: Vukušić, Z., ed., Zbornik radova III Simpozijuma o endemskoj nefropatiji, Beograd, Galenika, pp. 210–219

Petković, S., Skoro-Milić, A., Spasić, P. & Bojanić, N. (1985) Prilog studiji virusne etiologije tumora pijeloma i uretera u regionima endemske nefropatije. In: Petković, S., ed., III Simpozijum o endemskoj nefropatiji, Srpska Akademija Nauka i Umetnosti, Naućni Skupovi, Knjiga XXIII, Odelenje Medicinskih Nauka, Knjiga 3, Beograd, pp.279–288

Puhlev, A., Astrug, A., Popov, N. & Docev, O. (1960) Klinicki proucavanija vrhu endemicnija nefrit, Endemicnijat nefrit v Bulgaria, Sofia, Medicina i fizkultura, pp. 7–17

Suša, S. (1979) Endemska nefropatija (2. izdanje) Savremena administracija, Beograd, pp. 215–290

Mycotoxins, Endemic Nephropathy and Urinary Tract Tumours
Ed. M. Castegnaro, R. Pleština, G. Dirheimer, I.N. Chernozemsky & H. Bartsch
Lyon, International Agency for Research on Cancer, pp. 37-42

SOME PATHOMORPHOLOGICAL FEATURES OF BALKAN ENDEMIC NEPHROPATHY IN CROATIA

M. Vukelić[1], B. Šoštarić & R. Fuchs[2]

[1]Department of Pathology and Forensic Medicine, Slavonski Brod Medical Centre, Slavonski Brod; and [2]Department of Toxicology, Institute for Medical Research and Occupational Health, University of Zagreb, Zagreb, Yugoslavia

Summary

Over the last 16 years, 214 autopsies were done at the Department of Pathology and Forensic Medicine, Slavonski Brod Medical Centre, on people from an area recognized as endemic for Balkan nephropathy in the county. Balkan endemic nephropathy was diagnosed pathoanatomically and histopathologically in 94 of these cases, and in none of 1040 autopsies on people from a nonendemic area. The most striking pathological finding in all advanced cases of the disease was a marked reduction in kidney size and weight; in one extreme case, the organ weighed only 20 g. The process is invariably bilateral, but there are considerable differences in the degree of involvement in each pair. Although the pathoanatomical changes, including lung oedema and haemorrhage and fibrinous pericarditis, are seen in the majority of cases, they are not considered to be specific for Balkan endemic nephropathy, since they are well recognized signs of long-standing primary uraemia. The relevance of the peculiar finding of sulfurous yellow discoloration and hardened subcutaneous adipose tissue, seen frequently *post mortem*, is unknown and should be investigated in more detail. Histopathologically, fibrosis and atrophy of kidney cortex, with tubular degeneration, are the most consistent findings. Lesions characteristic of pyelonephritis were superimposed over the picture of 'pure' Balkan endemic nephropathy in a considerable number of cases.

Text

In the county of Slavonski Brod, in the north-east part of Croatia, several hyperendemic foci of Balkan endemic nephropathy have been recognized and thoroughly studied (Radonić, 1961; Radonić *et al.*, 1966; Vukelić & Belicza, 1977; Belicza *et al.*, 1979; Čeović *et al.*, 1979a,b; Vukelić *et al.*, 1979; Čeović *et al.*, 1983).

During the last 16 years at the Department of Pathology and Forensic Medicine, Slavonski Brod Medical Centre, 214 autopsies were carried out of people originating from a recognized endemic area within the county. Balkan endemic nephropathy was diagnosed in 94 of them, but in none of 1040 autopsies from a nonendemic area. The group of 94 recognized cases was composed mostly of older people with clinically well-recognized, long-standing kidney disease, and most deaths were due to uraemia. The other cases occurred in younger persons with no history of kidney disease; death was due to violent causes. The average age at the time of death was 60 in men and 66 in women. The purpose of this article is to summarize the gross morbid and histopathological findings in this group and to propose possible new directions in the study of this disease.

All autopsies were performed in accordance with widely accepted methods and techniques described elsewhere. Processing of tissue and preparation of histological slides was done using generally accepted methods (Lillie, 1954; Luna, 1979).

The most relevant autopsy findings are as follows. In almost all advanced cases, a characteristic, pale, dirty-yellow discoloration of the skin was seen, with a peculiar, yellowish coloration of adipose tissue throughout the body. The adipose tissue is harder then normal, with a granular texture. A number of lesions seen in several organs in cases with previous chronic kidney impairment were attributed to long-standing uraemia and are not described here. The most striking pathological finding in all advanced cases of the disease was a marked reduction in kidney size and weight (Figure 1); in one extreme case, one organ weighed only 20 g. This process of kidney shrinkage is progressive, and in the early stage of the disease the organs can be of normal or just slightly smaller size (Figure 2). The kidneys are pale grey and hard to cut. On the cut surface, the cortex is reduced to a thin rim, in some cases no thicker than 1–2 mm. Cysts are frequently seen within the cortex. Invariably, both kidneys are involved, but the process is by no means bilaterally symmetrical; and occasionally paired kidneys vary considerably in weight and shape.

The earliest histopathological changes were observed in cases with no organ reduction on gross examination. There was an increased amount of light eosinophilic to amphophilic, acellular material throughout the interstitium of the cortex. Some of the proximal tubular

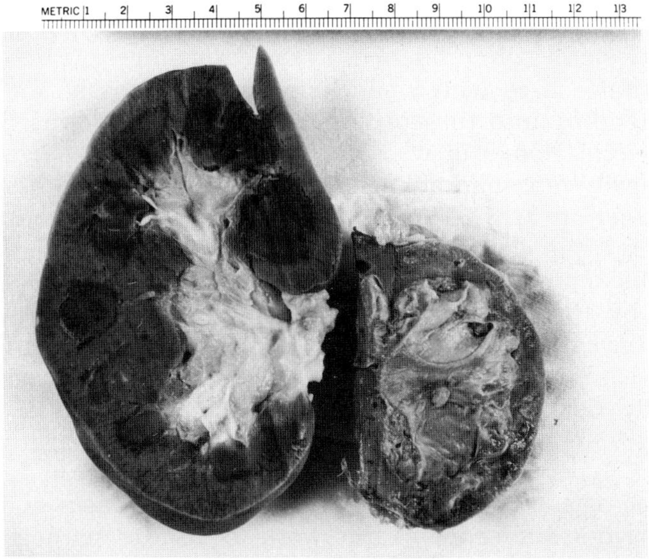

Figure 1. Dissected kidney (70 g) from a patient in an advanced stage of Balkan endemic nephropathy (right); a normal human kidney is shown on the left for comparison.

Both specimens fixed in formalin

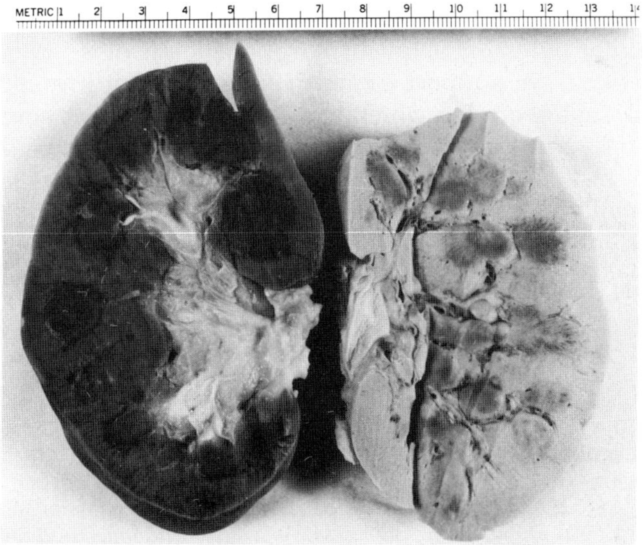

Figure 2. Dissected kidney from a patient in an early stage of Balkan endemic nephropathy (right); a normal human kidney is shown on the left for comparison.

Note the paleness and slight reduction in size of the diseased kidney. Both specimens fixed in formalin epithelial cells had undergone degenerative and regenerative processes, but neither inflammatory cell components nor evidence of primary glomerular lesions were prominent in the early stages.

In chronic cases with small kidneys seen grossly, diffuse cortical fibrosis extending into the corticomedullary junction was the principal finding. Several glomeruli were completely or partially hyalinized, and morphologically unremarkable glomeruli or glomeruli with small lesions could be seen nearby. The Bowman's capsule was usually slightly thickened, but the glomerular tufts appeared normal so it was assumed that the thickening of the capsule was due to external compression (Figure 3) by the increased amount of fibrous tissue (Figure 4). The epithelium was usually severely compromised, presenting several degenerative and necrotic changes. The inflammatory cell component was usually nonexistent or very scant, except in cases with superimposed pyelonephritis, mostly in patients with an associated tumour in the ureter. Hyperplastic arteriopathy of the onion-skin type, without fibrinoid necrosis, was seen frequently in advanced cases of the disease.

It is beyond the scope of this work to discuss pathogenesis of the lesions or to speculate about possible etiology. In the extensive medical literature on this disease, there is still no unanimous agreement about the basic pathomorphology of Balkan endemic nephropathy, and the different reports on this subject vary so widely that some authors (Sinđić, 1985), not without reason, wonder if they are all describing the same disease. A proper description of the morphological changes involved is imperative, as the understanding of any disease or reactive process usually relies on detailed morphologic studies. In order to fulfil this critical need, we support the proposal of several investigators to set up a central 'histoteque' to which centres would submit several sections from the same paraffin block of formalin-fixed tissues of interest. The collected slides could be evaluated by several pathologists and the results critically discussed, allowing a better understanding of basic pathological processes and facilitating the discovery of the etiology of the disease.

References

Belicza, M., Radonic, M. & Radošević, Z. (1979) Patoanatomski nalazi bubrega bolesnika umrlih od endemske nefropatije. In: II Simpozijum o endemskoj nefropatiji, Srpska Akademija Nauka i Umetnosti, Odelenje Medicinskih Nauka, Beograd, pp. 103–108

Čeović, S., Radonić, M., Hrabar, A., Radošević, Z., Bobinec, E., Pleština, R. & Habazin-Novak, V. (1979a) Endemic nephropathy in Brodska Posavina in a twenty year period. In: Strahinjić, S. & Stefanović, V., eds, *Endemic (Balkan) Nephropathy* (Proceedings of the 4th Symposium on Endemic (Balkan) Nephropathy), Niš, University of Niš, pp. 223–229

Čeović, S., Hrabar, A. & Radonić, M. (1979b) Endemic nephropathy in different ethnic groups in the area of Slavonski Brod. In: Strahinjić, S. & Stefanović, V., eds, *Endemic (Balkan) Nephropathy* (Proceedings of the 4th Symposium on Endemic (Balkan) Nephropathy, Niš, University of Niš, pp. 229–233

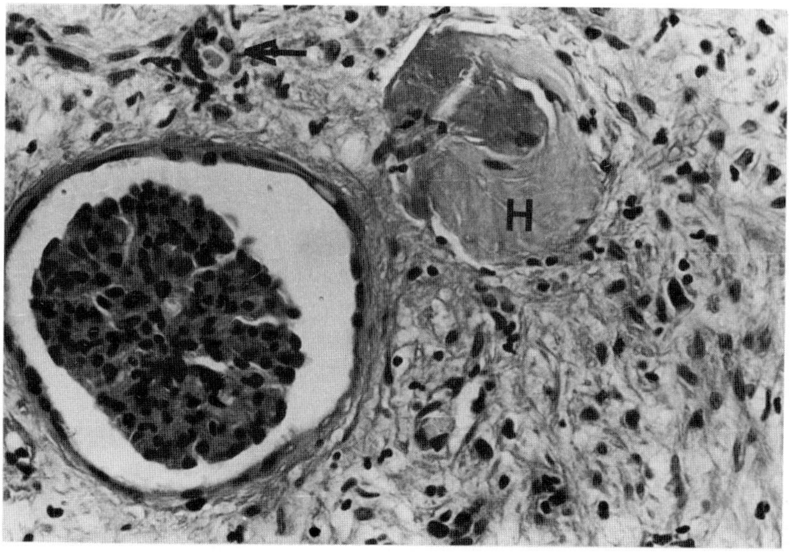

Figure 3. Photomicrograph of the kidney cortex of a patient with an advanced stage of Balkan endemic nephropathy

Note the incompletely hyalinized glomerulus (H) close to a relatively normal one (with absence of crescent formation). The arrow indicates a severely degenerated tubule containing proteinaceous material in the reduced lumen. Haematoxylin-eosin staining, 10 x 40 magnification, 7 μm section

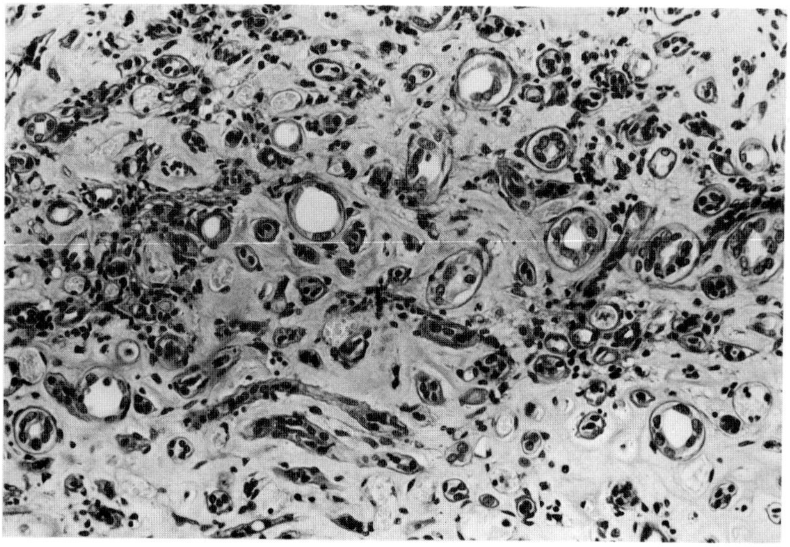

Figure 4. Photomicrograph from the same histological section as shown in Figure 3

Note the acellular, amorphous material compressing the tubuli. The majority of tubuli are undergoing degeneration. Haematoxylin-eosin staining, 10 x 20 magnification, 7 μm section

Čeović, S., Hrabar, A., Radonjić, M., Čeović, J., Mitar, I., Pleština, R., Bistrović, D. & Miletić-Medved, M. (1983) Distribution of Balkan endemic nephropathy in the region of Slavonski Brod, Croatia, Yugoslavia. In: Strahinjić, S. & Stefanović, V., eds, *Current Research in Endemic (Balkan) Nephropathy* (Proceedings of the 5th Symposium on Endemic (Balkan) Nephropathy, Nis, University Press, pp. 257–262

Lillie, R.D. (1954) *Histopathologic Technic and Practical Histochemistry*, New York, The Blakiston Co

Luna, L.G. (1979) *Manual of Histologic Staining Methods of the Armed Forces Institute of Pathology*, 3rd. ed., New York, McGraw-Hill

Radonić, M. (1961) *Proceedings of First International Congress of Nephropathy, Geneva, 1960*, Basel, Karger, pp. 528–537

Radonić, M., Radošević, Z. & Županić, V. (1966) Endemic nephropathy in Yugoslavia. In: *The Kidney* (International Academy of Pathology Monograph), Baltimore, Williams & Wilkins, pp. 503–522

Sinđić, M. (1985) Interpretacija morfoloskih promena u bubrezima kod endemske nefropatije. In: Petković, S., ed., III Simpozijum o endemskoj nefropatiji, Srpska Akademija Nauka i Umetnosti, Naučni Skupovi, Knjiga XXIII, Odelenje Medicinskih Nauka, Knjiga 3, Beograd, pp. 159–185

Vukelić, M. & Belicza, M. (1977) Analiza morfoloskih promjena bubrega u 21 slucaju endemske nefropatije. Zbornik radova VII Intersekcijskog sastanka patologa Jugoslavije,Kranjska Gora, 8–11 juna 1977, p. 38

Vukelić, M., Belicza, M. & Radonić, M. (1979) Patološkoanatomski nalazi bubrega bolesnika umrlih od endemske nefropatije i drugih bolesti na području zapadne brodske Posavine. In: Danilović, V., ed., II Simpozijum o endemskoj nefropatiji, Srpska Akademija Nauka i Umetnosti, Odelenje Medicinskih Nuka, Beograd, pp.109–115

Mycotoxins, Endemic Nephropathy and Urinary Tract Tumours
Ed. M. Castegnaro, R. Pleština, G. Dirheimer, I.N. Chernozemsky & H. Bartsch
Lyon, International Agency for Research on Cancer, pp. 43-46

HAEMATOLOGICAL FEATURES OF THE POPULATION OF THE AREA OF CROATIA, YUGOSLAVIA, ENDEMIC FOR BALKAN NEPHROPATHY

S. Pleština[1,2], A. Stavljenić[3], S. Čeović[4] & R. Fuchs[2]

[1]Zagreb University Hospital, [2]Institute for Medical Research and Occupational Health, University of Zagreb, [3]Medical Faculty, University of Zagreb, Zagreb, Yugoslavia; and [4]Department of Epidemiology, Slavonski Brod Medical Centre, Slavonski Brod, Yugoslavia

Summary

Normochromic, normocytic anaemia is a sign recognized as essential for the diagnosis of Balkan endemic nephropathy, although its relationship to the disease is still unclear. The aim of this study was to investigate whether a random sample of the population of a village endemic for nephropathy differed from the population of a village with no clinical case of nephropathy with respect to certain basic haematological parameters. During a screening campaign in 1984, 133 blood samples were collected from the endemic village and 40 from the nonendemic village and analysed for a number of haematological parameters; in 1991, 449 samples were collected in the endemic village and 156 in the nonendemic village and analysed for haemoglobin content and red clood cell count. Whereas in 1984 the haemoglobin content and red blood cell count were significantly lower in the endemic village ($p < 0.01$), in 1991 the erythrocyte count was much lower in the nonendemic village and there was no difference in haemoglobin concentration. Several controversial explanations, all of them speculative, are offered.

Introduction

Certain haematological disturbances occur commonly in cases of renal insufficiency. Anaemia is the most important of these changes. Its pathogenesis is complex: it appears to be caused both by depressed erythropoiesis, due to lack of erythropoietin, and by accelerated loss of red cells, due to mild haemolysis. Renal anaemias are usually normocytic and normochromic. Balkan endemic nephropathy is difficult to diagnose during its evolutive stages, before uraemia manifests. Because of its early onset, anaemia became one of the signs recognized as essential for epidemiological investigations and for diagnosis of the disease (Čeović *et al.*, 1979).

There is general agreement that the basic characteristics of anaemia in the advanced stages of Balkan endemic nephropathy are identical to those of anaemias associated with other chronic nephropathies. The interesting feature in patients suffering from Balkan endemic nephropathy, however, is the early onset of anaemia and the discrepancy between the degree of renal insufficiency and that of anaemia. No strong correlation could be established between the degree of anaemia and azotaemia (Radošević & Horvat, 1970). Furthermore, mild anaemia was present in patients without azotaemia and in people suspected of having nephropathy but with no subjective complaints or clinical evidence of disease (Radonić *et al.*, 1966).

These findings, which are contrary to current concepts of the pathogenesis of renal anaemias, raise the possibility that the unknown etiological agent in Balkan endemic nephropathy can damage both the kidney and the haematopoietic system. The aim of our study was to investigate whether an unselected sample of the population of a village endemic for nephropathy differed from the population of a village in a nonendemic area with respect to certain basic haematological parameters.

Subjects and Methods

During a screening campaign in the early spring of 1984, 133 blood samples were collected randomly from the population of the village of Kaniža, endemic for Balkan nephropathy, and 40 samples from the nonendemic village of Klakar. Erythrocyte, leukocyte and platelet counts, haematocrit, haemoglobin concentration, mean corpuscular volume and mean heamoglobin concentration were determined by routine methods at the Clinical Hospital Centre in Zagreb.

In 1991, 449 blood samples were collected in Kaniža and 156 in Klakar and analysed for haemoglobin content and erythrocyte count.

The results were evaluated statistically by calculating *p* values.

Results

The results are shown in Table 1. In 1984, the erythrocyte count, haematocrit and haemoglobin concentration were much lower in the population sample in the endemic village than in the nonendemic village ($p < 0.01$). In contrast, in 1991, the erythrocyte count was significantly lower in the nonendemic village than in the endemic one ($p < 0.01$), and there was no difference in haemoglobin concentration.

Table 1. Haematological parameters (mean ± SE) in villages endemic and nonendemic for Balkan nephropathy in 1984 and 1991

Village	Year	No. of subjects	Erythrocyte count (x 10^{12}/l)	Haemoglobin concentration (g/l)	Haematocrit (l/l)
Endemic	1984	133	4.46 ± 0.52	138 ± 16	0.41 ± 0.05
Nonendemic	1984	40	4.87 ± 0.52	151 ± 16	0.44 ± 0.04
Endemic	1991	449	4.65 ± 0.50	136 ± 13	–
Nonendemic	1991	156	4.40 ± 0.39	136 ± 13	–

The erythrocyte count in the endemic village was significantly higher in 1991 than in 1984 ($p < 0.01$), but it was still lower than that in the nonendemic village in 1984 ($p < 0.05$). The haemoglobin concentrations in the endemic village were similar in 1984 and 1991 and were much lower than those in the nonendemic village. The eryhtrocyte count and haemoglobin content were much lower in 1991 than in 1984 in the nonendemic village ($p < 0.01$).

Similar results were obtained in the endemic and the nonendemic villages for mean corpuscular volume (endemic, 92 fl; nonendemic, 92 fl), mean corpuscular haemoglobin (31.2 and 31.5 pg), mean corpuscular haemoglobin concentration (339 and 341 g/l) and leukocyte count (both 6.2 x 10^9/l). The platelet count was higher in the endemic village (309 x 10^9/l) than in the nonendemic village (279 x 10^9/l) ($p < 0.01$).

Discussion

The results obtained in 1984 indicate the presence of haematological disturbances in the endemic village; in contrast, in 1991 decreased red blood cell count and haemoglobin concentration were seen in the village in which there were no clinical cases of nephropathy. Moreover, more cases of proteinuria were found in Klakar in 1989 then in the endemic village (S. Čeović *et al.*, unpublished data). These findings raise the question of whether Klakar is still an appropriate control village or whether the hypothetical agent is now present in this village as well. Another possibility is that the agent has been leaving Kaniža, the endemic village. These results also raise the possibility that anaemia is not a valid indicator of the early stages of Balkan endemic nephropathy. The findings are, however, insufficient for drawing a firm conclusion, and further research should be undertaken.

References

Čeović, S., Radonić, M., Hrabar, A., Radošević, Z., Bobinac, E., Pleština, R. & Habazin-Novak, V. (1979) In: Strahinjić, S. & Stefanović, V., eds, *Endemic (Balkan) Nephropathy, Proceedings of the 4th Symposium on Endemic (Balkan) Nephropathy*, Niš, University of Niš, pp. 223–227

Radonić, M., Radošević, Z. & Županić, V. (1966) Endemic nephropathy in Yugoslavia. *Kidney*, **6**, 503–522

Radošević, Z. & Horvat, Z. (1970) Haematological changes in endemic nephropathy of south-east Europe. *Acta Med. Jug.*, **24**, 325–332

MYCOTOXINS AS ETIOLOGICAL AGENTS: ENVIRONMENTAL OCCURRENCE, ANIMAL MODELS AND ANALYSIS

Mycotoxins, Endemic Nephropathy and Urinary Tract Tumours
Ed. M. Castegnaro, R. Pleština, G. Dirheimer, I.N. Chernozemsky & H. Bartsch
Lyon, International Agency for Research on Cancer, pp. 49-56

PORCINE NEPHROPATHY IN EUROPE

B. Hald

*Department of Veterinary Microbiology,
Royal Veterinary and Agricultural University, Frederiksberg, Denmark*

Summary

Numerous surveys conducted in North America, Asia and Europe have revealed that ochratoxin A is a natural contaminant of plant products. Contamination frequencies of up to 40% have been encountered, at levels in the range of 5–500 µg/kg. Ochratoxin A is a major causal determinant of the disease porcine nephropathy; but other nephrotoxic mycotoxins, such as citrinin and the fungal quinones, may be involved. The disease is characterized clinically by polyuria and growth depression. Renal lesions in pigs include degeneration of the proximal tubules, interstitial fibrosis and hyalinization of the glomeruli. The disease is endemic, outbreaks being associated with bad weather conditions. A positive correlation has been observed between the prevalence rates of porcine nephropathy and the frequency of ochratoxin A in corresponding feed samples. Surveys for residues of ochratoxin A in kidneys from cases of porcine nephropathy in a number of European countries other than Denmark have demonstrated that 21–42% of samples contain ochratoxin A in the range of 1–100 µg/kg.

Introduction

Porcine nephropathy is a renal disease causally associated with feedborne ochratoxin A and other nephrotoxic mycotoxins. The disease was first seen more than 50 years ago in Denmark but has since been encountered in other European countries. This is not surprising, as ochratoxin A has been found to be a contaminant of cereals in all countries where surveys have been conducted.

Relationship between Porcine Nephropathy and Feed Contaminated with Ochratoxin A

Porcine nephropathy was recognized in Denmark as a disease entity in pigs in 1928 (Larsen, 1928). The disease was later reproduced by feeding rats and pigs barley cultures of an isolate of *Penicillium viridicatum* (i.e., by *P. verrucosum* (Frisvad, 1989)) from suspect Danish feed (Krogh & Hasselager, 1968). Examination of its metabolites revealed the presence of two nephrotoxins, oxalic acid and citrinin, in the cultures (Friis *et al.*, 1969). Kidney damage induced in pigs by citrinin, but not that induced by oxalate, resembled kidney damage observed in field cases (Krogh *et al.*, 1970). This observation led to the suggestion that citrinin, possibly in combination with other mycotoxins, was responible for porcine nephropathy. Van Walbeck *et al.* (1969) recognized ochratoxin A as another metabolite of *P. viridicatum*, and they therefore speculated that ochratoxin A might contribute to porcine nephropathy. Subsequent detection of ochratoxin A and citrinin in Danish feed (Krogh *et al.*, 1973) and of ochratoxin A in organs and tissues of pigs with nephropathy (Hald & Krogh, 1972) provided a solid basis for these assumptions.

It was concluded that ochratoxin A is the major disease determinant, since (i) porcine nephropathy can be reproduced in pigs by giving them feed containing ochratoxin A in the absence of citrinin (Krogh *et al.*, 1971), (ii) citrinin causes a lower incidence of nephrotoxicity than ochratoxin A, (iii) the levels of citrinin in Danish feed are lower than those of ochratoxin A and (iv) citrinin is less stable than ochratoxin A.

Natural Occurrence of Nephrotoxic Mycotoxins in Plant Products

Ochratoxins are a group of seven isocumarin derivates linked by an amide bond to the amino group of L-ß-phenylalanine. Ochratoxin A is the dominant and most toxic member of the family. Ochratoxins have been found to be produced by six species of *Penicillium* and seven species of *Aspergillus* (Krogh, 1978).

Ochratoxin A has been detected in a variety of foods and feeds, mostly from countries with temperate or continental climates (Krogh, 1978). The prevalence of ochratoxin A in grains used for food is generally low, and levels found were mostly lower than 500 µg/kg. Ochratoxin A has also been isolated from crops in warmer climatic zones, such as Yugoslavia and Australia (Krogh, 1987). In feed suspected of being contaminated, either because of mouldiness or because of its association with porcine nephropathy, levels up to 27 500 µg/kg have been reported, with ochratoxin A occurring in more than 50% of the samples (Krogh *et al.*, 1973).

Citrinin is a frequent co-contaminant with ochratoxin A. No satisfactory analytical method exists for quantifying citrinin; however, it has been found in samples analysed by unvalidated procedures. These results should be interpreted as indicating the widespread exposure to this mycotoxin (Pohland *et al.*, 1987).

Citrinin has been isolated from fungal strains of 14 species of *Penicillium* and three species of *Aspergillus* (Krogh, 1978). Of particular interest is *P. viridicatum*, because this species produces the nephrotoxic compounds xanthomegnin and viomellein (Hald *et al.*, 1983).

Nephrotoxic mycotoxins can be produced in most plant products, and production is governed by two factors: temperature and the water activity (A_w) of the substrate. The water content of cereals is dependent on a number of factors, such as climatic conditions during growth and harvest and harvesting and storage techniques. Since the requirements for the production of nephrotoxic mycotoxins are thus relatively nonspecific, moulds can produce toxins on almost any feedstuff that will support growth, as exemplified by the diverse types of commodities that become contaminated with nephrotoxic mycotoxins.

Regulatory control of nephrotoxic mycotoxins in animal feeds in European countries was summarized by van Egmond in 1987, at which time three countries had regulations for ochratoxin A levels. By 1990, at least nine countries had proposed or enforced official limits for ochratoxin A (see van Egmond, this volume). No guidelines exist in Denmark at present for nephrotoxic mycotoxins in agricultural commodities, but a control programme has been established. All kidneys that display macroscopic lesions indicating porcine nephropathy are analysed, and the entire carcass is condemned if the ochratoxin A concentration exceeds 25 μg/kg. This threshold value has been effective since 1979.

Porcine Nephropathy

Clinical aspects

Porcine nephropathy is a naturally occurring disease in Europe. In practically all reports on field cases of porcine nephropathy, the diagnosis was based on renal pathology and chemical analysis for ochratoxin A residues. The disease is characterized clinically by impairment of proximal renal function, glucosuria, proteinuria, and decreased maximal tubular clearance of *para*-aminohippurate, clearance of inulin, ability to concentrate urine and growth depression. When the nephrotoxic components are removed from the diet, the growth curve of the pigs returns to normal shape, but the polyuria continues.

Pathology

At slaughter, both kidneys show equally developed changes, and no change is found in other organs or tissues. The kidneys are enlarged and pale, the colour change ranging from slight to pronounced paleness, with white patches. The surface is smooth or may show development of vesicles. On the cut surface, large amounts of connective tissue may be observed in the cortex, and cysts up to 1 mm in diameter may be seen throughout the cortex. In advanced cases, the texture is increased.

Histologically, the disease is characterized by tubular degeneration and atrophy, affecting primarily the proximal tubules; interstitial, peritubular and periglomerular fibrosis and hyalinization of glomeruli are seen in advanced cases. These lesions are not specific for any single agent (Krogh, 1978).

Occurrence of Ochratoxin A in Animal Tissues

The transfer of ochratoxin A from feed to animal was elucidated in studies in which pigs were exposed to dietary levels of ochratoxin A. At slaughter, ochratoxin A was present in the animals, at decreasing levels in blood, kidney, lean meat, liver and fat.

The statistical association between ochratoxin A in kidney, lean meat, liver and fat was calculated on the basis of the ochratoxin A content in blood collected at slaughter from 284 pigs and is represented in Table 1 by regression analysis. No intercept was assumed. R^2 varied from 0.21 to 0.79 (Mortensen *et al.*, 1983).

In Denmark, both the occurrence of porcine nephropathy and condemnation rates due to ochratoxicosis declined steadily between 1980 and 1982 (Table 2). This positive development was attributed

Table 1. Regression of ochratoxin A in porcine kidney, lean meat, liver and fat on ochratoxin A in serum at slaughter (n = 284)

Kidney	ochratoxin A ppb = 0.0651 x serum ochratoxin A ppb S_b = 0.002
Lean meat	ochratoxin A ppb = 0.0346 x serum ochratoxin A ppb S_b = 0.001
Liver	ochratoxin A ppb = 0.0259 x serum ochratoxin A ppb S_b = 0.001
Fat	ochratoxin A ppb = 0.0181 x serum ochratoxin A ppb S_b = 0.001

S_b, standard error of slope

mainly to improved harvest and storage practices introduced by farmers, who suffer the full financial loss of condemnations. Therefore, it came as a surprise when in 1983 the incidence of porcine nephropathy increased sharply, resulting in almost 10 times as many condemnations as in the previous year.

Table 2. Prevalence of porcine nephropathy and condemnation rates due to ochratoxin A in Denmark, 1980–90[a]

Year	No. of pigs slaughtered (millions)	No. of nephropathic kidneys analysed	No. of kidneys with ochratoxin A (µg/kg)		Condemnations	
			< 10	10–25	No.	%
1980	14.35	6 845			1 725	25
1981	14.48	7 645			841	11
1982	14.28	2 336			229	10
1983	14.99	7 639	4 294	1 155	2 190	29
1984	14.65	1 298	1 116	54	128	10
1985	15.08	816	701	36	79	10
1986	15.98	5 264	3 120	842	1 302	25
1987	15.96	8 705	5 159	1 432	2 114	25
1988	16.08	41 288			7 807	19
1989	15.84	6 809	4 884	1 258	667	10
1990	15.93	3 138	3 010	97	128	4

[a]Quoted with the permission of the Danish National Veterinary Services

This increase was extremely well defined geographically. When the northern half of Jutland was compared with the rest of Denmark, the total number of pigs slaughtered was comparable (Table 3); however, the rate of nephropathy in northern Jutland was more than five times and the condemnation rate more than seven times higher than in the rest of the country. The spread of the outbreak reflects the unfavourable climatic conditions which prevailed in the northern part of Jutland during the harvest of 1982 (Bennetsen, 1982). Data for 1984–85 indicated that the general downward trend in occurrence was re-established; but in 1986, 1987, 1989 and particularly in 1988, the incidence of porcine nephropathy increased again. These outbreaks also reflected wet climatic conditions, which prevailed during the summer and harvest time in 1986–89. The data for 1990 show the lowest percentage condemnation rate in 10 years.

Surveys in a number of other European countries for residues of ochratoxin A in kidneys from cases of porcine nephropathy, based on meat inspection data (Table 4), revealed that 42% of samples contained ochratoxin A in the range of 0.1–100 µg/kg.

Table 3. Incidence of porcine nephropathy in Denmark, 1983

Area	No. of pigs slaughtered	No. of nephropathic kidneys analysed	Condemnations	
			No.	Per 100 000 slaughters
Northern Jutland	7 525 000	6426	1927	25.6
Rest of Denmark	7 470 000	1213	263	3.5

Table 4. Residues of ochratoxin A in kidneys from cases of porcine nephropathy, based on data from meat inspections

Country	No. of kidneys investigated	% containing ochratoxin	Ochratoxin A (μg/kg, range)	Reference
Belgium	385[a]	18	0.2–12	Rousseau *et al.* (1989)
Germany	104	21	0.1–1.8	Bauer *et al.* (1984)
Hungary	122	39	2–100	Sandor *et al.* (1982)
Poland	113	23	≤ 23	Goliński *et al.* (1984)
Poland	122	42	1–10	Goliński *et al.* (1985)
Sweden	129	25	2–104	Rutquist *et al.* (1977)
Sweden	90	27	2–88	Josefsson (1979)

[a] Suspected

The finding of ochratoxin A in renal tissue showing nephropathy lesions has been thought to indicate its causal role and to provide the basis for an etiological diagnosis. Strictly speaking, however, the presence of residues of ochratoxin A in the kidney merely indicates exposure to the compound immediately before analysis and does not necessarily demonstrate involvement of ochratoxin A in the pathogenesis of nephropathy.

Ochratoxin A is a short-lived metabolite in the pig, as demonstrated experimentally (Galtier, 1979). The data presented in Tables 2 and 4 illustrate further problems. In the case of a recent poisoning, the kidney appeared normal at slaughter and thus escaped analysis altogether, even though the carcass may well have contained a high level of ochratoxin A. During the later stages of porcine nephropathy, in cases in which farmers have held their pigs on an improved diet, the residue levels are low but the kidneys remain degenerated (Elling, 1983). This fact might explain the observation that ochratoxin A is detected in only 4–42% of cases of porcine nephropathy seen in the field. Another explanation may be that mycotoxins other than ochratoxin A, such as citrinin, viomellein, xanthomegnin, and other metabolites from *Penicillium* and *Aspergillus* are causally involved in porcine nephropathy.

Biochemical Effects

An etiological diagnosis of porcine nephropathy should be based on biochemical alterations rather than structural and functional changes: Although the target segment of the nephron is clearly identified, the impairment is not specific for ochratoxin A. Biochemical investigations have revealed changes in gluconeogenesis and decreased activity of renal cytosolic phosphoenolpyruvate carboxykinase and γ-glutamyl transpeptidase, a brush-border enzyme found in proximal tubules, as early signals of renal damage in pigs (Krogh *et al.*, 1988), and these enzymes can be considered to be sensitive indicators of ochratoxin A-induced porcine nephropathy. The activity of phosphoenolpyruvate carboxykinase is not affected by citrinin.

Endo (1983) showed with isolated renal nephron preparations that ochratoxin A promotes the release of alanine aminotransferase, followed by leucine aminopeptide and γ-glutamyl transpeptidase. His studies also indicate that ochratoxin A disturbs the membrane transport system of *para*-hippuric acid.

Conclusion

Ochratoxin A has nephrotoxic potential in all monogastric animals studied so far. It may similarly be causally involved in human renal disease, and ochratoxicosis is suspected. Other mycotoxins, such as citrinin and the fungal quinones, may also be involved as causal determinants.

Assuming that porcine nephropathy represents a valid model for human kidney disease, measurements of the activity of renal phosphoenolpyruvate carboxykinase, γ-glutamyl transpeptidase and perhaps alanine aminotransferase could provide the basis for a sensitive, selective diagnostic test for early stages of ochratoxin A-induced disease in humans.

References

Bauer, J., Gareis, M. & Gedek, B. (1984) Detection and occurrence of ochratoxin A in slaughter pigs. *Berlin München Tierärztl. Wochenschr.*, **97**, 279–283 (in German)

Bennetsen, A. (1982) *Results of field experiments in the Farmers' Organization (Aarhus)* Aarhus, p. 12 (in Danish)

van Egmond, H.P. (1987) Current situation on regulations for mycotoxins: overview of tolerances and status of standard methods for scanning and analysis. In: *Joint FAO/WHO/UNEP 2nd International Conference on Mycotoxins, Bangkok, Thailand*, Rome, Food and Ahricultural Organization, p. 35

Elling, F. (1983) Feeding experiments with ochratoxin A contaminated barley to bacon pigs. IV. Renal lesions. *Acta Agric. Scand.*, **33**, 153–159

Endo, H. (1983) Effect of ochratoxin A on enzyme release from isolated nephron segments in rats. *Proc. Jpn. Assoc. Mycotoxicol.*, **18**, 18 (in Japanese)

Friis, P., Hasselager, E. & Krogh, P. (1969) Isolation of citrinin and oxalic acid from Penicillium viridicatum Westling and their nephrotoxicity in rats and pigs. *Acta Pathol. Microbiol. Scand.*, **77**, 559–560

Frisvad, J.C. (1989) The connection between the Penicillia and Aspergilli and mycotoxins with special emphasis on misidentified isolates. *Arch. Environ. Contam. Toxicol.*, **18**, 452–467

Galtier, P. (1979) Toxicological and pharmacokinetic study of a mycotoxin, ochratoxin A. Thesis, University of Toulouse, France

Goliński, P., Hult, K., Grabarkiewicz-Szczęsna, J., Chełkowski, J., Kneblewski, P. & Szebiotko, K. (1984) Mycotoxic porcine nephropathy and spontaneous occurrence of ochratoxin A residues in kidneys and blood of Polish swine. *Appl. Environ. Microbiol.*, **47**, 1210–1212

Goliński, P., Hult, K., Grabarkiewicz-Szczęsna, J., Chełkowski, J. & Szebiotko, K. (1985) Spontaneous occurrence of ochratoxin A residue in porcine kidney and serum samples in Poland. *Appl. Environ. Microbiol.*, **49**, 1014–1015

Hald, B. & Krogh, P. (1972) Ochratoxin residues in bacon pigs. In: *IUPAC Symposium on the Control of Mycotoxins, Kungsälv, Sweden*, p. 18 (unpublished)

Hald, B., Christensen, D.H. & Krogh, P. (1983) Natural occurrence of the mycotoxin viomellein in barley and the associated quinone-producing Penicillia. *Appl. Environ. Microbiol.*, **46**, 1311–1317

Josefsson, E. (1979) Examination of ochratoxin A in pig kidneys. *Var Föda*, **31**, 415–420 (in Swedish)

Krogh, P. (1978) Causal associations of mycotoxic nephropathy. *Acta Pathol. Microbiol. Scand. A*, **Suppl. 269**, pp. 1–28

Krogh, P. (1987) Ochratoxins in food. In: Krogh, P., ed., *Mycotoxins in Food, Food Science and Technology*, New York, Academic Press, pp. 97–121

Krogh, P. & Hasselager, E. (1968) Studies on fungal nephrotoxicity. *R. Vet. Agric. Coll. Yearb.*, **116**, 198–214

Krogh, P., Hasselager, E. & Friis, P. (1970) Studies on fungal nephrotoxicity. II. Isolation of two nephrotoxic compounds from Penicillium viridicatum Westling: citrinin and oxalic acid. *Acta Pathol. Microbiol. Scand. B*, **78**, 401–413

Krogh, P., Axelsen, N.H., Elling, F., Gyrd-Hansen, N., Hald, B., Hyldgaard-Jensen, J., Larsen, A.E., Madsen, A., Mortensen, H.P. & Aalund, O. (1971) Experimental porcine nephropathy: changes of renal function and structure induced by ochratoxin A-contaminated feed. *Acta Pathol. Microbiol. Scand. A*, **Suppl. 246**, pp. 1–21

Krogh, P., Hald, B. & Pedersen, E.J. (1973) Occurrence of ochratoxin A and citrinin in cereals associated with mycotoxic porcine nephropathy. *Acta Pathol. Microbiol. Scand. B*, **81**, 689–695

Krogh, P., Gyrd-Hansen, N., Hald, B., Larsen, S., Nielsen, J.P., Smith, M., Ivanoff, C. & Meisner, H. (1988) Diagnostic potential of phosphoenolpyruvate, carboxykinase and gamma-glutamyl transpeptidase activity. *J. Toxicol. Environ. Health*, **23**, 1–14

Larsen, S. (1928) On chronic degeneration of the kidneys caused by mouldy rye. *Maanedsskr. Dyrl.*, **40**, 259–284, 289–300 (in Danish)

Mortensen, H.P., Hald, B. & Madsen, A. (1983) Feeding experiments with ochratoxin A contaminated barley for bacon pigs. 5. Ochratoxin A in pig blood. *Acta Agric. Scand.*, **33**, 235–239

Pohland, A.E. & Wood, G.E. (1987) Occurrence of mycotoxins in foods. In: Krogh, P., ed., *Mycotoxins in Food, Food Science and Technology*, New York, Academic Press, pp. 35–65

Rousseau, D.M. & van Peteghem, C.H. (1989) Spontaneous occurrence of ochratoxin A residues in porcine kidneys in Belgium. *Bull. Environ. Contam. Toxicol.*, **42**, 181–186

Rutquist, L., Björklund, N.E., Hult, K. & Gatenbeck, S. (1977) Spontaneous occurrence of ochratoxin residues in kidneys of fattening pigs. *Zentralbl. Vet. Med.*, **A24**, 402–408

Sandor, G., Glavits, R., Vajda, L., Ványi, A. & Krogh, P. (1982) Epidemiologic study of ochratoxin A-associated porcine nephropathy in Hungary. In: *Proceedings of the 5th IUPAC Symposium on Mycotoxins and Phycotoxins, Vienna, Austria*, pp. 349–352 (unpublished)

Van Walbeck, W., Scott, P.M., Harwig, J. & Lawrence, J.W. (1969) Penicillium viridicatum Westling: a new source of ochratoxin A. *Can. J. Microbiol.*, **15**, 1281–1285

Mycotoxins, Endemic Nephropathy and Urinary Tract Tumours
Ed. M. Castegnaro, R. Pleština, G. Dirheimer, I.N. Chernozemsky & H. Bartsch
Lyon, International Agency for Research on Cancer, pp. 57-70

METHODS FOR DETERMINING OCHRATOXIN A AND OTHER NEPHROTOXIC MYCOTOXINS

H.P. van Egmond

National Institute of Public Health and Environmental Protection, Laboratory for Residue Analysis, Bilthoven, The Netherlands

Summary

Chemical assays are of major importance for the determination of mycotoxins. Generally, all chemical methods for the analysis of mycotoxins include the basic steps of extraction, clean-up, separation, detection, quantification and confirmation of identity. The various approaches that exist for the determination of nephrotoxic mycotoxins, and in particular the ochratoxins, are discussed below. In conventional procedures, clean-up is usually achieved by liquid–liquid extraction or adsorption column chromatography, followed by thin-layer chromatography or high-performance liquid chromatography and ultraviolet or fluorescence detection. The recent introduction of methods based on immunochemical principles has had a large impact on analytical methodology for mycotoxins, including the ochratoxins, in both the purification and determination steps. The enzyme-linked immunosorbent assay approach for screening and (semi-)quantitative determination and the immunoaffinity column approach for rapid clean-up followed by conventional instrumental analysis are rapidly gaining ground. These techniques also offer possibilities for automated systems. To assist analysts in improving their data in mycotoxin research, the Community Bureau of Reference of the European Commission has produced several reference materials for mycotoxins; others, such as a reference material for ochratoxin A in grains, are in development.

Introduction

The availability of methods of analysis plays a key role in survey and research programmes for mycotoxins. In this paper, some approaches are discussed that are applied in practice to determine nephrotoxic mycotoxins (see Figure 1). Rather than attempt to cover exhaustively all of the methods that are available for analysing these toxins, the discussion is directed to general analytical aspects and examples of major conventional and novel analytical techniques that are used to determine nephrotoxins, and particularly ochratoxin A, in various matrices.

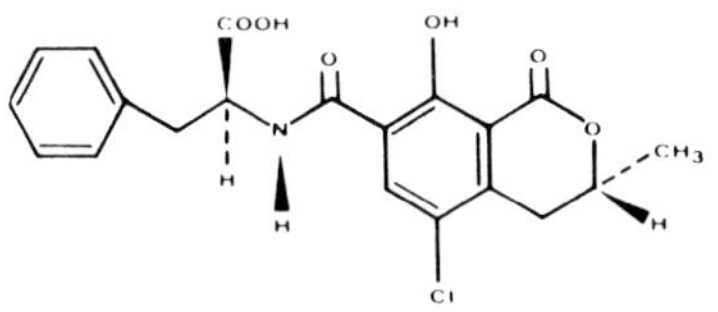

Ochratoxin A

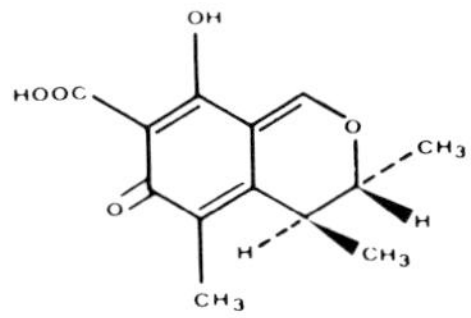

Citrinin

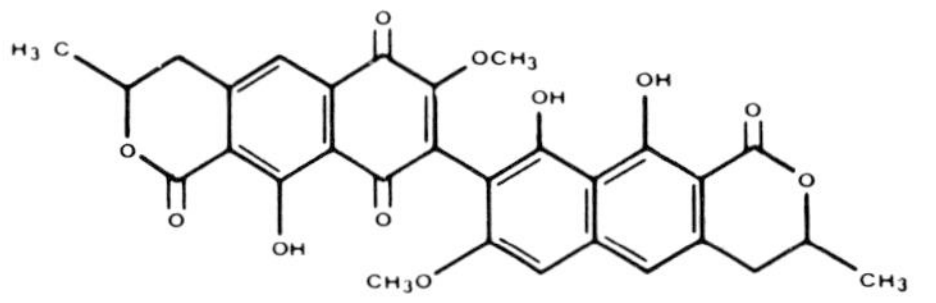

Viomellein

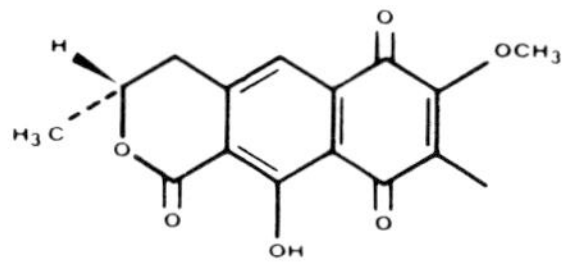

Xanthomegnin

Figure 1. Chemical structures of some nephrotoxic mycotoxins

Because mycotoxins are usually present in agricultural commodities and biological materials in concentrations ranging from (sub-) micrograms to milligrams per kilogram, the possibilities for determining them are limited to certain trace analytical methods. Both biological and chemical procedures have been developed. Biological methods can be useful in screening for mycotoxins, especially when the identity of the mycotoxin(s) is not known; if the identity is known, (bio-)chemical assays, if available, are preferable. These are generally much more specific, more rapid, more reproducible and have lower limits of detection. Hence, the biological methods are not further discussed in this article.

Analytical Procedures

Several factors determine and limit the analyst's choice of an appropriate analytical procedure for a certain mycotoxin. These include variations in the composition of the materials to be analysed and requirements for the utility and reliability of the method.

In spite of the differences, there are certain similarities in these analytical procedures. The basic steps are:

Sampling
|
Extraction
|
Clean-up
|
Concentration
|
Ultimate separation
|
Detection and quantification
|
Confirmation of identity

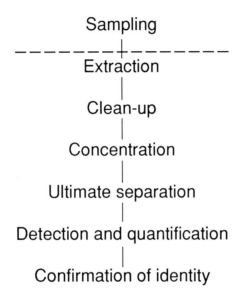

Although sampling is the most important, it is not discussed here because of the limited scope of this article. Test portions that are taken for analysis usually vary in weight from 25 to 100 g, a range that results from a compromise between homogeneity and practical requirements. For biological materials such as blood and milk, which may be analysed for ochratoxin A, smaller test portions are often taken—a few millilitres or even less.

Extraction

The first laboratory step in chemical analysis involves extraction of the test portion, to separate the mycotoxin from the bulk of the matrix components in a manageable form. The solvent and the solid substrate are put in contact either for a short period (1–3 min) in a high-speed blender or for a longer period (30 min) by shaking in a flask. Liquids can be extracted in a separatory funnel or by applying (prepacked) solid-phase extraction columns.

The choice of the solvent depends on the chemical properties of the matrix, the toxin and the choice of clean-up and determination steps. Often, mixtures of solvents or solvents with small amounts of water or acid are found to be the most effective. To improve extraction of ochratoxin A from animal tissues, enzymic digestion is also used (Hunt *et al.*, 1979).

Ochratoxin A present in acidified commodities is readily soluble in many organic solvents, and this characteristic has been used as the basis for extraction in several methods. In a method recommended by IARC (Egan *et al.*, 1982), ochratoxin A is extracted from ground samples with chloroform, after acidification with aqueous H_3PO_4. Chloroform is also often used to extract ochratoxin A from biological samples such as pig kidney (Paulsch *et al.*, 1982), human plasma (Breitholtz *et al.*, 1991) and milk (Gareis *et al.*, 1988). Acidified chloroform is used to extract the naphthoquinone mycotoxins xanthomegnin, viomellein and vioxanthin from cereals and animal feeds (Carman *et al.*, 1984; Scudamore *et al.*, 1986), and for citrinin (Hald & Krogh, 1975). For citrinin, acetonitrile is also often used as an extraction solvent, in combination with aqueous solvents to which sodium ethylenediaminetetraacetic acid (Trantham & Wilson, 1984) or glycolic acid (Gimeno, 1984) may be added to improve recovery. Analysis for citrinin is relatively difficult because it is often not efficiently extracted and is thermally unstable (Trantham & Wilson, 1984).

When immunoaffinity chromatography is used to purify extracts of ochratoxin A and when enzyme-linked immunosorbent assay (ELISA) is used to detect ochratoxin A, water-miscible extraction solvents are often used, such as methanol (Lee & Chu, 1984) and acetonitrile (Ramakrishna *et al.*, 1990). Extracts thus obtained are easily diluted with water, and the immunochemical reaction takes place in an aqueous environment.

Clean-up

After the test portion has been extracted, the usual next step is partial purification of the extract to remove lipids and other substances that may interfere in the determination. The purification step is omitted in some screening procedures, such as the flow injection method of Hult *et al.* (1984) for rapid detection of ochratoxin A in human sera, as well as in certain ELISA methods (see section on immunoassays). When chromatographic methods are used, column clean-up and liquid–liquid extraction are often used to clean extracts containing mycotoxins, including the nephrotoxins. In the method recommended by IARC for ochratoxin A (Egan *et al.*, 1982), ochratoxins are trapped in a laboratory-prepared column containing diatomaceous earth impregnated with $NaHCO_3$ solution. Extraneous compounds are washed off the column with hexane and chloroform, and the ochratoxins are eluted with benzene:acetic acid (98:2, v/v). In the procedure of Paulsch *et al.* (1982) for the determination of ochratoxin A in kidneys of swine, a liquid–liquid partitioning step is used instead. For the clean-up of citrinin-containing extracts of grains, both hydrophilic matrices impregnated with $NaHCO_3$ solution (Dick *et al.*, 1988) and liquid–liquid partitioning (Trantham & Wilson, 1984) are used.

More recently, solid-phase extraction and chromatography cartridges, such as SEP-PAK® and Baker®, have become popular. The sample extract is usually added to the cartridge in an appropriate solvent; then, the cartridge is washed with one or more solvents in which the toxins are insoluble or less soluble than the impurities. The solvent composition is subsequently changed in such a way that the toxins are selectively eluted from the cartridge, and the eluate is collected. Examples of analytical procedures in which solid-phase extraction columns are used in the clean-up step are those of Cohen and Lapointe (1986) for ochratoxin A (silicagel cartridge and cyano cartridge) and Carman *et al.* (1984) for xanthomegnin (silicagel cartridge).

The most recent advance in clean-up of extracts containing mycotoxins is use of immunoaffinity cartridges (Gilbert, 1991), and the first commercial prototypes for ochratoxin A have become available. These columns are composed of monoclonal antibodies, specific for ochratoxin A, which are immobilized on Sepharose® and packed into small plastic cartridges. The principle of the extraction is that the crude extract is forced through the column and ochratoxins are left bound to the recognition site of the immunoglobulin (see Figure 2). Extraneous material can be washed off the column with water or aqueous buffer, and the ochratoxins are finally eluted with acetonitrile. The volume of the acetonitrile eluate can be adjusted and the eluate injected directly onto a high-performance liquid chromatography (HPLC) system for final separation and determination of the toxin. Irrespective of the matrix to be analysed, the approach is much the same. At the time of writing, it was too early to judge the value of this new method for clean-up for ochratoxin A, but the first practical results may soon be expected from interlaboratory studies coordinated by the Community Bureau of Reference (see section on reference materials).

Both solid-phase extraction cartridges and immunoaffinity cartridges can be incorporated in fully automated sample preparation systems that take the sample from the extraction stage through to completion of HPLC in an unattended mode of action.

Ultimate separation, detection and quantification

Despite extraction and clean-up, the final extract may contain co-extracted substances that might interfere with the determination of mycotoxins. Several possibilities exist for separating the toxin(s) from the matrix components, to allow qualitative and quantitative determination.

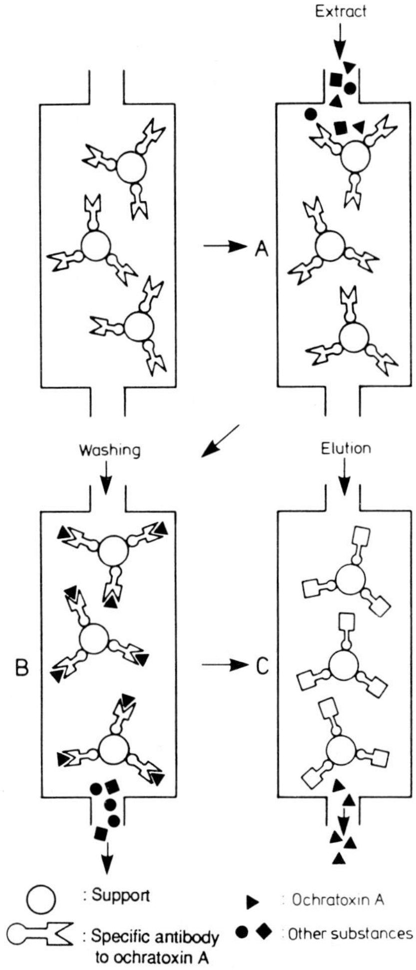

Figure 2. Schematic diagram of immunoaffinity chromatography for concentration and purification of ochratoxin A

Test extract containing ochratoxin A is first loaded onto the affinity gel column containing antibody against ochratoxin A (A). After washing to remove impurities (B), ochratoxin A is eluted from the column with acetonitrile (C).

Chromatographic procedures, which are physical separation techniques, are most often used for nephrotoxins, and they are used in combination with visual or instrumental determination of the mycotoxin(s) of interest. Another physicochemical technique that has been used for ochratoxin A involves spectrophotometry. Newer methods are the immunoassays. These biochemical separation techniques, which are used in combination with instrumental determination, are rapidly gaining ground for the determination of ochratoxin A.

Chromatographic techniques: The most important types of chromatography for the determination of nephrotoxic mycotoxins are thin-layer chromatography (TLC) and HPLC. Although methods have been published in which a mini-column was used in the determination step of ochratoxin A (Hald & Krogh, 1975; Holaday, 1976), these methods have not had wide application in practice.

In the first years of mycotoxin research, TLC was very common and popular for separating the components of extracts, and it still has many applications. Initially, separations were carried out in one dimension, using a single developing solvent. Later, two-dimensional TLC was introduced for the determination of mycotoxins, such as for ochratoxin A in pig kidneys (Paulsch *et al.*, 1982). It is a powerful separation technique, which provides much better separation than one-dimensional TLC and is useful especially when low levels must be detected in matrices that contain many interfering substances, such as feeds.

Ochratoxin A is readily detected after TLC by placing the developed TLC plate under longwave ultraviolet light, which brings out blue-green fluorescence on the ochratoxin spots (see Figure 3). On exposure to base (e.g., vapours of ammonia or methanol-ammonia) a more intense blue fluorescent spot is observed. In the method of Paulsch *et al.* (1982), a simple confirmatory test is included, which is based on the formation of ochratoxin A methylester on the TLC plate. With TLC, the limits of detection for ochratoxin A in plant and animal products are in the microgram per kilogram range.

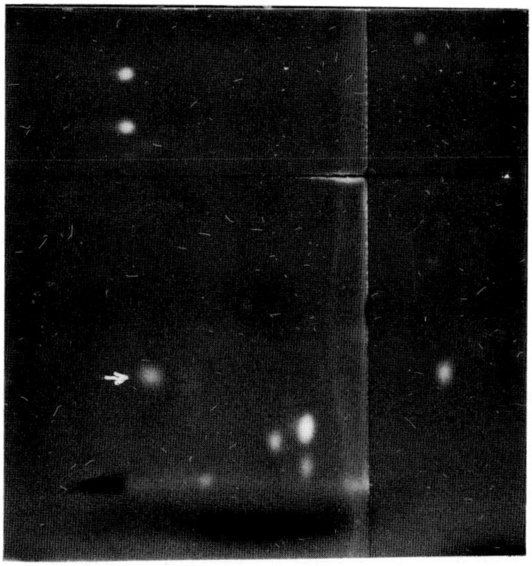

Figure 3. Result of two-dimensional thin-layer chromatography of an extract of kidney (spiked with ochratoxin A at 10 µg/kg), as applied by Paulsch *et al.* (1982)

TLC is also used for the detection of citrinin, which readily fluoresces yellow under longwave ultraviolet light. One problem is that spots of citrinin on TLC plates are unstable and show excessive streaking. Several techniques have been proposed to overcome this problem. In one, the TLC plate is impregnated with oxalic acid (Marti *et al.*, 1978); in another (Stubblefield, 1979), the plate is impregnated with ethylene diaminetetraacetic acid. Gimeno (1984) described a method for analysing citrinin in corn and barley, in which TLC was done with plates impregnated with glycolic acid in ethanol. Although these modifications tend to reduce streaking and tailing of the citrinin spots, none is completely satisfactory. Limits of detection for citrinin in agricultural commodities are around 50 µg/kg.

Use of TLC as a technique for separating (nephrotoxic) mycotoxins from matrix components has decreased in recent years in favour of liquid chromatography and immunoassays. Nevertheless, TLC remains a major separation technique in mycotoxin research. It is particularly recommended to scientists who cannot afford to purchase sophisticated analytical instrumentation.

Liquid chromatography separations are usually much better than those obtained with one-dimensional TLC. HPLC methods generally provide good quantitative information, and the equipment involved in HPLC systems can be automated relatively easily. To separate (nephrotoxic) mycotoxins from matrix components, bonded phase packings, e.g., octadecyl silane (ODS, C18) are most often used.

For ochratoxin A, an acidic mobile phase in conjunction with fluorescence detection with excitation and emission wavelengths at 330 and 460 nm, respectively (Cohen & Lapointe, 1986), usually offers limits of detection in the microgram per kilogram range or even less. The limit of detection may be improved by treating the column effluent with a 10% solution of ammonia (Hunt *et al.*, 1979). Thermospray quadrupole mass spectrometry has been described as an alternative to fluorescence detection (Rajakylae *et al.*, 1987). Although this advanced detection system is very selective and sensitive, its practical use is limited, however, to those laboratories which can afford to purchase these expensive instruments.

The carboxyl group in ochratoxin A makes it impossible to use reversed-phase chromatography under neutral or alkaline conditions with an ordinary liquid chromatographic technique: ochratoxin A would not be retained on the column under such conditions. Addition of a counter-ion to the mobile phase, however, allows formation of a neutral complex, which is retained on the column. This approach was followed by Breitholtz *et al.* (1991) for the liquid chromatographic determination of ochratoxin A in human plasma.

Reversed-phase ion-pair liquid chromatography is also used for the determination of citrinin in grains with ultraviolet detection at 254 nm, yielding a limit of detection of 20 µg/kg (Nakagawa *et al.*, 1982).

Liquid chromatography is also used to separate the naphthoquinone mycotoxins, usually, with C8 or C18 reverse-phase columns, whereas ultraviolet or electrochemical detection is used to detect the toxins (Carman *et al.*, 1984; Scudamore *et al.*, 1986). Figure 4 shows a chromatogram of an extract of barley obtained after separation on a C18 column and ultraviolet detection at 390 nm.

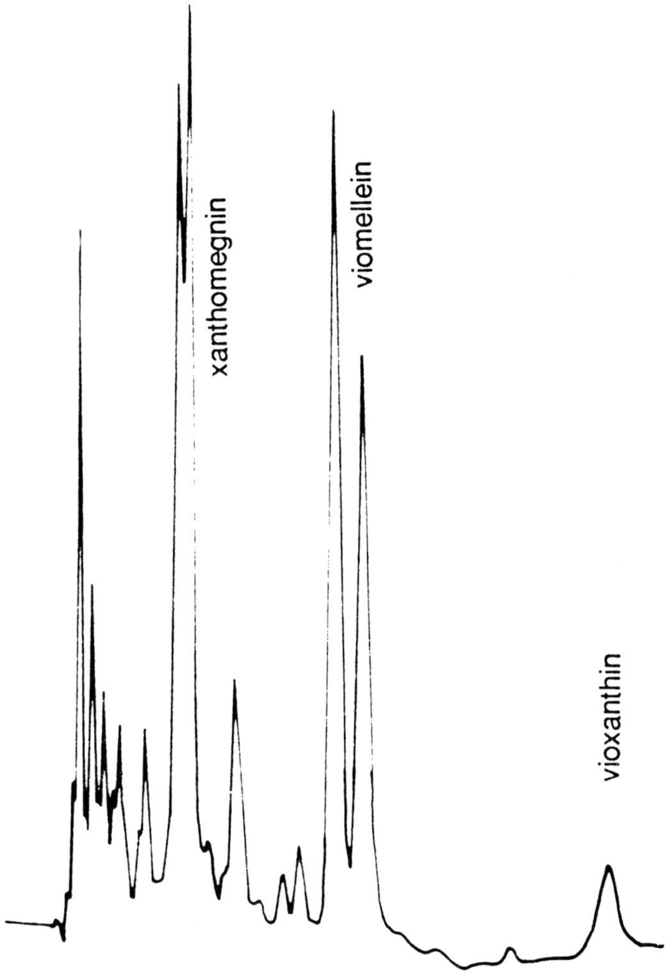

Figure 4. High-performance liquid chromatograph of an extract of barley, containing xanthomegnin, viomellein and vioxanthin at approximately 450, 600 and 90 µg/kg, respectively[a]

[a]From Scudamore *et al.* (1986), with permission

Spectrophotometry: Hult and Gatenbeck (1976) described a spectrophotometric method for the determination of ochratoxin A, in which this toxin is cleaved into ochratoxin α and phenylalanine, using the enzyme carboxypeptidase. The quantification of ochratoxin A is based on the loss of fluorescence intensity at 380 nm, at which ochratoxin A shows maximal excitation. The procedure has been used in practice for the determination of ochratoxin A in barley, pig kidneys and human blood. Detection levels range from 1 to 4 μg/kg, depending on the matrix.

Immunoassays: In the 1980s, there was a rapidly growing interest in the use of immunoassay techniques to determine mycotoxins. Initially, radioimmunoassays and ELISA were carried out only in specialist laboratories. Now, radioimmunoassays have been surpassed by ELISAs, which have found many applications in the determination of mycotoxins, including ochratoxin A (Morgan, 1989). The ELISAs described for determination of ochratoxin A all involve use of antibodies generated against conjugates made by linking the toxin to protein through the carboxylic acid function. Both polyclonal and monoclonal antibodies have been used, and various variants and formats of ELISA have been described for determination of ochratoxin A in cereals, pig kidneys and blood. Microtitre plates are often employed, which make possible full automation as well as use in the field. The principle of an indirect, double-antibody ELISA applied to determination of ochratoxin A in pig kidneys (Morgan *et al.*, 1986) is outlined in Figure 5. A fixed volume of specific antibody distributes itself between a constant amount of immobilized toxin and toxin in solution, according to how much of the

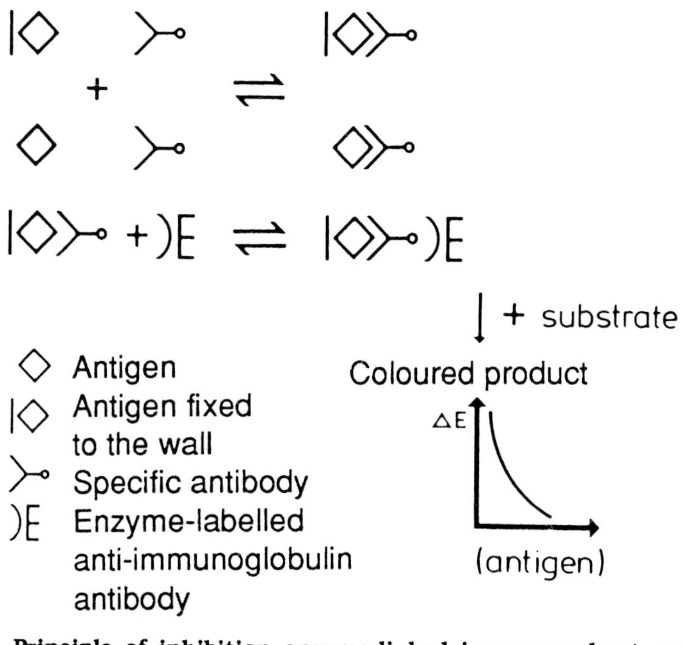

Figure 5. Principle of inhibition enzyme-linked immunosorbent assay for the determination of ochratoxin A, as applied by Morgan (1982)

latter is present. The bound antibody is then quantified by addition of a species-specific second antibody, labelled covalently with enzyme. Bound enzyme on the immobilized phase is detected after addition of substrate. The lower the colour intensity, the higher the concentration of ochratoxin A in the test portion.

Instead of microtitre plates, which are used for (semi-) quantitative determination of ochratoxin A, a commercial 'La Carte'® test can be used for rapid qualitative detection of ochratoxin A. In this procedure, a controlled amount of anti-ochratoxin A antibody is mounted onto each of two ports in a plastic card the size of a credit card. A drop of ochratoxin-free control solution is added to one port and a drop of the test solution to the other. Enzyme-labelled ochratoxin A is added to both card ports, followed by substrate solution. With increasing amounts of ochratoxin A, the colour in the port will appear lighter; conversely, if no ochratoxin A is present, a strong grey-blue dot will develop.

The simplicity of the ELISAs and the large number of samples that can be handled in one day (resulting in relatively low costs per analysis) have made these tests important, especially for screening. The limits of detection of ELISAs for ochratoxin A are comparable to those of chromatographic procedures. These assays are particularly useful for measuring ochratoxin A and its metabolites in body fluids, although their lack of selectivity might prevent their use as quantitative tools. Another point that requires attention is their specificity: although most of the antisera seem to be quite specific, the possibility of cross-reactions cannot be fully ruled out. It is therefore good laboratory practice to confirm positive findings obtained in immunoassays by using methods of analysis based on other principles.

Reference Materials

The availability of reliable methods of analysis is no guarantee of accurate results. Check sample programmes for ochratoxin A in wheat have shown that wide variability in results should be considered more the norm than the exception (Friesen, 1989). The large scatter of analytical results (Figure 6) is of little comfort to those people who either must either pay for the measurement or base potentially important decisions upon them. This situation prompted the Community Bureau of Reference of the European Commission to explore the possibilities for developing ochratoxin A reference materials.

Activities are undertaken by the Community Bureau of Reference in collaboration with a group of European laboratories experienced in analyses of mycotoxins. The objective is to improve the accuracy, and thereby the comparability, of ochratoxin measurements. Currently, efforts are being made to develop reference materials for ochratoxin A in wheat and in lyophilized pig kidney, following the success of previous

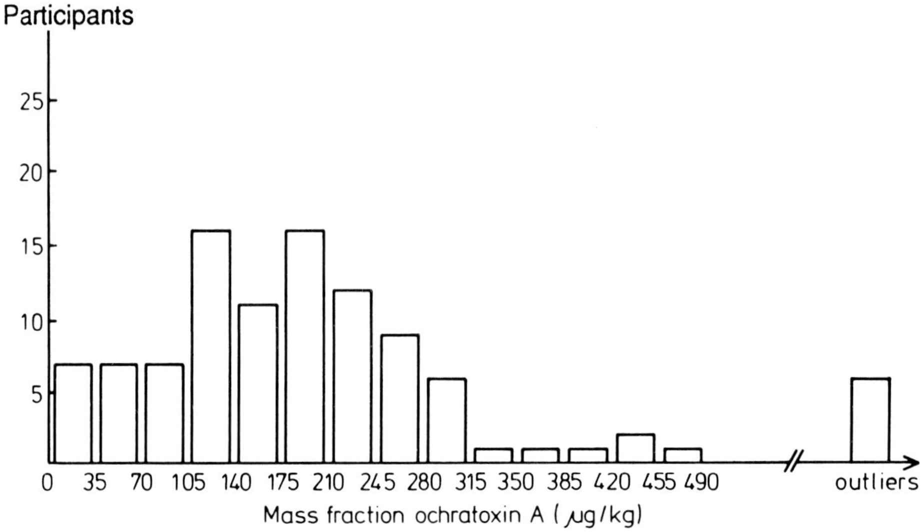

Figure 6. Frequency distribution of analytical results in the Check Sample Programme for ochratoxin A in wheat[a]

[a]From Friesen (1989)

projects for the development of mycotoxin reference materials (van Egmond, 1988). Experience with previous projects indicate that it will be a few more years before certified reference materials for ochratoxin A are available on the market.

Conclusion

The present state of methodology for determination of nephrotoxic mycotoxins, in particular ochratoxin A, may be summarized as follows:

—Chemical assays are of major importance in the determination of these mycotoxins. Solid-phase extraction columns and immunoaffinity columns offer attractive possibilities for efficient purification of extracts.

—Thin-layer chromatography, although a veteran in mycotoxin methodology, remains a reliable, practical and relatively simple separation technique with a broad field of application.

—High-performance liquid chromatography can be an attractive alternative to thin-layer chromatography, because it offers the possibility of automating the ultimate separation and quantification steps.

—The enzyme-linked immunosorbent assay is rapidly gaining ground. This technique is easy to perform and allows rapid screening and semi-quantitative determination.

—The development of (certified) reference materials for mycotoxins by the Community Bureau of Reference of the European Commission is an important activity for improving the accuracy and thereby the comparability of analytical measurements.

References

Breitholtz, A., Olsen, M., Dahlbäck, A. & Hult, K. (1991) Plasma ochratoxin A levels in three Swedish populations surveyed using an ion-pair HPLC technique. *Food Addit. Contam.*, **8**, 183–192

Carman, A.S., Kuan, S.S., Francis, O.J., Ware, G.M. & Luedtke, A.E. (1984) Determination of xanthomegnin in grains and animal feeds by liquid chromatography with electrochemical detection. *J. Assoc. Off. Anal. Chem.*, **67**, 1095–1098

Cohen, H. & Lapointe, M. (1986). Determination of ochratoxin A in animal feed and cereal grains by liquid chromatography with fluorescence detection. *J. Assoc. Off. Anal. Chem.*, **69**, 957–959

Dick, R., Baumann, U. & Zimmerli, B. (1988) Occurrence of citrinin in cereals. *Mitt. Geb. Lebensm. Hyg.*, **79**, 159–164 (in German)

Egan, H., Stoloff, L., Castegnaro, M., Scott, P., O'Neill, I.K. & Bartsch, H., eds (1982) *Environmental Carcinogens: Selected Methods of Analysis*, Vol. 5, *Some Mycotoxins*, Lyon, International Agency for Research on Cancer, pp. 255–270

van Egmond, H.P. (1988) Development of mycotoxin reference materials. *Fresenius Z. Anal. Chem.*, **332**, 598–601

Friesen, M.D. (1989) *Report on the Statistical Analysis of Results Obtained for the Analysis of Ochratoxin in Wheat Flour. Mycotoxin Check Sample Survey Programme* (International Agency for Research on Cancer Report EC/92/15-2), Lyon

Gareis, M., Märtlbauer, E., Bauer, J. & Gedek, B. (1988) Determination of ochratoxin A in breast milk. *Z. Lebensm. Unters. Forsch.*, **186**, 114–117 (in German)

Gilbert, J. (1991) Recent advances in analytical methods for mycotoxins. In: *Proceedings of the International Conference on Fungi and Mycotoxins in Stored Products, Bangkok, Thailand, 23-26 April 1991* (in press)

Gimeno, A. (1984) Determination of citrinin in corn and barley on thin layer chromatographic plates impregnated with glycolic acid. *J. Assoc. Off. Anal. Chem.*, **67**, 194–196

Hald, B. & Krogh, P. (1975) Detection of ochratoxin A in barley, using silica gel minicolumns. *J. Assoc. Off. Anal. Chem.*, **58**, 156–158

Holaday, C.E. (1976) A rapid screening method for the aflatoxins and ochratoxin A. *J. Am. Oil Chem. Soc.*, **53**, 603–605

Hult, K. & Gatenbeck, S. (1976) A spectrophotometic procedure, using carboxypeptidase A, for the quantitative measurement of ochratoxin A. *J. Assoc. Off. Anal. Chem.*, **59**, 128–129

Hult, K., Fuchs, R., Peraica, M., Pleština, R. & Čeović, S. (1984) Screening for ochratoxin A in blood by flow injection analysis. *J. Appl. Toxicol.*, **4**, 326–329

Hunt, D.C., Philip, L.A. & Crosby, N.T. (1979) Determination of ochratoxin A in pig's kidney using enzymic digestion, dialysis and high-performance liquid chromatography with post-column derivatization. *Analyst*, **104**, 1171–1175

Lee, S.C. & Chu, F.S. (1984) Enzyme-linked immunosorbent assay of ochratoxin A in wheat. *J. Assoc. Off. Anal. Chem.*, **67**, 45–49

Marti, L.R., Wilson, D.M. & Evans, B.D. (1978) Determination of citrinin in corn and barley. *J. Assoc. Off. Anal. Chem.*, **61**, 1353–1358

Morgan M.R.A. (1989) Mycotoxin immunoassays (with special reference to ELISAs). *Tetrahedron*, **45**, 2237–2249

Morgan, M.R.A., McNerney, R., Chan, H.W.-S. & Anderson, P.H. (1986) Ochratoxin A in pig kidney determined by enzyme-linked immunosorbent assay (ELISA). *J. Sci. Food Agric.*, **37**, 475–480

Nagakawa, T., Kawamura, T., Fujimoto, Y. & Tatsuno, T. (1982) Determination of citrinin in grain by reversed-phase ion-pair partition chromatography. *J. Food Hyg. Soc. Jpn*, **23**, 297–301 (in Japanese)

Paulsch, W.E., van Egmond, H.P. & Schuller, P.L. (1982) Thin layer chromatographic method for analysis and chemical confirmation of ochratoxin A in kidneys of pigs. In: *Proceedings, V International IUPAC Symposium on Mycotoxins and Phycotoxins, September 1-3, 1982, Vienna, Austria*, Vienna, Austrian Chemical Society, pp. 40–43

Rajakylae, E., Laasasenaho, K. & Sakkers, P.J. (1987) Determination of mycotoxins in grain by high-performance liquid chromatography and thermospray liquid chromatography-mass spectrometry. *J. Chromatogr.*, **384**, 391–402

Ramakrishna, N., Lacey, J., Candlish, A.A., Smith, J.E. & Goodbrand, I.A. (1990) Monoclonal antibody-based enzyme-linked immunosorbent assay of aflatoxin B_1, T-2 toxin, and ochratoxin A in barley. *J. Assoc. Off. Anal. Chem.*, **73**, 71–76

Scudamore, K.A., Clarke, J.H. & Atkin, P.M. (1986) Natural occurrence of fungal naphthoquinones in cereals and animal feedstuffs. In: Flannigan, B., ed., *Spoilage and Mycotoxins of Cereals and Other Stored Products* (International Biodeterioration Suppl. 22), Farnham Royal, CAB International, pp. 71–81

Stubblefield, R.D. (1979) Thin layer chromatographic determination of citrinin. *J. Assoc. Off. Anal. Chem.*, **62**, 201–202

Trantham, A.L. & Wilson, D.M. (1984) Fluorometric screening method for citrinin in corn, barley and peanuts. *J. Assoc. Off. Anal. Chem.*, **67**, 37–38

Mycotoxins, Endemic Nephropathy and Urinary Tract Tumours
Ed. M. Castegnaro, R. Pleština, G. Dirheimer, I.N. Chernozemsky & H. Bartsch
Lyon, International Agency for Research on Cancer, pp. 71-75

USE OF MONOCLONAL ANTIBODIES, ENZYME-LINKED IMMUNOSORBENT ASSAY AND IMMUNOAFFINITY COLUMN CHROMATOGRAPHY TO DETERMINE OCHRATOXIN A IN PORCINE SERA, COFFEE PRODUCTS AND TOXIN-PRODUCING FUNGI

Y. Ueno[1], O. Kawamura[1], Y. Sugiura[1], K. Horiguchi[1], M. Nakajima[2], K. Yamamoto[2] & S. Sato[3]

[1]Department of Toxicology and Microbial Chemistry, Faculty of Pharmaceutical Sciences, Science University of Tokyo, Ichigaya, Tokyo; [2]Division of Food Chemistry, Nagoya City Public Health Research Institute, Nagoya; and [3]Ueda Meat Inspection Station, Ueda, Nagano, Japan

Summary

Ochratoxin A, produced by a number of fungal species, has been found in many milieu, including porcine sera and coffee beans. It was therefore analysed by enzyme-linked immunosorbent assay (ELISA) in porcine sera, coffee products and fungal cultures, using monoclonal antibodies, a monoclonal antibody-linked immunoaffinity column (IAC) and high-performance liquid chromatography (HPLC). The chloroform extracts of acidified porcine sera were assayed directly by ELISA, with alkaline phosphatase and horseradish peroxidase as marker enzymes, at detection limits of 0.1 and 0.01 ng/ml, respectively. The presence of ochratoxin A in ELISA was confirmed by HPLC. The average contents in the five different lots tested were: 0.4 ng/ml in lot A (19 samples), 0.36 ng/ml in lot B (104 samples), 5.20 ng/ml in lot C (17 samples), 1.24 ng/ml in lot D (23 samples) and 0.22 ng/ml in lot E (24 samples). ELISA of methanol extracts of rice cultures showed the presence of more than 0.1 ng/g in 3 of 15 isolates of *Aspergillus*, in 16 of 67 isolates of *Penicillium* and in 7 of 17 isolates of *Eupenicillium*; none was found in an isolate of *Emericella*. IAC–HPLC analysis revealed that *P. foetidus*, which is similar to *A. niger* and is used for the production of a Japanese alcoholic drink (*shou-chuu*), also produced ochratoxin A. Use of IAC–HPLC to analyse coffee beans and instant coffee powder resulted in a sharp resolution of ochratoxin A without complicated clean-up steps. The IAC–HPLC technique could thus be used for mass surveys of ochratoxin A residues in biological specimens.

Introduction

Rapid, specific methods are needed in order to examine the relationship between exposure to ochratoxin A and the incidence of human diseases (WHO, 1990). We used monoclonal antibodies specific for ochratoxin A and an enzyme-linked immunosorbent assay (ELISA), developed by Kawamura *et al.* (1989), to detect ochratoxin A in porcine serum and fungal cultures. Furthermore, as ochratoxin A has been reported to occur naturally in coffee beans (Levis *et al.*, 1974; Tsubouchi *et al.*, 1984), an immunoaffinity column (IAC) was used to clean up coffee products, and ochratoxin A was estimated by high-performance liquid chromatography (HPLC). Some of the results reported here have been published elsewhere (Kawamura *et al.*, 1990; Nakajima *et al.*, 1991).

Materials and Methods

Anti-ochratoxin A monoclonal antibodies OTA.5 and OTA.7, their cross-reactivity to ochratoxin A and related compounds and the basic procedures for ELISA were reported previously (Kawamura *et al.*, 1989). Porcine sera, sampled randomly in a meat inspection station in the Nagano area over the past several years, were stored at –20 °C until required for analysis. At that time, 6 ml of each serum sample were acidified to pH 3–4, extracted with 6 ml chloroform and the residue, dissolved in methanol-phosphate-buffered saline/Tween (1:9, v/v), was assayed by ELISA with OTA.7.

To screen for fungi that produce ochratoxin A, they were inoculated onto rice grains and cultured for three weeks; then, methanol:water (3:1, v/v) extracts of the mouldy grains were assayed by ELISA.

An IAC was prepared by coupling OTA.5 with bromium cyanide-activated Sepharose 4B, and 0.5 ml of the gel was packed in a glass column (8 mm x 15 cm) containing 0.5 M NaCl-phosphate-buffered saline. Finely powdered coffee beans and instant coffee powder were suspended in 1% aqueous $NaHCO_3$, sonicated and filtered through a glass-fibre filter. Tinned coffee drink was filtered. These coffee sample solutions were charged onto the IAC, which was then washed with water, and ochratoxin A was eluted using 50% dimethyl sulfoxide–40 mM phosphate buffer (pH 5.0, 1:1, v/v). The content of ochratoxin A in the eluates was estimated by a HPLC–fluorodetector (Capcell Pak SG120 column 4.6 mm x 25 cm, acetonitrile-water-0.2 M phosphate buffer, pH 9.0, 64:33:3, v/v/v, 365 nm excitation and 450 nm emission).

Results and Discussion

ELISA revealed the presence of ochratoxin A residues at the nanogram per millilitre level (Table 1), which is far lower than those reported in the Balkans, Europe and Canada (WHO, 1990). Ochratoxin A was, however, detected in over 90% of samples. This is the first report of such residues in porcine sera in Japan. The ELISA reported here, based on OTA.7/horseradish peroxidase and a one-step extraction procedure, has a detection limit of 20 pg/ml and is thus suitable for use in mass surveys of ochratoxin A.

Of 100 isolates of fungi, 26 produced ochratoxin A at levels over the detection limit (0.1 ng/g rice grains) (Table 2). The species in which it was found are listed in Table 3. When the methanol extracts of these

Table 1. Ochratoxin A in porcine sera

Lot	Year	No. of samples	Ochratoxin A (ng/ml)	
			Mean	Range
A	1988	19	0.4	0.1–1.5[a]
B	1985	104	0.36	0.03–2.44[a]
C	1989	17	5.20	1.93–9.00[a]
D	1989	23	1.24	0.11–6.70
E	1990	24	0.22	0.04–0.57
Total		197	1.48	0.03–9.00

[a]From Kawamura *et al.* (1990)

Table 2. Distribution of ochratoxin A-producing fungi

Genus	No. of isolates examined	No. positive for ochratoxin A			
		10–100 ng/g	1–10 ng/g	0.1–1 ng/g	Not detected
Aspergillus	15	0	1	2	12
Penicillium	67	2	9	5	51
Eupenicillium	17	1	4	2	10
Emericella	1	0	0	0	1
Total	100	3	14	9	74

isolates were cleaned-up by IAC, HPLC analysis of the eluates revealed that two isolates of *A. foetidus*, *A. ochraceus* and others produced ochratoxin A. The results for *A. foetidus* were confirmed by chemical conversion of ochratoxin A into its methyl ester. This fungus, a variety of *A. niger* newly identified as a producer of ochratoxin A, is used in the production of a local alcoholic drink (*shou-chuu*).

Usual HPLC analysis of ochratoxin A in coffee products, particularly

Table 3. Ochratoxin A-producing fungi

Genus	Species
Aspergillus	*A. sydowii, A. terreus, A. ustus, A. foetidus*
Penicillium	*P. sclerotorum, P. implicatum, P. montanense, P. canescens, P. janczewskii, P. melinii, P. raistrickii, P. miczynskii, P. corylophilum, P. purpurogenum, P. verruculosum*
Eupenicillium	*E. cinnamopurpureum, E. hirayamae, E. pinetorum, E. javanicum*

roasted beans, gives rise to many fluorescent peaks which interfere with the estimation of this mycotoxin, and complicated clean-up steps are required. With our method, introduction of an IAC resulted in clear resolution of ochratoxin A in aqueous extracts of coffee beans and instant coffee powder, without interfering peaks (Figure 1). Tinned coffee drink was simply pasted over the IAC prior to HPLC. The detection

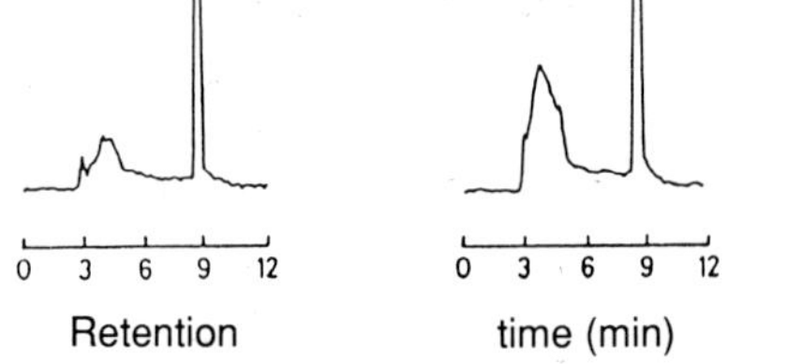

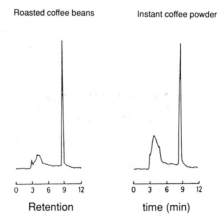

Figure 1. Immunoaffinity column (IAC)–high-performance liquid chromatography (HPLC) profiles of coffee products

5 g of each were spiked with 1 μg/kg ochratoxin A and sonicated in 80 ml aqueous 1% $NaHCO_3$ solution; the filtrates were charged on the IAC, and the column eluates with 50% dimethyl sulfoxide were analysed by HPLC (Nakajima *et al.*, 1991).

limits were 0.5 µg/kg for coffee beans and instant coffee powder and 0.025 µg/kg for the tinned coffee drink (Nakajima *et al.*, 1991). Using this simplified procedure, a mass survey of ochratoxin A in coffee products could be carried out.

References

Kawamura, O., Sato, S., Kajii, H., Nagayama, S., Ohtani, K., Chiba, J. & Ueno, Y. (1989) A sensitive enzyme-linked immunosorbent assay of ochratoxin A based on monoclonal antibodies. *Toxicon*, **27**, 887–897

Kawamura, O., Sato, S., Nagura, M., Kishimoto, S., Ueno, I., Sato, S., Uda, T., Ito, Y. & Ueno, Y. (1990) Enzyme-linked immunosorbent assay for detection and survey of ochratoxin A in livestock sera and mixed feed. *Food Agric. Immunol.*, **2**, 135–143

Levis, C.P., Trenk, H.L. & Mohr, H.K. (1974) Study of the occurrence of ochratoxin A in green coffee beans. *J. Assoc. Off. Anal. Chem.*, **57**, 866–870

Nakajima, M., Terada, H., Hisada, K., Tsubouchi, H., Yamamoto, K., Uda, T., Itoh, Y., Kawamura, O. & Ueno, Y. (1991) Determination of ochratoxin A in coffee beans and coffee products by monoclonal antobodies affinity chromatography. *Food Agric. Immunol.*, **2**, 189–195

Tsubouchi, H., Yamamoto, K., Hisada, K. & Sakabe, Y. (1984) A survey of occurrence of mycotoxins and toxigenic fungi in imported green coffee beans. *Proc. Jpn. Assoc. Mycotoxicol.*, **19**, 16–21 (in Japanese)

WHO (1990) *Selected Mycotoxins: Ochratoxins, Trichothecenes, Ergot* (Environmental Health Criteria No. 105), Geneva

Mycotoxins, Endemic Nephropathy and Urinary Tract Tumours
Ed. M. Castegnaro, R. Pleština, G. Dirheimer, I.N. Chernozemsky & H. Bartsch
Lyon, International Agency for Research on Cancer, pp. 77-81

FOOD CONTAMINATION BY OCHRATOXIN A IN GERMANY

H.K. Frank

Formerly: Federal Research Centre for Nutrition, Ettlingen, Germany

Summary

The results of studies reported in the literature on the occurrence of ochratoxin A in central Europe have been evaluated. Only data that were obtained from random samples were included, thus excluding those derived from samples suspected of being contaminated. Of 1100 samples of cereals (other than maize) and cereal products, 113 (10.3%) were contaminated with an average of 3.8 μg/kg. Of 325 samples of sausages containing porcine serum, 58 (17.8%) were contaminated, at an average of 0.15 μg/kg. The daily intake of ochratoxin A, calculated on the basis of information on consumption of such products, is 80 ng with cereals and 1.6 ng with sausages, corresponding to about 1 ng/kg body weight. People who consume maize products, corresponding to about 40 g maize per day, would take in a further 150 ng/day or 2 ng/kg body weight. These findings and the report that the biological half-life of ochratoxin A in *Macaca mulatta* is about 510 h, explain the fact that more than 50% of samples of human blood and serum tested contain ochratoxin A. As the levels of ochratoxin A in food are not subject to legal regulation in Germany, systematically collected data are not yet available.

Introduction

After the demonstration by Boorman (1989) of the carcinogenicity of ochratoxin A, a study was carried out at the request of the Deutsche Forschungsgemeinschaft, which should provide the basis for better estimation of the risk to consumers. The finding of ochratoxin A in the serum of about 50% of tested populations (Bauer *et al.*, 1986; Hald, 1989) suggests continuous uptake of this mycotoxin by consumers.

Natural Occurrence of Ochratoxin A in Foods of Vegetable Origin

In temperate climatic regions, ochratoxin A is unlikely to be produced in crops before harvest (Shotwell & Hesseltine, 1983). The conidia of ochratoxin A-producing moulds, however, adhere to grains and germinate under sufficient conditions during transport and storage, leading to contamination of the external layers of the grain. Infestation may also occur later; for instance, in flour stored under damp conditions, in oats or maize wetted before processing or in barley in germinating beds of malthouses.

Results reported between 1973 and 1988 based on data obtained from random samples were evaluated, eliminating those from samples suspected of being contaminated, in order to minimize the risk of having too high a rate of positive values. Early reports obtained using methods with detection limits of 20–50 μg/kg may have included false-negative results, and these also were considered to be of uncertain value.

A total of 1100 results were evaluated in the study for the Deutsche Forschungsgemeinschaft, of which 10.3% were found to contain ochratoxin A; the mean level was 3.8 μg/kg. In 1991, annual reports of various official institutions for 1989 and 1990 were consulted. All of the data reported had been obtained using analytical methods with limits of detection of 0.5–0.1 μg/kg. The results influenced the proportion of positive samples (Table 1), but the mean values were clearly lower than those in 1973–88.

Table 1. Frequency of occurrence and concentration of ochratoxin A in grains and grain products in Germany

Product	No. of samples			Concentration of ochratoxin A (μg/kg)
	Total	No. positive	% positive	
1973–88				
Cereals	765	24	3.1	11.8
Bran	84	9	10.7	6.8
Flour	93	17	18.2	2.2
Miscellaneous[a]	158	63	39.8	0.9
1989–90				
Wheat	64	1	1.6	0.4
Wheat flour	38	7	18.0	0.5
Semolina	4	2	50.0	0.5
Rye	30	7	23.3	1.5
Flaked oats	21	8	38.0	1.2
Miscellaneous	316	37	11.7	1.0
Total	1573	175	11.1	2.9

[a]Including bread, bakeries, pastries, semolina, flaked oats

Evaluation of studies of maize and maize products, for a total of 1539 samples, showed that 5.1% contained ochratoxin A, with a mean concentration of 77.6 µg/kg. Other potential carriers of ochratoxin A, such as beer (Majerus & Woller, 1983; Payen *et al.*, 1983), coffee beans (Tsubouchi *et al.*, 1988; Micco *et al.*, 1989), dried figs, ground hazelnuts, olive oil and bean sprouts were not evaluated, since they are not part of the basic German diet; mouldy products, such as bread and flour, were also not included, as spoilt food is usually not consumed.

Worth mention is a study in Switzerland in which citrinin, a toxin that frequently occurs concomitantly with ochratoxin A (Frank, this volume), was found in 50% of flour and semolina samples, at a concentration of about 0.5 µg/kg (Dick *et al.*, 1988).

Occurrence of Ochratoxin A in Foods of Animal Origin

Domestic animals take up ochratoxin A with fodder. The toxin is excreted slowly from the organism, the rate depending on the species, and residues are found only in monogastric animals. In pigs, however, ochratoxin A binds differently to various matrices: for a concentration in blood plasma of 100, that in blood is 44, that in kidneys, 9, and that in lean meat, 4.3.

In Germany, 13.3% of 354 pig kidneys were found to contain ochratoxin A, at a mean value of 1.4 µg/kg (Hadlok, 1989a). Consumption of pig kidneys in this country is, however, insignificant. Pigs' blood and plasma, which are essential ingredients of sausages, are of greater importance. Of 1169 blood samples, 49% contained ochratoxin A, at about 1 µg/kg. Of 283 plasma samples, 52% were positive, containing 1.2 µg/kg (Hadlok, 1989b; Majerus *et al.*, 1989). Dried plasma, which is used far less, contained no ochratoxin A (Scheuer & Leistner, 1986). Sausages were also found to contain the toxin: it was found in 16% of 125 black pudding samples, in 19% of 100 liver sausage samples and in 19% of 100 boiling sausage samples. The mean ochratoxin A content was 0.1–0.2 µg/kg, and the maximum was 3.4 µg/kg (Scheuer & Leistner, 1985).

Daily Uptake by Consumers

The uptake of ochratoxin A with basic food was calculated on the basis of information about the consumption of various foods. Each person consumes an average of 205 g of grain products daily; of this, 11.2%, corresponding to 23 g, is contaminated with 66 ng of ochratoxin A. Consumers also eat 37 g of sausages per day, to give an uptake of 1.6 ng. A total of 68 ng ochratoxin A are therefore consumed daily, or about 1 ng/kg body weight. People who regularly eat maize products, such as cornflakes (in Germany, mainly children and adolescents), corresponding to 40 g/day, take up another 154 ng of ochratoxin A per day.

The biological half-time of ochratoxin A in humans is not known, but it can be assumed to be the same as or higher than that in monkeys, which is 510 h (Hagelberg *et al.*, 1989). The finding that 50% of human blood samples contain ochratoxin A can be explained on this basis. Most of the negative findings in people on a 'normal' diet can be assumed to be below the analytical detection limit and not to correspond to 'zero'. Thus, there is continual contamination with small quantities of ochratoxin A.

Discussion

The values given above provide clues as to the order magnitude of contamination by ochratoxin A. They should not be regarded as 'absolute' values, as they are not derived from the comprehensive, systematic studies that would be necessary if the level of ochratoxin A in food were subject to legal regulation. As there are regional and annual fluctuations in such levels, depending on climatic conditions, establishment of 'safe' levels would require studies lasting three to four years.

On the basis of the risk estimation of Kuiper-Goodman and Scott (1989) and of data from tests in experimental animals, the estimated tolerable daily intake of ochratoxin A for humans is 0.2–4.2 ng/kg body weight. Bauer and Gareis (1987) in Germany, who used a similar basis, regarded 1.2 ng/kg body weight as tolerable. In contrast, the Joint FAO/WHO Expert Committee on Food Additives has proposed a provisional tolerable weekly intake of 112 ng/kg body weight, corresponding to 16 ng/kg body weight per day (Herrman, this volume).

References

Bauer, J. & Gareis, M. (1987) Ochratoxin A in the foodchain. *Z. Veterinärmed. B*, **34**, 613–627 (in German)

Bauer, J., Gareis, M. & Gedek, B. (1986) Incidence of ochratoxin A in blood serum and kidneys of man and animals. In: *Proceedings of the 2nd World Congress on Foodborne Infections and Intoxications*, Berlin, Robert von Ostertag Institute, pp. 903–908

Boorman, G., ed. (1989) *NTP Technical Report on the Toxicology and Carcinogenesis Studies of Ochratoxin A (CAS No. 303-47-9) in F344/N Rats (Gavage Studies)* (NTP TR 358, NIH Publication No. 89-2813), Research Triangle Park, NC, National Toxicology Program, US Department of Health and Human Services, National Institutes of Health

Deutsche Forschungsgemeinschaft (1990) *Ochratoxin A*, Weinheim, VCH Verlag (in German)

Dick, R., Baumann, U. & Zimmerli, B. (1988) Occurrence of citrinin in cereal grains. *Mitt. Geb. Lenbensmittelhyg.*, **79**, 159–164 (in German)

Hadlok, R.M. (1989a) Mycotoxins and their toxicological-hygienic assessment. In: Heeschen, W., ed., *Pathogenic Microorganisms and their Toxins in Foodstuffs of Animal Origin*, Hamburg, Behr's Verlag, pp. (in German)

Hadlok, R.M. (1989b) Occurrence of ochratoxin A in blood of slaughter pigs and human beings. In: *11th Mykotoxin-Workshop, 22–24 May, Berlin* (in German)

Hagelberg, S., Hult, H. & Fuchs, R. (1989) Toxicokinetics of ochratoxin A in several species and its plasma-binding properties. *J. Appl. Toxicol.*, **9**, 91–96
Hald, B. (1989) Human exposure to ochratoxin A. In: Natori, S., Hashimoto, K. & Ueno, Y., eds, *Mycotoxins and Phytotoxins '88*, Amsterdam, Elsevier, pp. 57–67
Kuiper-Goodman, T. & Scott, P.M. (1989) Risk assessment of the mycotoxin ochratoxin A. *Biomed. Environ. Sci.*, **2**, 179–248
Majerus, P. & Woller, R. (1983) Mycotoxins in beer. II. Ochratoxin A and citrinin. *Monatsschr. Brauwiss.*, **36**, 335–336 (in German)
Majerus, P., Otteneder, H. & Hower, C. (1989) Occurrence of ochratoxin A in blood serum of swine. *Dtsch. Lebensm. Rundsch.*, **85**, 307–313 (in German)
Micco, C., Grossi, M., Miraglia, M. & Brera, C. (1989) A study of the contamination by ochratoxin A of green and roasted coffee beans. *Food Addit. Contam.*, **6**, 333–339
Payen, J., Girard, T., Gaillardin, M. & Lafont, P. (1983) About contamination of beers by mycotoxins. *Microbiol. Aliment. Nutr.*, **1**, 143–146
Scheuer, R. & Leistner, L. (1985) Occurrence of ochratoxin A in slaughter pigs and meat products. *Mitt. Bundesanst. Fleischforsch.*, **88**, 6436–6439 (in German)
Scheuer, R. & Leistner, L. (1986) Occurrence of ochratoxin A in pork and pork products. In: *Proceedings of the 32nd European Meeting of Meat Research Workers*, Vol. 1, p. 2
Shotwell, O.L. & Hesseltine, C.W. (1983) Five-year study of mycotoxins in Virginia wheat and dent corn. *J. Assoc. Off. Anal. Chem.*, **66**, 1466–1469
Tsubouchi, H., Terada, H., Yamamoto, K., Hisada, K. & Sakabe, Y. (1988) Ochratoxin A found in commercial roast coffee. *J. Agric. Food Chem.*, **36**, 540–542

Mycotoxins, Endemic Nephropathy and Urinary Tract Tumours
Ed. M. Castegnaro, R. Pleština, G. Dirheimer, I.N. Chernozemsky & H. Bartsch
Lyon, International Agency for Research on Cancer, pp. 83-87

OCHRATOXIN A AND OTHER MYCOTOXINS IN CEREALS FROM AN AREA OF BALKAN ENDEMIC NEPHROPATHY AND URINARY TRACT TUMOURS IN BULGARIA

T. Petkova-Bocharova[1], M. Castegnaro[2], J. Michelon[2] & V. Maru[2,3]

[1]National Oncological Centre, Sofia, Bulgaria; and [2]International Agency for Research on Cancer, Lyon, France

Summary

The etiology of Balkan endemic nephropathy and urinary tract tumours in the rural population of the endemic regions remains unknown. As one hypothesis involves mycotoxins, a survey was carried out to investigate the possible involvement of the nephrotoxic mycotoxins ochratoxin A and citrinin. Recently, this survey was extended to screening for the presence of other mycotoxins—aflatoxins, citrinin, sterigmatocystin and zearalenone. A total of 524 samples of home-produced and home-stored beans and maize from the harvests of 1984, 1985, 1986, 1989 and 1990 were analysed. Ochratoxin A was found in samples from both endemic and nonendemic areas, but more of the samples from affected families were contaminated, and at higher levels, than those from unaffected households. Citrinin and aflatoxins B_1 and G_1 were also found more frequently in samples from endemic areas. These results support the theory that mycotoxins are involved in the etiology of Balkan endemic nephropathy and urinary tract tumours.

---

[3]Present address: Biochemistry Section, Pathology Department, Jaslok Hospital and Research Centre, Bombay 40026, India

Introduction

Ochratoxin A is a natural contaminant of food and feed. Because of certain reported similarities between Balkan endemic nephropathy and ochratoxin-induced porcine nephropathy (Krogh, 1974), it has been suggested that ochratoxin A plays a causal role in the induction of Balkan endemic nephropathy. It may also play a role in the induction of the urinary tract tumours associated with this disease (Castegnaro *et al.*, 1990). Previous studies have shown that ochratoxin A is frequently present in food in areas of Yugoslavia (Krogh *et al.*, 1977; Pavlovic *et al.*, 1979; Pepeljnjak & Blažević, 1982) and Bulgaria (Petkova-Bocharova & Castegnaro, 1985), where nephropathy is endemic.

We report here results of analyses for ochratoxin A in staple foods from Vratza, a district in which there is a high incidence of Balkan endemic nephropathy and renal tumours. Samples were also examined for contamination by aflatoxins B_1, B_2, G_1 and G_2, citrinin, sterigmatocystin and zearalenone.

Materials and Methods

A total of 524 samples of home-produced, home-stored beans and maize from the harvests of 1984, 1985, 1986, 1989 and 1990 were analysed for ochratoxin A; 466 samples from 1987, 1988, 1989 and 1990 were analysed for citrinin; and 244 samples from 1989 and 1990 were examined for the presence of aflatoxins, sterigmatocystin and zearalenone. Most of the samples came from stores that had been kept in cellars for 5–8 months; none was visibly mouldy.

Ochratoxin A was determined partly by thin-layer chromatography, using the method of the Association of Official Analytical Chemists (Stoloff *et al.*, 1982), and partly by the high-pressure liquid chromatography method described by the same authors, with modification of the chromatographic conditions. The limit of detection of the first method is approximately 10 µg/kg and that of the second, 0.1 µg/kg.

Citrinin was determined by the method of Gimeno (1984), with a limit of detection of 15–20 µg/kg. Contamination with aflatoxins, sterigmatocystin and zearalenone was analysed using the multi-toxin thin-layer chromatography method of Soares and Rodriguez-Amaya (1989). The limits of detection were 2 µg/kg for aflatoxins, 15 µg/kg for sterigmatocystin and 55 µg/kg for zearalenone.

All data were evaluated statistically, using Student's *t* test with transformation for differences between values.

Results and Discussion

Ochratoxin A was found in samples of beans and maize from both endemic and nonendemic areas for Balkan nephropathy (Table 1). Many more samples from endemic areas than from control areas were contaminated, and the differences were significant ($p < 0.01$) for products

Table 1. Prevalence of ochratoxin A in bean and maize samples from different harvests collected in an area with a high incidence of endemic nephropathy and in a nonendemic area in Bulgaria

Harvest (year)	Food	Area	No. of samples	No. of samples containing ochratoxin A			
				No.	%	95% Confidence interval	Range (µg/kg)
1984	Beans	Endemic	24	9	37.5	19.3–57.7	0.5–94.0
		Control	20	3	15.0	3.0–33.7	0.3–5.0
	Maize	Endemic	24	15	62.5	42.2–80.7	5.0–24.7
		Control	20	4	20.0	5.7–40.2	0.7–5.9
1985	Beans	Endemic	34	30	88.2	75.1–96.8	0.05–264.0
		Control	4	4	100.0		0.2–284.6
	Maize	Endemic	36	35	97.2	89.2–99.9	0.2–1417.5
		Control	4	4	100.0		0.2–138.8
1986	Beans	Endemic	34	13	38.2	22.4–55.1	25–200
		Control	24	2	8.3	0.77–22.6	20–150
	Maize	Endemic	34	14	41.7	25.6–58.7	25–250
		Control	24	2	8.3	0.77–22.6	20–180
1989	Beans	Endemic	30	11	36.6	25.6–58.7	25–240
		Control	25	2	8.0	0.76–21.9	25–200
	Maize	Endemic	32	14	43.7	26.9–61.3	25–900
		Control	25	2	8.0	0.76–21.9	10–230
1990	Beans	Endemic	25	10	40.0	21.7–59.8	85–260
		Control	40	2	5.0	0.45–14.0	10–220
	Maize	Endemic	25	11	44.0	25.1–63.8	25–890
		Control	40	2	5.0	0.45–14.0	20–235

collected in 1984, 1986, 1989 and 1990. No statistical evaluation could be made for products collected in 1985, because the number of samples from control areas was too small and because ochratoxin A was found in most samples, owing to the low limit of detection of the high-pressure liquid chromatographic method used. Analysis of samples from the 1985 harvest consumed by families affected by Balkan nephropathy and those not affected, however, showed a statistically significant difference ($p < 0.05$) in ochratoxin A contamination (Table 2); using a similar analysis, highly significant differences were also found for the 1984 ($p < 0.05$), 1986 ($p < 0.05$), 1989 ($p < 0.001$) and 1990 ($p < 0.001$) harvests.

Citrinin was found in samples from both endemic and nonendemic areas. Although the difference was not statistically significant, it was detected more frequently and at higher concentrations in samples from endemic areas (Table 3). Of the other mycotoxins, only aflatoxins B_1 and G_1 were detected; they were found in samples of beans and maize from the endemic area and in one sample of beans and one of maize from the

Table 2. Ochratoxin A contamination of bean and maize samples from the 1985 harvest collected from families affected with Balkan nephropathy and from unaffected families within and outside the endemic area

Families	Beans		Maize	
	Range (µg/kg)	% samples with > 10 µg/kg	Range (µg/kg)	% samples with > 25 µg/kg
Affected	ND–250	53	0.2–1200	70
Control	ND–280	23	ND–1400	37

ND, not detected

1990 harvest in a control area. Owing to the small number of samples, however, no statistical evaluation could be made. The other mycotoxins were not detected.

The population of this area of Bulgaria where Balkan nephropathy is endemic is thus exposed to ochratoxin A, citrinin and probably aflatoxins. The reported synergism between citrinin and ochratoxin A (Kanisawa, 1984) and that between aflatoxins and ochratoxin A (Huff & Doerr, 1981) suggest that the risk in this area for Balkan nephropathy and urinary tract cancer may be increased. The finding of ochratoxin A, citrinin and aflatoxins in staple foods in the area provides support for the hypothesis that these mycotoxins are involved in the etiology of these diseases.

References

Castegnaro, M., Chernozemsky, I.N., Hietanen, E. & Bartsch, H. (1990) Are mycotoxins risk factors for endemic nephropathy and associated urothelial cancers? *Arch. Geschwulstforsch.*, **60**, 295–303

Gimeno, A. (1984) Determination of citrinin in corn and barley on thin layer chromatographic plates impregnated with glycolic acid. *J. Assoc. Off. Anal. Chem.*, **67**, 194–196

Huff, W.E. & Doerr, J.A. (1981) Synergism between aflatoxin and ochratoxin A in broiler chickens. *Poult. Sci.*, **60**, 550–555

Kanisawa, M. (1984) Synergistic effects of citrinin on hepatorenal carcinogenesis of ochratoxin A in mice. *Dev. Food Sci.*, **7**, 245–254

Krogh, P. (1974) Mycotoxin porcine nephropathy: a possible model for Balkan endemic nephropathy. In: Puchlev, A., Dinev, I., Milev, B. & Doichinov, D., eds, *Endemic Nephropathy*, Sofia, Bulgarian Academy of Sciences, pp. 266–270

Krogh, P., Hald, B., Pleština, R. & Čeović, S. (1977) Balkan (endemic) nephropathy and foodborn ochratoxin A: preliminary results of a survey in foodstuffs. *Acta Pathol. Microbiol. Scand. Sect. B.*, **85**, 238–270

Pavlović, M., Pleština, R. & Krogh, P. (1979) Ochratoxin A contamination of foodstuffs in an area with Balkan (endemic) nephropathy. *Acta Pathol. Microbiol. Scand. Sect. B.*, **87**, 243–246

Pepeljnjak, S. & Blažević, N. (1982) Contamination with moulds and occurrence of ochratoxin A in smoked meat products from endemic regions of Yugoslavia. In: *Fifth International IUPAC Symposium on Mycotoxins and Phytotoxins*, Vienna, Austrian Chemical Society, pp. 102–105

Table 3. Prevalence of mycotoxins other than ochratoxin A in bean and maize samples from different harvests collected in an area with a high incidence of endemic nephropathy and in nonendemic areas in Bulgaria

Harvest (year)	Food	Area	No. of samples	No. of samples containing mycotoxin			
				No.	%	95% Confidence interval	Range (μg/kg)
Citrinin							
1987	Beans	Endemic	24	8	33.3	15.5–53.1	50–1000
		Control	20	2	10.0	0.97–26.9	100–120
	Maize	Endemic	24	10	41.6	22.6–61.9	120–1500
		Control	20	3	15.0	3.0–33.9	100–180
1988	Beans	Endemic	48	13	27.0	15.4–40.6	20–380
		Control	20	2	10.0	0.97–26.9	120–200
	Maize	Endemic	48	13	27.0	15.4–40.6	50–1000
		Control	20	2	10.0	0.97–26.9	80–100
1989	Beans	Endemic	30	12	40.0	23.2–58.1	50–800
		Control	25	3	12.0	2.4–27.6	50–120
	Maize	Endemic	32	14	43.7	26.9–61.2	50–1100
		Control	25	3	12.0	2.4–27.6	150–380
1990	Beans	Endemic	25	9	36.0	18.4–55.8	30–800
		Control	40	4	10.0	2.7–21.4	20–200
	Maize	Endemic	25	10	40.0	21.7–59.8	50–1000
		Control	40	4	10.0	2.7–21.4	50–140
Aflatoxin B_1							
1989	Beans	Endemic	30	2	6.3	0.63–18.5	20–30
		Control	25	ND	–		–
	Maize	Endemic	32	3	9.3	1.8–21.8	20–50
		Control	25	ND	–		–
1990	Beans	Endemic	25	2	8.0	0.76–21.9	30
		Control	40	1	2.5	0–9.7	18
	Maize	Endemic	25	2	8.0	0.76–21.9	30–70
		Control	40	1	2.5	0–9.7	25
Aflatoxin G_1							
1989	Beans	Endemic	30	2	6.7	0.63–18.5	10–20
		Control	25	ND	–		–
	Maize	Endemic	32	3	9.3	1.8–21.8	20–30
		Control	25	ND	–		–
1990	Beans	Endemic	25	1	4.0	0–15.3	20
		Control	40	1	2.5	0–9.7	8
	Maize	Endemic	25	2	8.0	0.76–21.9	20–30
		Control	40	1	2.5	0–9.7	18

Petkova-Bocharova, T. & Castegnaro, M. (1985) Ochratoxin A contamination of cereals in an area of high incidence of Balkan endemic nephropathy in Bulgaria. *Food Addit. Contam.*, **2**, 267–270

Soares, L.M.V. & Rodriguez-Amaya, D.B. (1989) Survey of aflatoxins, ochratoxin A, zearalenone and sterigmatocystin in some Brazilian foods by using multi-toxin thin layer chromatographic method. *J. Assoc. Off. Anal. Chem.*, **72**, 22–25

Stoloff, L., Castegnaro, M., Scott, P., O'Neill, I.K. & Bartsch, H., eds (1982) *Experimental Carcinogens. Selected Methods of Analysis*, Vol. 5, *Some Mycotoxins* (IARC Scientific Publications No. 44), Lyon, IARC, pp. 255–269, 271–282

Mycotoxins, Endemic Nephropathy and Urinary Tract Tumours
Ed. M. Castegnaro, R. Pleština, G. Dirheimer, I.N. Chernozemsky & H. Bartsch
Lyon, International Agency for Research on Cancer, pp. 89-92

ENZYME-LINKED IMMUNOSORBENT ASSAY FOR ANALYSIS OF OCHRATOXIN A IN HUMAN PLASMA SAMPLES USING ANTIBODIES RAISED AGAINST A NEW OCHRATOXIN–PROTEIN CONJUGATE

A. Breitholtz Emanuelsson & K. Hult[1]

Department of Biochemistry and Biotechnology, Royal Institute of Technology, Stockholm, Sweden

Summary

A derivative of ochratoxin A was linked to bovine serum albumin in such a way that the carboxylic group of the toxin was left free. Injection of the conjugate into rabbits resulted in sensitive antibodies towards ochratoxin A, which were used in an enzyme-linked immunosorbent assay.

Text

As recent results have shown an increased frequency of renal cancer in male rats fed ochratoxin A (Boorman, 1989), the daily intake of this toxin should be monitored. Several ways have been described for producing antibodies towards ochratoxin A after linking it through its carboxylic group to a carrier protein. In this study, a derivative of ochratoxin A was linked to bovine serum albumin in a new way. Lysine was substituted for phenylalanine, and the ε-amino group was linked to bovine serum albumin, leaving the carboxylic group free (Figure 1). The free carboxylic group is important for the charge distribution, as it is dissociated and charged at the pH used for analysis.

[1]To whom correspondence should be addressed

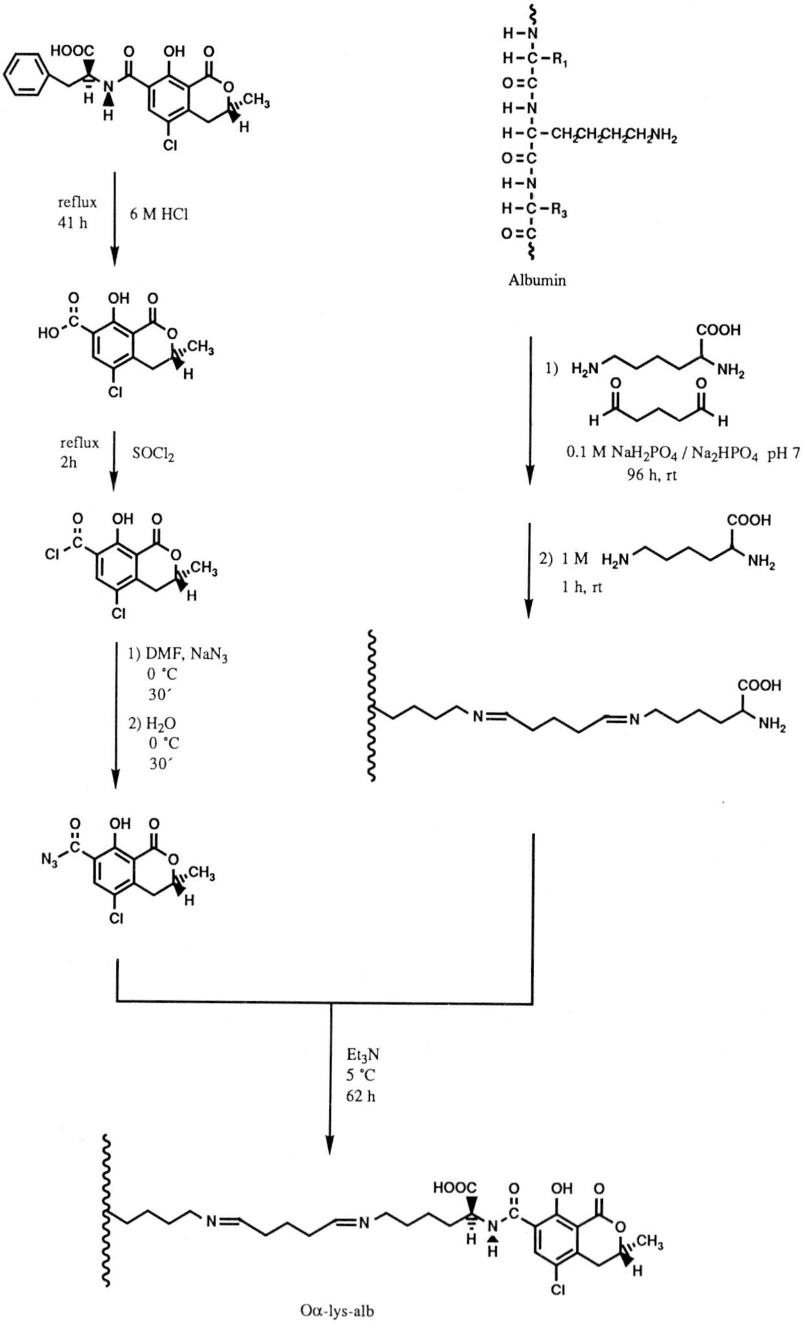

Figure 1. Linkage of a derivative of ochratoxin A to bovine serum albumin

The calculated electropotential surface of this modified ochratoxin A–bovine serum albumin conjugate (Oα-lys-alb) looks almost the same as the electropotential surface of ochratoxin A. The charge distribution pattern of the ochratoxin A–protein conjugate with ochratoxin A linked through its carboxylic group (OA-alb) is very different from that of ochratoxin A. Since the charge distribution is very important in determining the affinity between antigen and antibody, this difference in distribution can explain why the antibodies towards ochratoxin A produced in our earlier attempt, using a conjugate with ochratoxin A linked through its carboxylic group, did not have the expected affinity. In this study, antibodies were produced by injecting the Oα-lys-alb conjugate into rabbits. The polyclonal antibodies produced were purified from antibodies with an affinity to bovine serum albumin by running the serum through a column with bovine serum albumin linked to sepharose.

The purified polyclonal antibodies were used in a competitive enzyme-linked immunosorbent assay (ELISA), using an α-rabbit immunoglobulin G alkaline phosphatase conjugate as secondary antibody. The ELISA plates were coated with the Oα-lys-alb conjugate (0.01 μg/ml; 9.5 Oα-lys/bovine serum albumin). With this coating, the detection limit of ochratoxin A in phosphate buffer was 0.03 ng/ml (1.5 pg/well).

Human plasma samples, calf, rat and pig sera samples and milk samples were analysed successfully using this method. Since, however, the proteins in the samples must be removed before use in the ELISA, a new, easy purification method was developed: To a 300-μl sample in an Eppendorf tube, 400 μl acetone are added, and the solution is mixed on a vortex and subsequently centrifuged. The supernatant is removed from the precipitate to a new tube; buffer (100 μl 1 M KH_2PO_4/K_2HPO_4 at pH 8) and chloroform (500 μl) are added, and the mixture is stirred on a vortex. The tube is then centrifuged, and the water phase is used in the ELISA. The recovery using this purification method is about 65%, and the limit of detection of ochratoxin A in the samples is 0.1 ng/ml.

A total of 294 human plasma samples collected in Yugoslavia and in Sweden have been analysed using this method. In 46% of these samples, the ochratoxin A contamination level was less than 0.5 ng/ml; 30% of the samples were contaminated in the concentration range, 0.5–1 ng/ml; and 24% of the samples had an ochratoxin A content of more than 1 ng/ml.

We also analysed 129 of the samples by the high-performance liquid chromatography method described by Breitholtz *et al.* (1991). The agreement between the two methods was very good. The ELISA technique is much faster, however; and if a large number of samples is to be run daily, for example in an epidemiological study, the ELISA technique is preferable. Data collected from such analyses of serum samples can be used subsequently to calculate the daily intake of ochratoxin A (Breitholtz *et al.*, 1991) from the equation

$$k_0 = Cl_p C_p / A$$

(Klaassen, 1986), which shows the relation between continous intake (k_0, ng/kg body weight per day), the plasma clearance (Cl_p, ml/kg body weight per day), the plasma concentration of ochratoxin A (C_p, ng/ml) and the bioavailability (A, fraction of toxin taken up).

References

Boorman, G., ed. (1989) *NTP Technical Report on the Toxicology and Carcinogenesis Studies of Ochratoxin A (CAS No. 303-47-9) in F344/N Rats (Gavage Studies)* (NTP TR358; NIH Publication No. 89-2813), Research Triangle Park, NC, National Toxicology Program, US Department of Health and Human Services, National Institutes of Health, pp. 1–44

Breitholtz, A., Olsen, M., Dahlbäck, Å. & Hult, K. (1991) Plasma ochratoxin A levels in three Swedish populations surveyed using an ion-pair HPLC technique. *Food Addit. Contam.* , **8**, 183–192

Klaassen, C.D. (1986) Distribution, excretion and absorption of toxicants. In: Klaassen, C.D., Amdur, M.O. & Doull, J., eds, *Casarett and Doull's Toxicology. The Basic Science of Poisons*, 3rd ed., New York, Macmillan, pp. 33–63

Mycotoxins, Endemic Nephropathy and Urinary Tract Tumours
Ed. M. Castegnaro, R. Pleština, G. Dirheimer, I.N. Chernozemsky & H. Bartsch
Lyon, International Agency for Research on Cancer, pp. 93-96

NATURAL OCCURRENCE OF OCHRATOXIN A IN FOOD AND FEED IN SENEGAL

A. Kane, N. Diop & T.S. Diack

Institut de Technologie Alimentaire, Dakar, Senegal

Summary

A retrospective study covering 1984–88 showed that urogenital diseases were the third most frequent cause of death in Senegal. The purpose of our study was to see whether ochratoxin A is involved in the etiology of these nephropathies. A total of 166 samples of eight principal types of food and feed consumed in this country were obtained randomly at markets in Dakar and Kaolack and analysed for ochratoxin A using the method of analysis of the Association of Official Analytical Chemists. Only cowpea (*Vigna unguiculata*), from a legume widely distributed in Senegal, was contaminated (16% of samples), at an average level of 34 µg/kg. All other samples were free of ochratoxin A, indicating that it is not directly correlated with the renal diseases observed. Aflatoxin B_1 was detected in almost all of the samples, and a competition between moulds producing the two toxins is suggested.

Introduction

Studies carried out in Senegalese hospitals have shown that renal diseases are very frequent. Samb (1988) found that between 1984 and 1988, urogenital diseases accounted for 11.47% of all morbidity, with a ratio of 1.9 for men:women, and 14.65% of all mortality in the busiest hospital in Dakar. They represented the third most frequent cause of death, after hepatic and cardiovascular diseases. The conditions involved were mainly chronic and acute renal failure, chronic nephritis and pyelonephritis.

Chronic renal failure accounted for 78.6% of all deaths from urogenital diseases and constituted the second most frequent cause of death after primary liver cancer. Ngaide (1991) reported two cases of tubular necrosis and 16 cases of oedema among a series of patients with acute renal failure. Kane (1981) found that, of 183 patients with chronic renal failure, 73 were women and 110 were men, and 21.3% were aged between 20 and 29 years. The main nephropathy was glomerulonephropathy. Touré *et al.* (1984) found that 46.9% of cases of renal failure were glomerulonephritis, 20.1% interstitial nephritis, 18.7% unclassified nephropathies, 8.9% vascular nephropathies, 4.5% acute renal failure and 0.7% polycystic disease.

This evidence that nephropathies constitute a serious health problem in Senegal prompted us to investigate their cause. Ochratoxin A is found naturally in many foods and feeds (Harwig, 1974) and has been associated with nephropathy in swine and with Balkan endemic nephropathy in humans (Krogh *et al.*, 1977). Although the occurrence of aflatoxins in foods in Africa is well documented, no data are available concerning the natural occurrence of ochratoxin A in comestibles in Senegal and neighbouring countries. We therefore analysed the foods and feeds consumed most frequently in Senegal by humans and animals to determine whether they are contaminated by this mycotoxin.

Materials and Methods

Ochratoxin A was a gift from B.D. Jones (Overseas Development Natural Research Instsitute, London, United Kingdom). Its concentration was determined in methanol at 333 nm with an extinction coefficient of 5500, using a Beckman spectrophotometer. All chemicals used were of the purest grade available.

Samples of food and feed were collected randomly in different markets in Dakar and Kaolack, two of the most heavily populated towns in Senegal. They were analysed rapidly or kept in a refrigerator until analysis. Ochratoxin A was determined by the method of the Association of Official Analytical Chemists (1980), which has a limit of detection of 10 µg/kg. Aflatoxins B_1, B_2, G_1 and G_2 were evaluated qualitatively using the 'best food' method for groundnut and the 'control branch' method for mixed feed and groundnut cake (Association of Official Analytical Chemists, 1980) and the method of Siraj *et al.* (1981) with thin-layer chromatographic determination for all other commodities.

Results and Discussion

Ochratoxin A was found only in samples of cowpea (*Vigna unguiculata*), the bean of a legume that is distributed widely in Senegal and other African countries, such as Niger and Mali. Rice, millet and maize, the commodities consumed most frequently by humans, and animal feeds were not contaminated with ochratoxin A. All of the samples, however, contained aflatoxins at various levels.

Ochratoxin A does not therefore appear to be a common contaminant of food and feed in Senegal. The levels found in 16% of cowpea samples are not alarming, as this is not a staple food. Since the route of human exposure to this mycotoxin is oral, through consumption of infected foods or animal products, ochratoxin A is therefore unlikely to be involved significantly in the nephropathies reported there. Investigations of human and animal blood will be needed to confirm this conclusion.

Two hypotheses could be put forward to explain the relative absence of ochratoxin A in food and feed in Senegal. First, the finding of detectable amounts of aflatoxins in almost all the samples analysed, and especially maize and groundnut products (with more than 100 and up to 2 µg/kg, respectively), might suggest that there is competition between moulds producing aflatoxins and those producing ochratoxin A. Another possibility is that ochratoxin-producing moulds are present but that they do not have the optimal conditions of temperature and relative humidity to produce ochratoxin A.

Table 1. Occurrence of ochratoxin A in food and feed in Senegal

Sample	Number analysed	Contaminated samples		Average concentration (µg/kg)
		Number	Per cent	
Maize	32	0	–	–
Cowpea	31	5	16	34
Millet	20	0	–	–
White sorghum	10	0	–	–
Rice	20	0	–	–
Mixed feed	10	0	–	–
Groundnut cake	23	0	–	–
Groundnut kernels	20	0	–	–

References

Association of Official Analytical Chemists (1980) *Official Methods of Analysis*, 13th ed., Washington DC, pp. 26.026–26.036

Harwig, J. (1974) Ochratoxin A related metabolites. In: Purchase, I.F.H., ed., *Mycotoxins*, Amsterdam, Elsevier, pp. 345–367

Kane, B. (1981) *Notre Experience de l'Insuffisance rénale chronique 'A Propos de 232 Cas'*, MD Thesis, University of Dakar

Krogh, P., Hald, B., Pleština, R. & Čeović, S. (1977) Balkan endemic nephropathy and foodborn ochratoxin A: preliminary results of survey of foodstuffs. *Acta Pathol. Microbiol. Scand. Sect. B*, **85**, 238–240

Ngaide, S.A. (1991) *Insuffisance rénale aigüe—Bilan du Centre d'Hémodialyse de l'Hôpital Aristide Le Dantec du 10 Octobre 1987 au 9 Janvier 1990*, MD Thesis, University of Dakar

Samb, B. (1988) *Mortalité et Morbidité dans le Service de Médecine interne de l'Hôpital Aristide Le Dantec. Etude rétrospective du 1er Janvier 1984 au 31 Décembre 1988*, MD Thesis, University of Dakar

Siraj, M.Y., Hayes, A.W., Unger, P.D., Hogan, G.R., Ryan, N.J. & Wray, B.B. (1981) Analysis of aflatoxin B_1 in human tissues with high-pressure liquid chromatography. *Toxicol. Appl. Pharmacol.*, **58**, 422–430

Toure, Y.I., Abdallah, M. & Diop, B. (1984) Place de la néphrologie dans la morbidité dans un service de médecine interne pour adultes noirs africains à Dakar, à propos de 7379 cas. *Dakar Méd.*, **29**, 213–220

Mycotoxins, Endemic Nephropathy and Urinary Tract Tumours
Ed. M. Castegnaro, R. Pleština, G. Dirheimer, I.N. Chernozemsky & H. Bartsch
Lyon, International Agency for Research on Cancer, pp. 97-103

IMMUNOASSAY OF OCHRATOXIN AND OTHER MYCOTOXINS FROM A SINGLE EXTRACT OF CEREAL GRAINS UTILIZING MONOCLONAL ANTIBODIES

J. Lacey[1], N. Ramakrishna[1], A.A.G. Candlish[2] & J. E. Smith[3]

[1]AFRC Institute of Arable Crops Research, Rothamsted Experimental Station, Harpenden, Herts; [2]Rhône Poulenc Diagnostics Ltd, Glasgow, Scotland; and [3]University of Strathclyde, Department of Bioscience and Biotechnology, Glasgow, Scotland, United Kingdom

Summary

Immunoassays provide rapid, specific, sensitive and inexpensive methods for analysing mycotoxins but have generally been tested individually. Mycotoxigenic fungi rarely occur in pure culture in nature, and different mycotoxins may occur together. It would therefore be advantageous if several immunoassays for different mycotoxins could utilize a single extract. Monoclonal antibodies specific for ochratoxin A, aflatoxin B_1 and T-2 toxin, raised at the University of Strathclyde, allow detection limits of 1 ng/ml ochratoxin A, 0.1 ng/ml aflatoxin B_1 and 10 ng/ml T-2 toxin, when used in competitive enzyme-linked immunosorbent assays. These antibodies have been used to assay ochratoxin A, aflatoxin B_1 and T-2 toxin in a single acetonitrile: 0.5% KCl:6% H_2SO_4 (89:10:1) extract of cereal grain. Extracts were either diluted 1:10 for direct assay or subjected, before assay, to a simple liquid-liquid clean-up procedure, which removed interfering substances and resulted in a 5:1 concentration. Recoveries from barley to which mycotoxins had been added averaged 95.8% for ochratoxin A, 93.8%

for aflatoxin B_1 and 80.6% for T-2 toxin, and the detection limits were 5 ng/g for ochratoxin A, 4 ng/g for aflatoxin B_1 and 50 ng/g for T-2 toxin. Mean coefficients of variation within and between assays and between subsamples were < 12% for ochratoxin A and aflatoxin B_1 but up to 17% for T-2 toxin in assays of barley inoculated with toxigenic fungi. These assays were used to study the production of mycotoxins in mixed cultures of toxigenic fungi and show that toxin production depends on water availability, temperature, presence of interacting species and period of incubation and may be enhanced or decreased by the interaction.

Introduction

Traditional chemical methods for analysing mycotoxins in cereal grains are time consuming, require extensive sample clean-up and are expensive. By contrast, enzyme-linked immunosorbent assays (ELISA) based on antigen–antibody reactions are fast, specific, sensitive and inexpensive, especially if monoclonal antibodies are used. ELISA methods using polyclonal antibodies have been described for ochratoxin A, aflatoxin B_1 and T-2 toxin, but these can be produced in only limited amounts and may vary in affinity and specificity between bleedings. Monoclonal antibodies can potentially be produced in unlimited quantities with constant affinity and specificity and have been described for ochratoxins, aflatoxins, trichothecenes and zearalenone. Candlish, Goodbrand and their colleagues (Candlish *et al.*, 1985, 1986, 1988; Goodbrand *et al.*, 1987) recently developed monoclonal antibodies to ochratoxin A, aflatoxin B_1 and T-2 toxin at the University of Strathclyde, and this paper describes their use to assay these toxins in single extracts of barley grain.

Materials and Methods

Extraction of grain samples

Grain samples were extracted by blending 10 g with 50 ml acetonitrile, 0.5% KCl and 6% H_2SO_4 (89:10:1) for 2 min in a Waring blendor. The extract was filtered through Whatman No. 41 filter paper and diluted 1:10 with Tris/HCl buffer (0.02 M Tris, 0.05% Tween-20, 0.15 M NaCl and HCl to pH 7.4).

Monoclonal antibodies

Monoclonal antibodies (MAb) specific for aflatoxin B_1 and T-2 toxin (Candlish *et al.*, 1985; Goodbrand *et al.*, 1987) were conjugated to horse-radish peroxidase (HRP; type-IV, Sigma) following the method of Kurstak (1985), but ochratoxin A MAb (Candlish *et al.*, 1988) were used unconjugated. Antibodies were stored at –20 °C.

Enzyme-linked immunosorbent assay

Microtitre plates (Nunc MaxiSorp) were coated by adding to each well 150 µl coating buffer (0.02 M Tris/HCl buffer, pH 9) containing 1–2 µg/ml of ochratoxin A-bovine serum albumin (BSA) conjugate (prepared as described by Chu *et al.*, 1976), or aflatoxin B_1-oxime-BSA or T-2 toxin-hemisuccinate-BSA conjugates (Sigma), and incubating either at 4 °C overnight or at 32 °C for 3 h. Plates were then washed four times with Tris/HCl buffer, treated with 2 mg BSA/ml coating buffer (150 µl/well) for 30 min at 32 °C and again washed four times with Tris/HCl buffer. After they had been dried, sensitized plates could be stored at 4 °C for up to six months before use.

Ochratoxin A was assayed by indirect competitive ELISA with 100 µl grain extract and 50 µl ochratoxin A-MAb (diluted 1:5000 in Tris/HCl buffer and 25% normal sheep serum) added to each well of ochratoxin A BSA-sensitized plates and incubated for 1 h at 32 °C. Plates were washed four times with Tris/HCl buffer, incubated with 150 µl goat anti-mouse immunoglobulin G-HRP conjugate (diluted 1:1000)/well for 1 h at 32 °C and again washed four times. Bound enzyme activity was measured after 30 min incubation with 150 µl substrate (1 ml 1% 3,3′,5,5′-tetramethylbenzidine in dimethylsulfoxide and 16 µl H_2O_2 in 100 ml substrate buffer (0.1 M sodium acetate and citric acid to pH 5.0), stopping the reaction with 50 µl of 10% H_2SO_4/well and then measuring absorbance at 450 nm. Two replicate wells were prepared with each ochratoxin A standard (prepared in Tris/HCl buffer, pH 7.4) and three with each grain extract on each microtitre plate.

Aflatoxin B_1 and T-2 toxin were assayed by direct competitive ELISA with 100 µl grain extract and 50 µl aflatoxin B_1-MAb-HRP or T-2 toxin-MAb-HRP conjugate added to each well of microtitre plates sensitized, respectively, with aflatoxin B_1-oxime-BSA or T-2 toxin-hemisuccinate-BSA and incubated for 1 h at 32 °C. Aflatoxin B_1-MAb-HRP was routinely diluted to 1:2000 in Tris/HCl buffer and 25% normal sheep serum, while T-2 toxin-MAb-HRP was diluted to 1:5000. Neutralized antibody was removed by washing four times with Tris/HCl buffer, and bound enzyme activity was measured as described previously.

Standard curves were obtained by plotting $\log_{10}$ concentration against absorbance.

Effect of barley extract on ELISA

Extracts of barley were tested for their interference in ELISA by macerating 10 g of finely ground, healthy grain for 2 min in a Waring blendor with 50 ml acetonitrile or methanol, both containing 0.5% KCl and 6% H_2SO_4 (89:10:1). Extracts were filtered (Whatman No. 41 filter paper), and samples were diluted in Tris/HCl buffer before use in testing for interference in ELISA. Extracts were cleaned up by (i) removal of lipid-soluble components, by partitioning acetonitrile extracts with hexane (25 ml) before assaying the acetonitrile phase, diluted with Tris/HCl buffer, or (ii) removal of polar constituents by partitioning into water in an acetonitrile-water-chloroform mixture. The chloroform phase was evaporated and then reconstituted with Tris/HCl buffer before ELISA. Methanol and acetonitrile extracts in Tris/HCl buffer (100 µl) (with and without clean-up) were incubated with ochratoxin A-MAb or with aflatoxin B_1- or T-2 toxin-MAb-HRP conjugates in sensitized microtitre plates with no free ochratoxin A, aflatoxin B_1 or T-2 toxin. Percent inhibition of binding of MAb or MAb-HRP conjugates to mycotoxin-BSA was calculated.

Subsequently, with concentrations of less than 50 ng ochratoxin A, 10 ng aflatoxin B_1 or 500 ng T-2 toxin/g, the extract was concentrated using a clean-up procedure in which 10 ml acetonitrile extract was shaken in a separating funnel with 10 ml distilled water and 10 ml chloroform for 2 min. The chloroform layer was collected, evaporated to near dryness and redissolved in 2 ml Tris/HCl buffer containing 9% acetonitrile to give a 5:1 concentration of the extract for ELISA.

Recovery of ochratoxin A, aflatoxin B_1 and T-2 toxin from artificially contaminated barley

Finely ground barley samples (10 g) were spiked, one day before extraction, with pure mycotoxins to give 5–500 ng/g ochratoxin A, 4–60 ng/g aflatoxin B_1 and 50–5000 ng/g T-2 toxin. Samples were extracted, filtered and, as appropriate, either diluted 1:10 with Tris/HCl buffer or cleaned up and concentrated prior to assay in ELISA.

Estimation of ochratoxin A, aflatoxin B_1 and T-2 toxin in barley colonized with toxigenic fungi

Barley grain on which toxigenic *Penicillium verrucosum*, *Aspergillus flavus* or *Fusarium poae* had been grown and uninoculated grain were mixed in different proportions to give contaminated samples containing a range of mycotoxin concentrations. Samples (100 g) were finely ground and well mixed, and 10-g subsamples were extracted for assay. Within-assay, inter-assay and subsample coefficients of variation were calculated. Other samples examined were grain naturally contaminated during storage in underground Iron-Age-type pits at Butser Hill Ancient Farm, Hampshire, United Kingdom (B.P. Brenton, N. Ramakrishna & J. Lacey, unpublished data).

Results and Discussion

Tests showed that MAb assay of mycotoxins in grain is fast and specific and that immunoassays for different mycotoxins can be successfully applied to a single extract of grain. Sample preparation of only 5 min was required for direct assay and 11 min when clean-up was necessary. Up to 36 samples could be analysed in duplicate on a single microtitre plate within 2.5 h for ochratoxin A and within 1.5 h for aflatoxin B_1 and T-2 toxin. When the mycotoxin contamination is erratic, however, much time and work could be saved by initially using a mixture of antibodies and then assaying individual mycotoxins only in positive samples. Alternatively, affinity columns, containing a mixture of antibodies, might be used to clean-up extracts before thin-layer or liquid chromatography.

Ochratoxin A, aflatoxin B_1 and T-2 toxin differ greatly in their physical and chemical properties: ochratoxin A is acidic and, for efficient extraction, requires a solvent with pH < 3 (Chu, 1974). The addition of 1 ml 6% H_2SO_4 in 100 ml extraction solvent gave a final pH of 0.9. Extraction of healthy barley with acetonitrile or methanol (with 10 ml 0.5% KCl and 1 ml 6% H_2SO_4 added to each) showed that methanol extracts, diluted 1:8 with Tris/HCl buffer, were more turbid and caused significantly more interference in ELISA than acetonitrile extracts, even after the solvent was evaporated and the sample redissolved in Tris/HCl buffer. Even 90% acetonitrile or methanol did not inhibit MAb binding to the ELISA solid phase. Interference was avoided by diluting both methanol and acetonitrile extracts 1:10 with Tris/HCl buffer; as T-2 toxin is unstable in methanol at room temperature, however, acetonitrile was preferred for extraction.

Among the simple mycotoxin clean-up procedures tested, defatting with hexane had no effect on interference, although this procedure may be advantageous with seeds containing more lipids than barley. Mixing the acetonitrile extract with water, to remove water-soluble interference, and re-extracting with chloroform allowed concentration of the mycotoxins without affecting the standard curves, except that ELISA of T-2 toxin was highly sensitive to slight changes in acetonitrile concentration.

Figure 1 shows the standard curves for ochratoxin A with different dilutions of MAb. Similar curves were obtained for aflatoxin B_1 and T-2 toxin (Ramakrishna *et al.*, 1990). The assays were sensitive to 1–250 ng/ml for ochratoxin A, 0.1–5 ng/ml for aflatoxin B_1 and 10–2000 ng/ml for T-2 toxin, using dilutions of 1:5000 for ochratoxin A-MAb, 1:2000 for aflatoxin B_1-MAb-HRP and 1:5000 for T-2 toxin-MAb-HRP conjugates. Coefficients of variation, both within and between assays of all mycotoxins, were comparable to those reported previously for other antibodies and mycotoxins, ranging between 4 and 12%. Ochratoxin A-MAb gave 8.2% cross-reactivity in a competitive ELISA with ethylester ochratoxin C but no cross-reactivity with ochratoxin α, coumarin or phenylalanine (Candlish *et al.*, 1988). Aflatoxin B_1-MAb gave 12.6%, 14.3%, 1.2%, 7.3% and 4.4% cross-reactivities, respectively, with aflatoxin B_2, aflatoxin G_1, aflatoxin G_2, aflatoxin M_1 and aflatoxin M_2

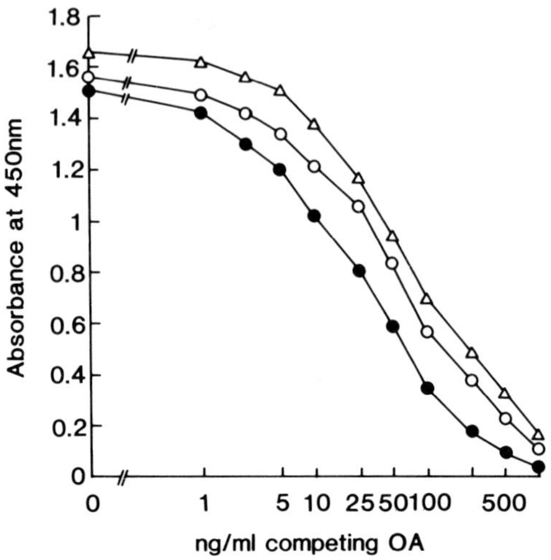

Figure 1. Standard curves from indirect competitive ELISA for ochratoxin A with different dilutions of MAb: 1:1000, filled circles; 1:2000, open circles; 1:5000, triangles

(Candlish *et al.*, 1985); T-2 toxin-MAb showed 25% cross-reactivity with HT-2 toxin but none with T-2 tetraol, T-2 triol, diacetoxyscirpenol or verrucarol (Goodbrand *et al.*, 1987).

Barley spiked with ochratoxin A at 50–500 ng/g, aflatoxin B_1 at 10–60 ng/g and T-2 toxin at 500–5000 ng/g allowed recovery of 98%, 98% and 78% of the respective toxins after 1:10 dilution of acetonitrile extracts. The minimal concentrations that can be determined from their standard curves are 1 ng/ml for ochratoxin A, 0.2 ng/ml for aflatoxin B_1 and 10 ng/ml for T-2 toxin. If extracts are diluted 1:10, the minimal contamination of barley that can be detected is 50 ng/g for ochratoxin A, 10 ng/g for aflatoxin B_1 and 500 ng/g for T-2 toxin. To assay smaller concentrations of ochratoxin A and T-2 toxin, sample extracts must be concentrated before ELISA. After clean-up and concentration at 5:1 prior to ELISA, mean recoveries from barley spiked with ochratoxin A at 5–10 ng/g, aflatoxin B_1 at 4–8 ng/g and T-2 toxin at 50–100 ng/g were 92%, 88% and 84%, respectively. Minimum detection limits for ochratoxin A and aflatoxin B_1, with and without clean-up, were comparable with those reported previously, but clean-up was particularly useful for improving the minimum detection limit for T-2 toxin in barley.

Ochratoxin A, aflatoxin B_1 and T-2 toxin concentrations in barley grain colonized with the toxigenic fungi *P. verrucosum*, *A. flavus* and *F. poae* were 42–5911, 3–559 and 528–21056 ng/g, respectively. Mean within-assay, inter-assay and subsample coefficients of variation were smaller for ochratoxin A and aflatoxin B_1 (8.9–10.0% for ochratoxin A; 8.1–11.9% for aflatoxin B_1) than for T-2 toxin (14.7–17.0%).

Ochratoxin was detected in 77% of barley samples after winter storage in Iron-Age underground pits. Concentrations ranged from 1.3 to 385 ng/g, the largest concentrations occurring in samples containing 24–30% water.

Immunoassays could be invaluable in epidemiological studies of mycotoxin-related diseases. Not only could they be used to screen cereal grains and other foods but they might also be used to screen fungi for mycotoxin-producing ability and to detect toxins in airborne dust that might be inhaled, in body fluids and, utilizing immuno-gold techniques, in biopsy and necropsy specimens.

References

Candlish, A.A.G., Stimson, W.H. & Smith, J.E. (1985) A monoclonal antibody to aflatoxin B1: detection of the mycotoxin by enzyme immunoassay. *Lett. Appl. Microbiol.*, **1**, 57–61

Candlish, A.A.G., Stimson, W.H. & Smith, J.E., (1986) A monoclonal antibody to ochratoxin A. *Lett. Appl. Microbiol.*, **3**, 9–11

Candlish, A.A.G., Stimson, W.H. & Smith, J.E. (1988) Determination of ochratoxin A by monoclonal antibody-based enzyme immunoassay. *J. Assoc. Off. Anal. Chem.*, **71**, 961–964

Chu, F.S. (1974) Studies on ochratoxins. *CRC Crit. Rev. Toxicol.*, **2**, 499–524

Chu, F.S., Chang, F.C.C. & Hinsdill, R.D. (1976) Production of antibody against ochratoxin A. *Appl. Environ. Microbiol.*, **31**, 831–835

Goodbrand, I.A., Stimson, W.H. & Smith, J.E. (1987) A monoclonal antibody to T-2 toxin. *Lett. Appl. Microbiol.*, **5**, 97–99

Kurstak, E. (1985) Progress in enzyme immunoassays: production of reagents, experimental design and interpretation. *Bull. World Health Organ.*, **63**, 793–811

Ramakrishna, N., Lacey, J., Candlish, A.A.G., Smith, J.E. & Goodbrand, I.A. (1990) Monoclonal antibody-based enzyme linked immunosorbent assay of aflatoxin B1, T-2 toxin and ochratoxin A in barley. *J. Assoc. Off. Anal. Chem.*, **73**, 71–76

Mycotoxins, Endemic Nephropathy and Urinary Tract Tumours
Ed. M. Castegnaro, R. Pleština, G. Dirheimer, I.N. Chernozemsky & H. Bartsch
Lyon, International Agency for Research on Cancer, pp. 105-108

CONTAMINATION OF HUMAN MILK WITH OCHRATOXIN A

C. Micco[1], M.A. Ambruzzi[2], M. Miraglia[1], C. Brera[1], R. Onori[1] & L. Benelli[1]

[1]*Istituto Superiore di Sanità, Laboratorio Alimenti and* [2]*Ospedale Pediatrico Bambino Gesù, Servizio Dietologia Clinica, Rome, Italy*

Summary

Ochratoxin A is a mycotoxin frequently found as a contaminant both in food and in animal feed. It can reach humans through the food chain and can then be excreted in biological fluids, one of which is human milk; it can therefore be transmitted from mother to child during breast-feeding.

This fact prompted us to carry out the present study, aimed at the determination of ochratoxin A in human milk in Italy, as done elsewhere. Fifty samples of human milk were collected randomly over one year and analysed by a high-performance liquid chromatorgaphy method. Nine samples were found to contain levels in the range of 1.7–6.6 ng/ml.

Introduction

Ochratoxin A is found in foods of plant origin and in edible animal tissues. The occurrence of ochratoxin A in blood and human milk in various European countries demonstrates the real risk of human exposure to this toxin. In Yugoslavia, 42 out of 639 human blood serum samples contained ochratoxin A at 1–57 ng/ml (Hult *et al.*,1982); in Poland, 77 out of 1065 human blood serum samples contained ochratoxin A at a mean value of 0.27 ng/ml and a maximum value of 40 ng/ml) (Goliński, 1987); in Germany, 173 out of 306 serum samples contained a mean level of 0.6 ng/ml (range, 0.1–14.4 ng/ml) (Bauer & Gareis, 1987); in Denmark, 46 out of 96 blood plasma samples had ochratoxin A at 0.1–9.2 ng/g (Hald, 1989); and in Bulgaria, 45 out of 312 blood samples contained a mean of approximately 14 ng/ml (Petkova-Bocharova *et al.*, 1988). Ochratoxin A was found in 4 out of 36 samples of human milk in Germany (range, 0.017–0.03 ng/ml) (Gareis *et al.*, 1988).

The nephrotoxic potential of ochratoxin A has been recognized for many years, but recently it was also shown to be a renal carcinogen in mice and rats (IARC,1987).

The preliminary study reported here is the first step in a broad monitoring programme of human milk samples collected randomly in different areas of Italy, in order to take into consideration the wide variation in daily diets across the country. Further studies will be necessary to highlight national areas with higher risks of exposure to ochratoxin A contamination in order to establish a potential correlation between the incidence of renal pathology in Italy and the presence of this toxin in human beings through the food chain. This study provides the first information on the presence of ochratoxin A in human milk samples in Italy.

Materials and Methods

Fifty samples of human milk taken at one feeding were collected randomly in 1989–90 from different areas of Italy and frozen at –20 °C until analysed by a high-performance liquid chromatography method (Gareis *et al.*, 1988).

The apparatus used was a Varian 5000 delivery system pump equipped with a Lichrospher 100 RP 18 column (Merck) and a C 18 Lichrospher precolumn (Merck). Detection was performed on a Perkin-Elmer LS 4 fluorescence spectrometer (excitation wavelength, 333 nm; emission, 470 nm). A precolumn derivatization step was performed to convert ochratoxin A to its methyl ester derivative for confirmation of identity. The limit of detection in milk was 0.2 ng/ml. Typical chromatograms of human milk extracts are shown in Figure 1.

For recovery experiments, a stock solution of ochratoxin A in methanol at a concentration of 0.5 mg/ml was prepared, and suitable volumes were drawn successively to prepare working standard solutions at concentrations of 0.01, 0.05, 0.10 and 0.20 ng/ml. These solutions were used both for spiking human milk samples and for establishing calibration curves. Recovery values are presented in Table 1; the average recovery of 87% was found for levels of 1–10 ng/ml.

Results and Discussion

Ochratoxin A was detected in nine out of 50 samples at 1.2–6.6 ng/ml (Table 2). These results confirm the possibility that ochratoxin A contamination occurs in human biological fluids, even in Italy and other

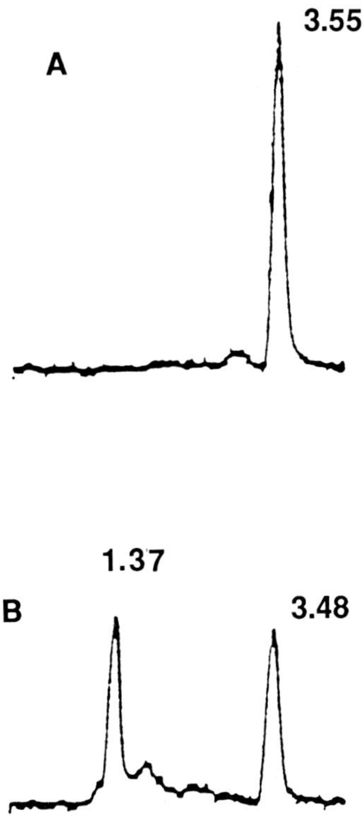

Figure 1. High-performance liquid chromatograms of (A) ochratoxin A standard and (B) ochratoxin A-contaminated human milk

Table 1. Average recovery from ochratoxin A-spiked human milk

Added (ng/ml)	Found (ng/ml)	Recovery (%)
1	0.8	86
1	0.9	90
5	4.2	84
5	4.1	82
10	9.0	90
10	9.0	90

Standard deviation, 3.5

countries where nephropathy is not endemic. The milk samples were drawn from healthy women in different geographical regions with varying daily diets. The reported levels of ochratoxin A in human milk are similar to those found in other European countries in other biological fluids, such as serum.

Table 2. Ochratoxin A levels in human milk samples

Sample	Ochratoxin A (ng/ml)
1	2.5
2	4.5
3	1.8
4	6.6
5	4.9
6	6.5
7	4.6
8	4.6
9	1.7

50 samples analysed; limit of detection, 0.2 ng/ml

No definitive conclusion can be drawn because the number of samples analysed was small and because no clear interrelation was found between routine diet of subjects and ochratoxin A levels in milk samples.

Conclusions

The aim of this study was to verify the correlation between direct consumption of ochratoxin A-contaminated food and its presence in biological fluids such as human milk. These findings should be considered as a starting point for future studies in Italy for the evaluation of human risk, for the monitoring of the presence of ochratoxin A in feed and food and for a potential correlation between contamination and renal pathology.

References

Bauer, J. & Gareis, M. (1987) Ochratoxin A in the food chain. *J. Vet. Med. B*, **34**, 613–627

Gareis, M., Märtlbauer, E., Bauer, J., & Gedek, B. (1988) Determination of ochratoxin A in breast milk. *Z. Lebensm. Unters. Forsch.*, **186**, 114–117

Goliński, P. (1987) *Ochratoxin A in Humans as a Result of Food and Feed Contamination*, Thesis, Poznań, Rocznalion Akademii Rolniczej

Hald, B. (1989) Human exposure to ochratoxin A. In: Natori, S., Hashimoto, K. & Ueno, Y., eds, *Mycotoxins and Phycotoxins*, Amsterdam, Elsevier Science Publishers

Hult, K., Pleština, R., Habazin-Novak, V., Radić, B. & Čeović, S. (1982) Ochratoxin A in human blood and Balkan endemic nephropathy. *Arch. Toxicol.*, **51**, 313–321

IARC (1987) *IARC Monographs on the Evaluation of Carcinogenic Risks to Humans*, Suppl. 7, *Overall Evaluations of Carcinogenicity: An Updating of IARC Monographs Volumes 1 to 42*, Lyon, pp. 271–272

Petkova-Bocharova, T., Chernozemsky, I. N. & Castegnaro, M. (1988) Ochratoxin A in human blood in relation to Balkan endemic nephropathy and urinary system tumours in Bulgaria. *Food Addit. Contam.*, **5**, 299–301

Mycotoxins, Endemic Nephropathy and Urinary Tract Tumours
Ed. M. Castegnaro, R. Pleština, G. Dirheimer, I.N. Chernozemsky & H. Bartsch
Lyon, International Agency for Research on Cancer, pp. 109-111

DECOMPOSITION OF OCHRATOXIN A BY HEAT AND γ-IRRADIATION

M. Kostecki[1], P. Goliński[1,3], W. Uchman[2] & J. Grabarkiewicz-Szczęsna[1]

[1]Department of Chemistry and [2]Institute of Food Technology of Animal Origin, Agricultural University of Poznań, Poland

Summary

Up to 50% of initial amounts of 2 and 100 ng/ml ochratoxin A were decomposed after γ-irradiation of solutions in water, in 2% aqeous NaCl or an aqueous solution of 2% NaCl and $NaNO_2$. Ochratoxin A in these solutions was not decomposed, however, after heating at 20, 121 or 135 °C for 15 min.

Introduction

Widely spread microflora and toxic secondary metabolites formed by fungi deteriorate or decrease the nutritive value of about 20% of cereals, decreasing animal production and presenting a public health problem (for review, see Goliński, 1987).

Several methods have been described for preventing the formation of mycotoxins and for the detoxification of secondary metabolites (Chełkowski *et al.*, 1982). The fate of mycotoxins during processes such as alcoholic fermentation, beer production, cleaning of kernels, fractionated cereal milling and bread baking has been investigated, but the techniques used were unsatisfactory (Chełkowski *et al.*, 1981). The effect of γ-irradiation on toxinogenic fungi and mycotoxins has also been investigated (Maxcy, 1982; Sommer & Mitchel, 1986).

---

[3]To whom correspondence should be addressed

Since ochratoxin A residues have been found in porcine tissue and blood in Poland (Goliński *et al.*, this volume) and Poles consume large amounts of pork and products made with pork blood, we decided to study the behaviour of ochratoxin A under the conditions encountered during meat processing—heating and γ-irradiation and dissolution in aqueous solutions of the food preservatives NaCl, $NaNO_2$ and ascorbic acid.

Materials and Methods

Ochratoxin A was purchased as crystalline standard from Sigma (St Louis, MO, USA) and dissolved to a final volume of 2.5 ml for a concentration of 100 ng/ml, and to a final volume of 100 ml for a concentration of 2 ng/ml, in one of the following: a 2% aqueous solution of NaCl; $NaNO_2$ at 0.15 mg/ml in 2% NaCl; ascorbic acid at 0.4 mg/ml in 2% NaCl; $NaNO_2$ plus ascorbic acid at the above concentrations in 2% NaCl; or water.

To test for the effects of temperature, these samples were heated for 15 min to 20, 121 or 135 °C in glass ampoules (100 ng/ml) or in glass bottles (2 ng/ml). To test for the effects of γ-irradiation, samples in NaCl, $NaNO_2$ or water were irradiated at 0.1–50 kGy with a ^{60}Co source type RCHM γ 20.

Ochratoxin A was determined by spectrofluorodensitometry according to the method of Nesheim (1973), with our modifications (Goliński *et al.*, 1984). The results reported are averages of six replicate measurements at each dose; standard deviations are also given.

Results and Discussion

Ochratoxin A was not affected by heating in any of the solutions, confirming our earlier observations of the stability of the mycotoxin during such hydrothermic processes as baking (Goliński, 1987).

Up to about 50% of the toxin was decomposed by γ-irradiation (Table 1). The efficiency of decomposition was comparable in the two solutions of food preservatives, indicating that this method could be of use in meat processing. Even approximately 1 min of irradiation with 1.0 kGy resulted in a significant reduction in the concentration of toxin. This radiation dose is higher than those used normally (0.1–0.2 kGy) but is still in the range allowed in food technology. The structure and toxicity of the decomposition product of ochratoxin A are being investigated.

In view of the fact that about 37% of porcine serum samples in Poland had residues of ochratoxin A, with an average level of 5 ng/ml (Goliński *et al.*, this volume), application of this technique in the meat industry would appear to be desirable.

Table 1. Decomposition of ochratoxin A in aqueous solutions during γ-irradiation

Concentration of ochratoxin A (ng/ml)	Radiation dose (kGy)	Decomposition (% ± SD)		
		NaCl	$NaNO_2$	Water
2	0.1	5	10	Not studied
	1.0	10	20	Not studied
	10.0	25	40	Not studied
	50.0	45	45	Not studied
100	0.1	6 ± 0.4	10 ± 1.3	22 ± 4.9
	1.0	32 ± 4.0	12 ± 1.2	36 ± 5.5
	5.0	30 ± 6.8	32 ± 7.2	50 ± 7.9
	10.0	40 ± 7.4	36 ± 6.4	45 ± 5.8
	20.0	48 ± 8.8	52 ± 9.1	50 ± 7.1
	50.0	44 ± 9.2	44 ± 7.1	42 ± 5.0

References

Chełkowski, J., Goliński, P. & Szebiotko, K. (1981) Mycotoxins in cereal grain. Part II. The fate of ochratoxin A after processing of wheat and barley grain. *Nahrung*, **25**, 423–426

Chełkowski, J., Szebiotko, K., Goliński, P., Buchowski, M., Godlewska, B., Radomysk W. & Wiewiówska, M. (1982) Mycotoxins in cereal grain. Part V. Changes of cereal grain biological value after ammoniation and mycotoxins (ochratoxins) inactivation. *Nahrung*, **26**, 1–7

Goliński, P. (1987) Ochratoxin A in human organisms as the result of food and feed contamination. *Rocz. AR Poznaniu*, **168** (in Polish)

Goliński, P. & Grabarkiewcz-Szczęsna, J. (984) Chemical confirmatory test for ochratoxin A, citrinin, penicillic acid, sterigmatocystin and zearalenone, performed directly on thin layer chromatographic plates. *J. Assoc. Off. Anal. Chem.*, **67**, 1108–1110

Maxcy, R.B. (1982) Irradiation of food for public health protection. *J. Food Prot.*, **45**, 363–366

Nesheim, S. (1973) Analysis of ochratoxin A and barley using partition and thin layer chromatography. I. Development of the method. *J. Assoc. Off. Anal. Chem.*, **56**, 817–821

Sommer, N.F. & Mitchel, F.G. (1986) Gamma irradiation—a quarantine treatment for fresh fruits and vegetables. *Hort. Sci.*, **21**, 356–360

Mycotoxins, Endemic Nephropathy and Urinary Tract Tumours
Ed. M. Castegnaro, R. Pleština, G. Dirheimer, I.N. Chernozemsky & H. Bartsch
Lyon, International Agency for Research on Cancer, pp. 113-117

COMPARATIVE ACUTE NEPHROTOXICITY OF *PENICILLIUM AURANTIOGRISEUM* IN RATS AND HAMSTERS

G.C. Hard[1] & J.B. Greig

Medical Research Council Toxicology Unit, Carshalton, Surrey, United Kingdom

Summary

Air-dried mycelium of *Penicillium aurantiogriseum*[2], grown as a surface culture on yeast extract-sucrose medium, was incorporated into powdered diet and fed to rats and hamsters for different periods up to 28 days. At intervals, animals were anaesthetized and the kidneys fixed *in situ* by perfusion. In rats, the fungus produced scattered exfoliation of pyknotic cells and an increased frequency of mitotic figures involving the pars recta segment of proximal tubular epithelium. This lesion was detectable as early as three days after beginning of treatment and was well developed by 14 days. No degenerative tubular change or mitogenic effect was observed in hamsters, even after feeding for 35 days; and there was no apparent renal pelvic or interstitial lesion in either species.

[1]Present address: American Health Foundation, Valhalla, New York, USA

[2]The original Balkan isolate of a nephrotoxic *Penicillium* species was classified as *P. verrucosum* var. *cyclopium* (Barnes *et al.*, 1977). Subsequently, a revision of the taxonomy of the *Penicillia* has meant that the isolate should be classified as *P. aurantiogriseum* (Yeulet *et al.*, 1988). It does not produce ochratoxin A or citrinin.

Introduction

Although attention has recently been focused on ochratoxin A as a possible causative factor of Balkan endemic nephropathy, involvement of unidentified mycotoxins cannot be excluded. *Penicillium aurantiogriseum* is one example of a food-contaminating mould that is commonly found in the endemic regions (Austwick, 1981; Pepeljnjak & Cvetnić, 1984; Macgeorge & Mantle, 1991). Previous work showed that a mycelial extract of this fungus fed for one month to rats produced electron microscopic evidence of degenerative change in the pars recta of proximal tubules, with marked cellular and nuclear enlargement (Peristianis *et al.*, 1978). After 12 months of feeding, there was extensive karyomegaly and multinucleate cells in the same tubular segment (Austwick *et al.*, 1979). As renal pelvic carcinoma occurs frequently in conjunction with Balkan endemic nephropathy (Radovanović *et al.*, 1984, 1985), and a common link is believed to exist between the cortical nephropathy and urothelial cancer (Sattler *et al.*, 1977), the experimental screening of candidate toxins should include the renal pelvis as well as the outer kidney zones, preferably in more than one laboratory animal species. In addition, knowledge of the early histopathological change produced by *P. aurantiogriseum* would be useful for devising a short-term assay in rodents for screening fungal contaminants from different regions in order to investigate the distribution and seasonal fluctuation of toxigenic varieties.

Methods

An isolate (M2) of *P. aurantiogriseum* collected from contaminated grain by Dr P.G. Mantle in a Yugoslav area of Balkan endemic nephropathy (Yeulet *et al.*, 1988) was grown on moist, sterile bran flakes. These cultures were used to inoculate sterile aqueous medium (200 ml) containing yeast extract (2% w/v, Difco Laboratories, Detroit, MI, USA) and sucrose (15% w/v, FSA Laboratory Supplies, Loughborough, United Kingdom) in 1-litre conical flasks. Cultures were grown for 13–16 days at 22 ± 2 °C and were then killed by the addition of chloroform. After the cultures had stood overnight, the mycelium was harvested by filtration through a nylon mesh; it was then air-dried in a fume cupboard until it was free of the odour of chloroform, whereupon residual medium was expressed and the mycelium finely minced. The experimental diet was prepared by incorporating the minced mycelium (15% w/w) into powdered rat diet 41(B) or hamster diet (Special Diets Services, Witham, Essex, United Kingdom), along with arachis oil (2% w/w) to prevent dust formation. The animals used were 10 eight-week-old male rats (180–200 g) of the outbred Lac:PAW strain, bred on-site at Carshalton, and 35 male Syrian hamsters aged 8–9 weeks (85–135 g) obtained from the National Institute for Medical Research (Mill Hill, London). After random allocation into groups, the rats were fed the test diet for 3, 7 and

14 days and the hamsters for 5, 15 and 28 days. Some hamsters were kept for a further 41–43 weeks on a normal pelletted diet. At the end of each period of exposure, animals were anaesthetized with intraperitoneal chloral hydrate (7% w/v; 1.0–2.0 ml/animal) and prepared for intravascular perfusion and fixation of the kidneys with buffered formalin or 2% glutaraldehyde. The perfusate was introduced into the abdominal aorta in rats and *via* the left ventricle in hamsters. After perfusion, the kidneys were removed and immersed in buffered formalin. Haematoxylin and eosin-stained sections were prepared by conventional means. Hamsters maintained for the longest observation period were not perfused but were killed with carbon dioxide, and their kidneys were fixed by immersion.

Results and Discussion

All rats exposed to the mycelium showed cellular changes involving the pars recta of the proximal tubule. The lesions were most severe in rats fed the test diet for 14 days, but those killed after three days of exposure were also affected. The fungus produced scattered pyknosis, exfoliation of degenerate cells into tubular lumens (Figure 1) and an increase in the frequency of mitotic figures of tubular epithelium (Figure 2). These changes appeared to be located mainly in the outer stripe of the outer medulla and did not extend into the cortical pars recta tubules comprising medullary rays. Mitotic figures were particularly frequent at the point where pars recta became the thin limb of Henle, involving junctional cells in both segments. Some mitotic figures were abnormal in appearance (Figure 3). Sparsely scattered pars recta cells had enlarged nuclei (karyomegaly) at 14 days (Figure 4), indicating mitotic division without cytokinesis, leading to increased ploidy (Jackson, 1974). Karyomegaly was first noted, but only occasionally, at the seven-day exposure stage.

None of the hamsters showed any cellular change in the kidney and were clearly refractory to the necrotizing and mitogenic effects observed in rats. Furthermore, the lining cells of the renal pelvis and papilla showed no degenerative or regenerative change in either species. Perfusion fixation enhances visualization of alterations that affect cells of the interstitium, but none was seen in rat or hamster.

The lesions produced in rats by *P. aurantiogriseum* mycelium were consistent with the nuclear changes in pars recta cells observed previously after longer periods of exposure (Barnes *et al.*, 1977; Peristianis *et al.*, 1978; Austwick *et al.*, 1979). Since our acute study was conducted, Adatia *et al.* (1991) have demonstrated identical cellular changes in rats administered a partially purified ethanol-extracted fraction of *P. aurantiogriseum* both orally and intraperitoneally, indicating that the various investigations have been dealing with the same toxic component.

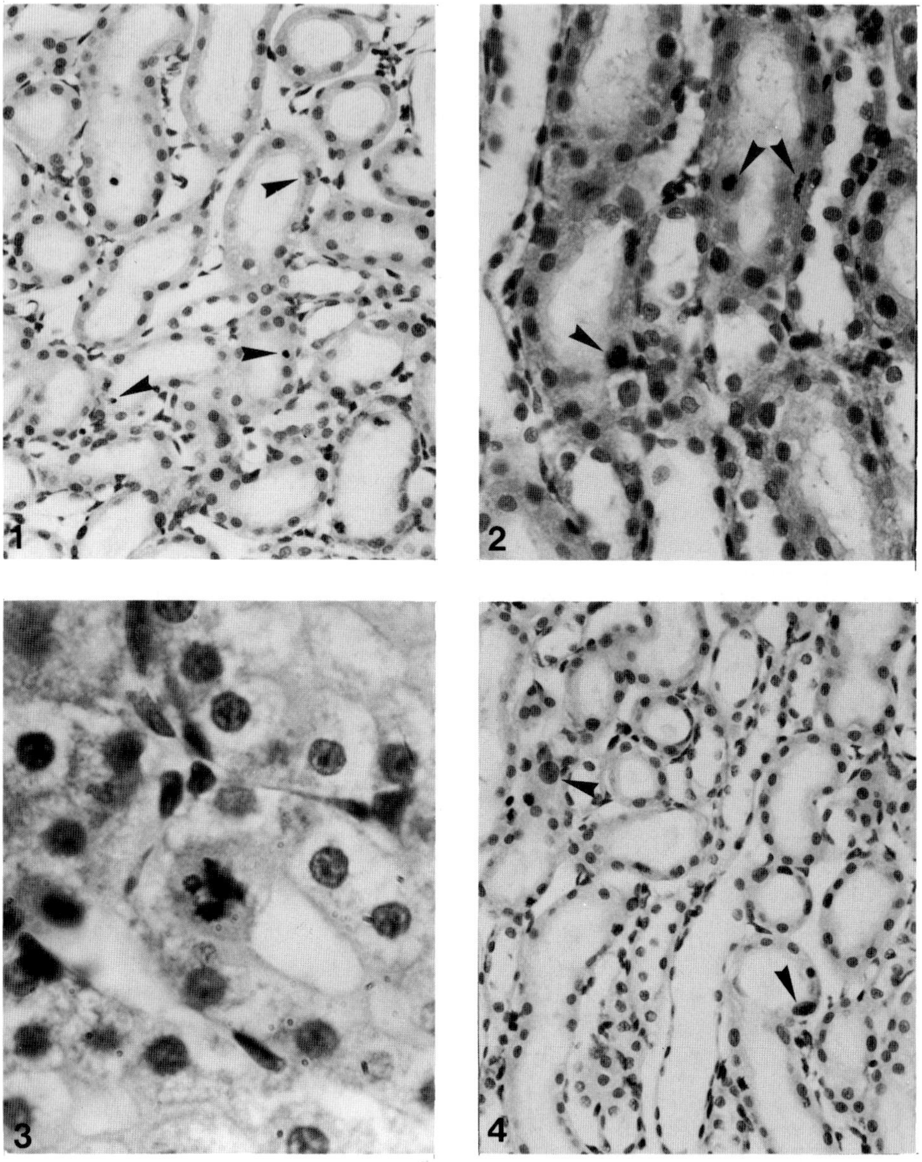

Fig. 1. Pyknotic lining cells (arrows) in pars recta tubules of rat kidney 7 days after exposure to mycelium and exfoliation of a degenerating cell into tubular lumen. H & E x 365

Fig. 2. Frequent mitotic figures (arrows), indicative of increased cell turnover, in rat pars recta cells after 14 days of exposure to mycelium. H & E x 590

Fig. 3. Abnormal mitotic figure suggestive of increased ploidy in rat pars recta cell after 14 days of exposure to mycelium. H & E x 1440

Fig. 4. Karyomegaly (arrows) in a pars recta cell of rat kidney after 14 days of exposure to mycelium. H & E x 365

Our study has indicated some species difference in susceptibility to the toxicity of this fungus, as well as showing that renal pelvic or interstitial cell populations are not involved in the acute response. More importantly, the investigation underlines the feasibility of screening field samples of the fungus obtained directly from contaminated foodstuffs in endemic areas for nephropathic activity. The basis of a short-term assay could lie in the oral administration by gavage of suspensions of fungus harvested from grain or other food to small groups of rats for a limited period of about five days, followed by evaluation of renal histopathology.

Acknowledgements

The authors are grateful to Dr P.G. Mantle of the Department of Biochemistry, Imperial College of Science, Technology and Medicine, London, for providing the starting culture of *P. aurantiogriseum.*

References

Adatia, R., Heaton, J.M., Macgeorge, K.M. & Mantle, P.G. (1991) Acute histopathological changes produced by *Penicillium aurantiogriseum* nephrotoxin in the rat. *Int. J. Exp. Pathol.*, **72**, 47–53

Austwick, P.K.C. (1981) Balkan nephropathy. *Practitioner*, **225**, 1031–1038

Austwick, P.K.C., Carter, R.L., Greig, J.B. & Peristianis, G.C. (1979) Mycotoxic activity in isolates of *Penicillium verrucosum* var. cyclopium from the endemic areas of Balkan nephropathy. In: Strahinjic, S. & Stefanovic, V., eds, *Proceedings of the 4th Symposium on Endemic (Balkan) Nephropathy*, Niš, University of Niš, p. 263 (Abstract)

Barnes, J.M., Carter, R.L., Peristianis, G.C., Austwick, P.K.C., Flynn, F.V. & Aldridge, W.N. (1977) Balkan (endemic) nephropathy and a toxin-producing strain of *Penicillium verrucosum* var. cyclopium: an experimental model in rats. *Lancet*, **i**, 671–675

Jackson, M.R. (1974) The nature of dimethylnitrosamine induced enlargement of rat hepatocyte nuclei. *J. Pathol.*, **113**, 197–208

Macgeorge, K.M. & Mantle, P.G. (1991) Nephrotoxic fungi in a Yugoslavian community in which Balkan nephropathy is hyperendemic. *Mycol. Res.*, (in press)

Pepeljnjak, S. & Cvetnić, Z. (1984) Distribution of moulds on stored grains in households in an area affected by endemic nephropathy in Yugoslavia. *Mycopathologia*, **86**, 83–88

Peristianis, G.C., Austwick, P.K.C. & Carter, R.L. (1978) An ultrastructural investigation of nephrotoxicity in rats induced by feeding cultures of *Penicillium verrucosum* var. cyclopium. *Virchows Arch. B Cell. Pathol.*, **28**, 321–337

Radovanović, Z., Gledovićić, Z. & Janković, S. (1984) Balkan nephropathy and malignant tumours. *Neoplasma*, **32**, 225–229

Radovanović, Z., Krajinović, S., Janković, S., Hall, P.W. & Petković, S. (1985) Family history of cancer among cases of upper urothelial tumours in a Balkan nephropathy area. *Cancer Res. Clin. Oncol.*, **110**, 181–183

Sattler, T.A., Dimitrov, T. & Hall, P.W. (1977) Relation between endemic (Balkan) nephropathy and urinary-tract tumours. *Lancet*, **i**, 278–280

Yeulet, S.E., Mantle, P.G., Rudge, M.S. & Greig, J.B. (1988) Nephrotoxicity of *Penicillium aurantiogriseum*, a possible factor in the aetiology of Balkan endemic nephropathy. *Mycopathologia*, **102**, 21–30

Mycotoxins, Endemic Nephropathy and Urinary Tract Tumours
Ed. M. Castegnaro, R. Pleština, G. Dirheimer, I.N. Chernozemsky & H. Bartsch
Lyon, International Agency for Research on Cancer, pp. 119-127

PENICILLIUM AURANTIOGRISEUM-INDUCED, PERSISTENT RENAL HISTOPATHOLOGICAL CHANGES IN RATS; AN EXPERIMENTAL MODEL FOR BALKAN ENDEMIC NEPHROPATHY COMPETITIVE WITH OCHRATOXIN A

P.G. Mantle[1], K.M. McHugh[1], R. Adatia[1], J.M. Heaton[2], T. Gray[3] & D.R. Turner[3]

[1]Department of Biochemistry, Imperial College of Science, Technology and Medicine, London; [2]Department of Histopathology, Victoria Hospital, Worksop, Notts; and [3]Department of Pathology, Queen's Medical Centre, Nottingham, United Kingdom

Summary

Renal histopathological changes in rats, caused by food partially moulded by a common fungus isolated from an area of nephropathy in Yugoslavia, were differentiated into acute and chronic responses. The acute response to a few days on the diet specifically involved necrosis and concomitant mitosis in proximal tubule cells. More protracted, continuous or intermittent administration of nephrotoxic mould led to a marked karyomegaly in the same corticomedullary region. The phenomenon is more promiment than that induced by treatment with ochratoxin A over a similar period, raising the question of the putative role of such mycotoxins in the etiology of chronic human renal disease.

Text

We have sought to differentiate the renal histopathological changes caused in rats given food moulded by Yugoslav *Penicillium aurantio-griseum* into acute and chronic responses. The striking karyomegaly which developed extensively in proximal convoluted tubules during several weeks of dosing with this common nephrotoxic fungus was compared with that induced by ochratoxin A, also given in food. The findings form a basis for discussing the potential role of these mycotoxins in Balkan endemic nephropathy and its association with a high frequency of renal tumours.

The circumstances in which clinical expression of a particular type of disease of the renal interstitium occurs, as seen in the advanced atrophy associated with Balkan nephropathy, are of rural agricultural communities which have traditionally produced and stored much of their own food. The locations at low altitude in regions of high rainfall near rivers at temperate latitudes (43–45 °) provide the physical conditions of high humidity and temperature that favour fungal spoilage of food which is not stored in a controlled environment. Consequently, in the past, mouldy commodities have been an inevitable part of the human diet. Growth of common fungi of the genera *Penicillium* and *Aspergillus* on carbohydrate-rich agricultural produce such as cereals (mainly maize and wheat) and legumes, and on smoked meats, need not necessarily be overtly luxuriant for spoilage, in the form of contributed toxic metabolites, to have occurred. After potently nephrotoxic mycotoxins were recognized in the 1970s, they soon became obvious candidates as etiological agents in the mysterious Balkan nephropathy which still defies satisfactory explanation.

The only nephrotoxic mould specifically sought and uniquely isolated from food in a hyperendemic area (the Vratza region of Bulgaria) and offered as a potential necrotizing agent active rather specifically against rat renal proximal tubules conformed to the description of *Penicillium verrucosum* var. *cyclopium* (Barnes *et al.*, 1977). The nephrotoxin(s) has still not been fully characterized (although it is known that it is neither citrinin nor ochratoxin) but is produced readily on damp shredded wheat by a wide range of penicillia conforming to the more modern species concepts of *P. aurantiogriseum* and *P. commune* (Pitt, 1988). Such fungi were among the commonest isolated from agricultural produce in a hyperendemic area of Yugoslavia (Macgeorge & Mantle, 1990, 1991). The effect of forced feeding laboratory-cultured fungal biomass to experimental rats on 20 days during a month was first described by Barnes *et al.* (1977) as a complex picture of necrosis, regeneration and some enlarged nuclei at the renal corticomedullary junction. Recent systematic reappraisal of this phenomenon has shown that an acute response seen as tubular necrosis and concomitant mitoses is already apparent after one day's consumption of a mouldy diet, although the effect becomes progressively more prominent (Figure 1) over a five-day period (Adatia *et al.*, 1991). These histopathological changes are apparently only temporary, the tissue becoming indistinguishable from normal after a week or two on uncontaminated diet. They are interpreted as necrosis in epithelial cells of the P3 (straight) segment of proximal tubules which have absorbed indiscriminately from the glomerular filtrate an unusual compound which is directly, or is metabolized to become, cytotoxic. Whether solely as a normal reaction mechanism, or also by some direct stimulus from another unusual component of the glomerular filtrate, adjacent cells respond to the necrosis by a regeneration mechanism seen as mitoses.

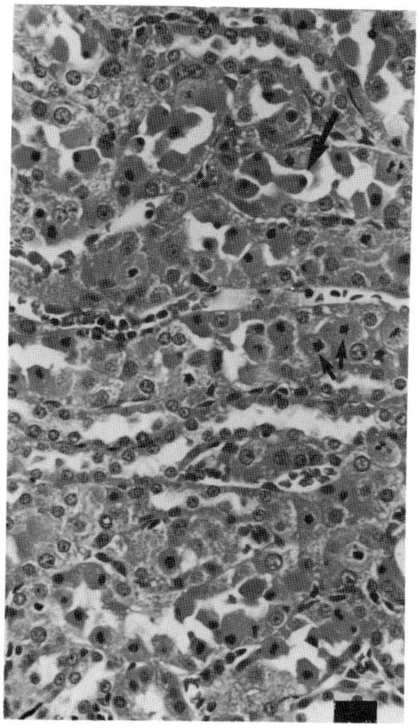

Figure 1. An area of acute renal tubular necrosis at the corticomedullary junction of a rat dosed for 5 days with shredded wheat moulded by *Penicillium aurantiogriseum*, showing degenerating cells (big arrow) and regenerating cells (small arrows) attributed to the P3 segment of nephrons

Haematoxylin and eosin; bar, 30 μm

In contrast, rats maintained for several weeks on a diet containing a 20% component of shredded wheat moulded by *P. aurantiogriseum* develop histologically striking cellular changes in the same region as that in which acute responses are evident (Figure 2). Many proximal tubular cells enlarge and develop either giant nuclei, up to 3.5 times the diameter of normal nuclei (a volume increase of about 40-fold), or multiple nuclei (Mantle *et al.*, 1991). These changes persist, as demonstrated in individual animals in which the effect of a chronic dosing regime is evidenced in renal tissues after unilateral nephrectomy,

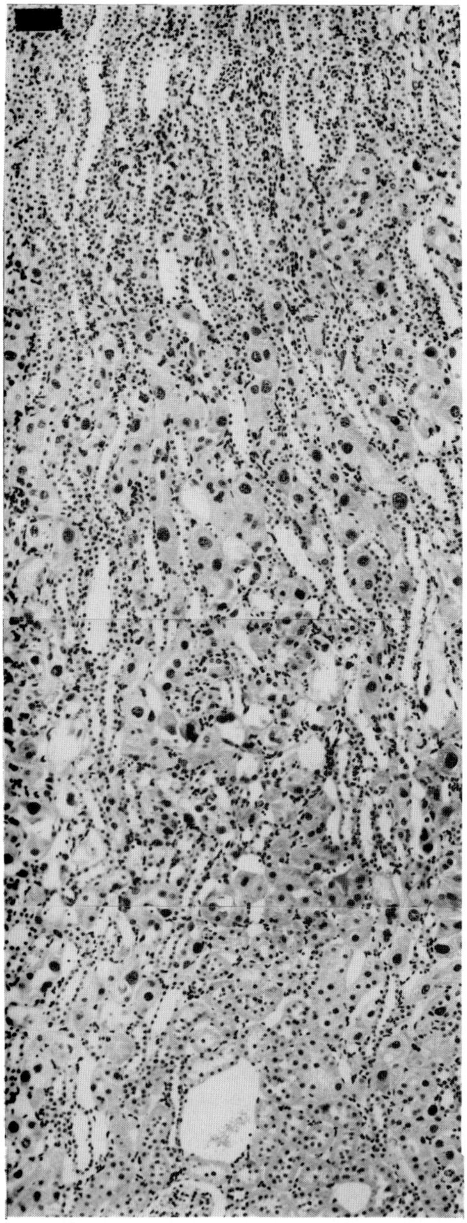

Figure 2. Section of rat kidney spanning the corticomedullary junction, showing the frequency and extent of distribution of karyomegaly 15 weeks after commencing a diet containing *P. aurantiogriseum*-moulded shredded wheat

The extremes of the section show (top) part of an innermost glomerulus immediately surrounded by unchanged tissues and (bottom) the medullary region including descending thin limbs of renal tubules in which the cells also appear normal. Haematoxylin and eosin; bar, 130 µm for at least 8 months after transfer to a diet free from the nephrotoxin. Throughout, the treated animals show no clinical effect of a nephrotoxic diet. Even a crude nephrotoxic extract of moulded shredded wheat given in food over 10 days at a dose about 100 times that which gave prominent acute renal histopathological changes was accepted without any general effect on the animal. Typical extensive necrosis and mitosis occurred at the corticomedullary junction of a nephrectomized kidney but, after a three-week period on normal diet, during which the histopathology could be expected to have resolved, the other kidney then showed a bizarre array of giant-nucleated tubular cells. Alternating weekly periods on and off a nephrotoxic diet, over eight cycles, also transformed proximal tubular cells into the bizarre karyomegaly. Nephrotoxic *P. aurantiogriseum* seems therefore to have a latent factor which after short-term high dosage or long-term low dosage stimulates nuclear division, which is at least abnormal in its polyploidy, and a concomitant failure of cells to divide.

Electron microscopy of karyomegalic cells showed that giant nuclei frequently had irregular outlines and that the cytoplasm contained many discrete regions rich in smooth endoplasmic reticulum (Figures 3 and 4). It is possible that this reflects enhanced metabolic potential in karyomegalic cells so that they survive in spite of the acutely cytotoxic principle(s) produced by the nephrotoxic mould. While, following protracted nephrotoxic insult, the extensive karyomegaly is also seen to be associated with some cells apparently in mitosis, this may be only a continuation of the nuclear replicating process and is probably not a response that reflects persistent susceptibility to a necrotic aspect of the toxicosis. Prominently thickened tubular basement membranes also occur.

Anticipating that some of the histopathological changes caused by *P. aurantiogriseum* in rats might be similar to those evoked by the best-known fungal nephrotoxin ochratoxin A, this compound was included in the study so that direct comparisons could be made. Since so little pure ochratoxin A is necessary to be nephrotoxic, and following some previous protocols in the extensive literature on this compound, it was given initially by gastric gavage. Consequent inappetance contraindicated this mode of administration, so a dilute solution of the compound in bicarbonate was incorporated into powdered diet in a subsequent study. The dose (about 25 mg/kg of diet or 1–1.5 mg/kg body weight) was selected according to Munro *et al.* (1974) as likely to produce renal histopathological changes, including enlarged nuclei, without seriously affecting other organs; this dose was only about one-tenth of the acute oral LD_{50} value.

Diet containing ochratoxin A alone for eight weeks clearly caused changes at the corticomedullary junction, but these were much less striking under light microscopy than those elicited by *P. aurantiogriseum.*

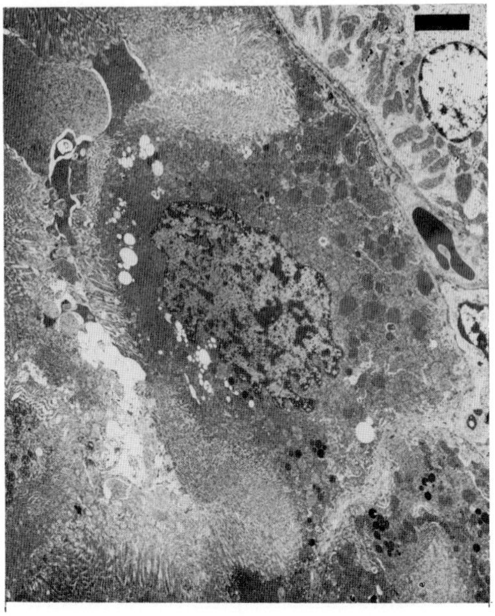

Figure 3. Electron micrograph of an area of the straight segment of a proximal tubule of a rat dosed for only 5 weeks with *P. aurantiogriseum*-moulded shredded wheat, showing a giant nucleus with a convoluted nuclear membrane

Bar, 4 μm

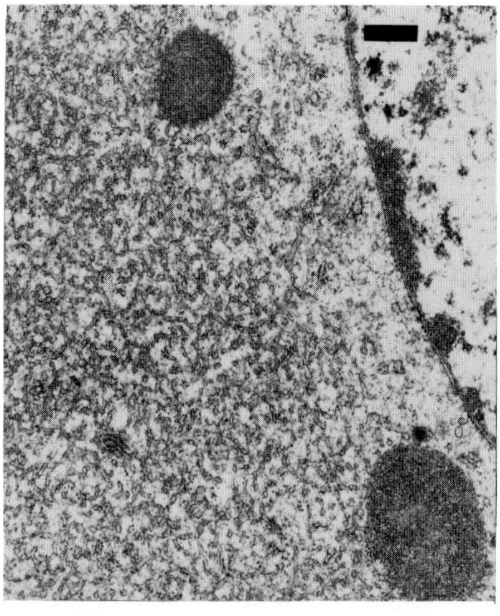

Figure 4. Electron micrograph showing the diffuse proliferation of smooth endoplasmic reticulum within the cytoplasm of karyomegalic proximal tubular cells illustrated in Figure 3

Bar, 0.4 μm

Nevertheless, electron micrographs showed significantly enlarged nuclei and thickened tubular basement membranes. Regimes combining ochratoxin A with *P. aurantiogriseum* gave an intermediate response with respect to the prominence of karyomegaly; there was no evidence of synergism. Thus, although it is not yet possible to quantify the *P. aurantiogriseum* nephrotoxin that induces karyomegaly, and thereby to compare its specific activity in this respect to that of ochratoxin A, it is probable that *P. aurantiogriseum* is much more potent than ochratoxin A in inducing karyomegaly and is certainly much more target-specific. Also, whereas *P. verrucosum* is as nephrotoxic as the ochratoxin A that it is shown to contain, it would be premature to assume that the principal nephrotoxin of *P. aurantiogriseum*, traced so far to a water-soluble amphoteric fraction that co-chromatographs (high-performance liquid and thin-layer chromatography) with leucine and valine and some as yet uncharacterized neutral peptides, accounts for all the activity of the whole fungus. Another component co-chromatographs (high-performance liquid chromatography) with threonine and serine and, when separated from the toxin that affects the P3 segment of rat nephrons, causes nuclear pyknosis further down the proximal tubules.

Considering the difficult application of limited pieces of evidence on mycotoxic nephropathy in the experimental rat to Balkan endemic nephropathy, several features of the human disease must be recognized. First, Balkan nephropathy seems to require residence of at least 10–15 years in an endemic village, and most people are not diagnosed clinically until well into middle age. Indeed, the disease probably became recognized only after the Second World War because control of infectious diseases was extending the life span and allowed this chronic renal disease to become more apparent. Secondly, in spite of several claims to the contrary, no clear picture of the early histopathology of the disease has been described. Of course, the end-stage extensive, bilateral renal atrophy goes through early stages in which excessive interstitial fibrosis is evident, but such evidence is still probably much too advanced to be helpful in defining the histological condition of tubules around the point at which interstitial fibrosis first occurs. This could be in childhood or adolescence; thus, in practice, evidence linked to eventual development of the nephropathy may never be obtainable. Thirdly, whatever causes Balkan nephropathy seems to be highly specific to the kidney. Fourthly, the high prevalence of renal tumours in people suffering from Balkan nephropathy is manifested by an increased incidence mainly of the usual range of urothelial tumour types.

It was reasonable, therefore, to suggest (Krogh, 1978) that ochratoxin A might be involved in the etiology of Balkan nephropathy. Ochratoxin A is well established as a causal factor in mycotoxic porcine nephropathy, and in the pig experimentally reproduces the histopathology of interstitial fibrosis seen in the natural cases recognized in the Danish bacon industry and attributed to barley moulded by what is now known as *P. verrucosum* (Pitt, 1987). Sensitive analytical techniques have been used to detect small amounts of the toxin in food and in human sera, not at high frequency in nephropathy areas but nevertheless somewhat above that in non-nephropathy areas. The presence of ochratoxin has not been monitored throughout the year, so the pattern of human exposure is unclear. It might be fair to conclude that either ochratoxin A is an exceptionally potent cumulative nephrotoxin in susceptible individuals or that its spasmodic trace occurrence in food cannot constitute a risk of causing the renal disease expressed by the clustering of cases of Balkan nephropathy. Even porcine nephropathy is not apparent in the endemic area; surely the poorest quality food would rather go to the pigs which many or most of the farmers keep.

Experimental evidence in rats suggests that ochratoxin A is too generally toxic to avoid causing only renal histopathology in humans subject to spasmodic ingestion of contaminated food. The renal carcinogenicity of ochratoxin A in rats (Boorman, 1989), however, has added a new, interesting dimension to Balkan nephropathy, particularly with respect to the induction of urothelial tumours, although the renal tubular sites of experimental rat renal tumours (carcinomas and adenomas) did not quite fit this category. A striking, consistent histopathological feature in rats was karyomegaly in proximal tubules, and the question arises as to whether this is a significant feature in the etiology of the tumorigenesis. Mandal *et al.* (1987) recently reanalysed biopsy specimens of kidney from Yugoslav nephropathy patients, some of whom had renal tumours, and reported that proximal convoluted tubules showed increased apparently nonspecific epithelial degeneration and regeneration, with enlarged nuclei and nucleoli.

Thus, the abnormal effects on nuclear division of the *P. aurantiogriseum* nephrotoxin, and also of ochratoxin A, deserve further exploration with respect to the long-term health of nephrons in which persistent epithelial cell changes have been caused. If ochratoxin A is seen by many as probably involved in Balkan nephropathy, then the ubiquitous *P. aurantiogriseum* and its nephrotoxins also deserve a high profile. To date, in our chronic studies with *P. aurantiogriseum* we have naturally sought its effect after more or less continuous exposure. It may be that much more intermittent insult could gradually necrose nephrons without instituting protective metabolic potential, as may be expressed by proliferation of smooth endoplasmic reticulum. Such experiments are perhaps less exciting but could be revealing. The rat model, however, may not allow sufficiently protracted experiments, before the natural onset of renal disease, to be an adequate mimic of the human situation. Experimental animals are usually maintained in such equable environments that imposition of the probable circumstantial complexities of the Balkan syndrome may present an insuperable challenge.

References

Adatia, R., Heaton, J.M., Macgeorge, K.M. & Mantle, P.G. (1991) Acute histopathological changes produced by *Penicillium aurantio griseum* nephrotoxin in the rat. *Int. J. Exp. Pathol.*, **72**, 47–53

Barnes, J.M., Carter, R.L., Peristianis, G.C., Austwick, P.K.C., Flynn, F.V. & Aldridge, W.N. (1977) Balkan (endemic) nephropathy and a toxin-producing strain of *Penicillium verrucosum* var. *cyclopium*: an experimental model in rats. *Lancet*, **i**, 671–675

Boorman, G. (1989) *Toxicology and Carcinogenesis Studies of Ochratoxin A* (NIH Publication No. 89-2813), Research Triangle Park, NC, National Toxicology Program

Krogh, P. (1978) Causal association of mycotoxic nephropathy. *Acta Pathol. Microbiol. Scand. A*, **Suppl. 269**

Macgeorge, K.M. & Mantle, P.G. (1990) Nephrotoxicity of *Penicillium aurantiogriseum* and *P. commune* from an endemic nephropathy area of Yugoslavia. *Mycopathologia*, **112**, 139–145

Macgeorge, K.M. & Mantle, P.G. (1991) Nephrotoxic fungi in a Yugoslavian community in which Balkan nephropathy is hyperendemic. *Mycol. Res.*, **95**, 660–664

Mandal, A.K., Sindjic, M. & Sommers, S.C. (1987) Kidney pathology in endemic nephropathy. *Clin. Nephropathy*, **27**, 304–308

Mantle, P.G., McHugh, K.M., Adatia, R., Gray, T. & Turner, D.R. (1991) Persistent karyomegaly caused by *Penicillium* nephrotoxins in the rat. *Proc. R. Soc. B* (in press)

Munro, I.C., Moodie, C.A., Kuiper-Goodman, T., Scott, P.M. & Grice, H.C. (1974) Toxicologic changes in rats fed graded dietary levels of ochratoxin A. *Toxicol. Appl. Pharmacol.*, **28**, 180–188

Pitt, J.I. (1987) *Penicillium viridicatum*, *Penicillium verruco sum*, and production of ochratoxin A. *Appl. Environ. Microbiol.*, **53**, 266–269

Pitt, J.I. (1988) *A Laboratory Guide to Common* Penicillium *Species*, North Ryde, NSW, Australia, CSIRO Food Research Laboratory

BIOLOGICAL MONITORING OF OCHRATOXIN A, PHARMACOKINETICS AND TOXICITY

Mycotoxins, Endemic Nephropathy and Urinary Tract Tumours
Ed. M. Castegnaro, R. Pleština, G. Dirheimer, I.N. Chernozemsky & H. Bartsch
Lyon, International Agency for Research on Cancer, pp. 131-135

HUMAN EXPOSURE TO OCHRATOXIN A

R. Fuchs[1], B. Radić[1], S. Čeović[2], B. Šoštarić[1] & K. Hult[3]

[1]Department of Toxicology, Institute for Medical Research and Occupational Health, University of Zagreb, Zagreb, Yugoslavia; [2]Department of Epidemiology, Medical Centre, Slavonski Brod, Yugoslavia; and [3]Department of Biochemistry and Biotechnology, Royal Institute of Technology, Stockholm, Sweden

Summary

Over a nine-year period during screening campaigns in villages where Balkan nephropathy is endemic, human blood samples were collected and analysed for ochratoxin A. The incidence of positive samples was 0.5–2.5%. Dried beans were found to be more frequently contaminated with the toxin than other food commodities. In view of a specific accumulation of ochratoxin A observed in eggs, more attention should be paid to contamination of this food with ochratoxin A.

Text

Since ochratoxin A was first suggested as a possible disease determinant of Balkan endemic nephropathy, considerable efforts have been made to determine the extent of human exposure to this toxin. A number of studies and food analyses have been performed to estimate possible human risk. In the 1970s, ochratoxin A was found in 9.38% of food samples in villages in the county of Slavonski Brod where Balkan nephropathy is endemic, at concentrations of 7.5–140 µg/kg (Pavlović *et al.*, 1979).

The development of analytical methods for determining ochratoxin A in biological materials like whole blood and serum has resulted in proof that humans are exposed to this nephrotoxic substance of natural origin and that it is absorbed from the gastrointestinal tract (Hult *et al.*, 1979a, 1982, 1984). To establish whether there is an association between exposure to ochratoxin A and Balkan endemic nephropathy, 17 175 human blood samples were collected in the field from 1981 to 1989 and analysed at the Institute for Medical Research and Occupational Health, University of Zagreb. Analyses were performed by the enzymatic method with carboxypeptidase A (Hult *et al.*, 1979b). The detection limit of the method in the laboratory was 5 ng/ml.

The results are summarized in Table 1. Ochratoxin A was detected each year during the nine-year observation period in 0.5–2.5% of blood specimens from the endemic area. In nonendemic villages, the toxin was detected at comparable incidence in four of the years, except in 1985 when the incidence in one control village was 3.7%. In this particular village, the samples were obtained during the summer, whereas all other samples were collected during winter. Ochratoxin A was not detected in human blood between 1986 and 1989, but the levels may have been below the limit of detection of the method. Furthermore, the same control villages were not always used, and some of them are located more than 100 km from the area of nephropathy.

Table 1. Prevalence of ochratoxin A in human blood collected in endemic and nonendemic villages for Balkan nephropathy in the county of Slavonski Brod, 1981–89

Year	Endemic villages			Nonendemic villages		
	No. of samples	% positive	Range (ng/ml)	No. of samples	% positive	Range (ng/ml)
1981	694	1.8	5–15	242	1.6	5–10
1982	1049	2.5	5–50	242	2.1	5–30
1983	1872	2.4	5–30	738	2.7	10–15
1984	2165	1.3	5–50	227	1.7	5–10
1985	1490	0.4	10–15	375[a]	3.7	5–50
1986	1887	1.7	5–35	401	0	
1987	2073	0.5	5–50	156	0	
1988	1554	1.3	5–100	570	0	
1989	1013	1.4	5–20	427	0	

[a]Samples collected during summer

The finding of ochratoxin A in human blood raised the question of which food commodity is the most important source of exposure. Cereals have been reported to be highly contaminated with ochratoxin A in several countries (WHO, 1979; Steyn, 1984). Our studies in villages in Croatia showed that dried beans are frequently contaminated with ochratoxin A. In five out of 10 selected households in which ochratoxin A was identified in the sera of family members, dried beans were found to contain ochratoxin A in a concentration range of 10–50 µg/kg. It is interesting to note that all of the beans did not look damaged or mouldy and that contamination was discovered by chemical analysis only. As dried beans are used frequently in dishes in some regions of Yugoslavia, they may be an important source of human exposure to the toxin.

Residues of ochratoxin A have been identified in pigs in a number of studies (Krogh *et al.*, 1976a; Mortensen *et al.*, 1983), but data on other farm animals, such as poultry, are less clear. Our results on the accumulation of ochratoxin A in organs and tissues of Japanese quail, studied by whole-body autoradiography (Fuchs *et al.*, 1988), are in agreement with those of other studies of poultry (Frye & Chu, 1977; Prior *et al.*, 1980; Reichman *et al.*, 1982), except in respect of toxin residues in eggs. Ochratoxin A was not detected in eggs in feeding experiments (Krogh *et al.*, 1976b) or was found only when hens were fed large amounts of the toxin (10 mg/kg) (Juszkiewicz *et al.*, 1982). Our study demonstrated the presence of ochratoxin A in the eggs of birds given low doses of ochratoxin A (Fuchs *et al.*, 1988) (Figure 1). As eggs are consumed widely, the presence of such residues may be of toxicological importance, especially if other sources of exposure are likely to be present.

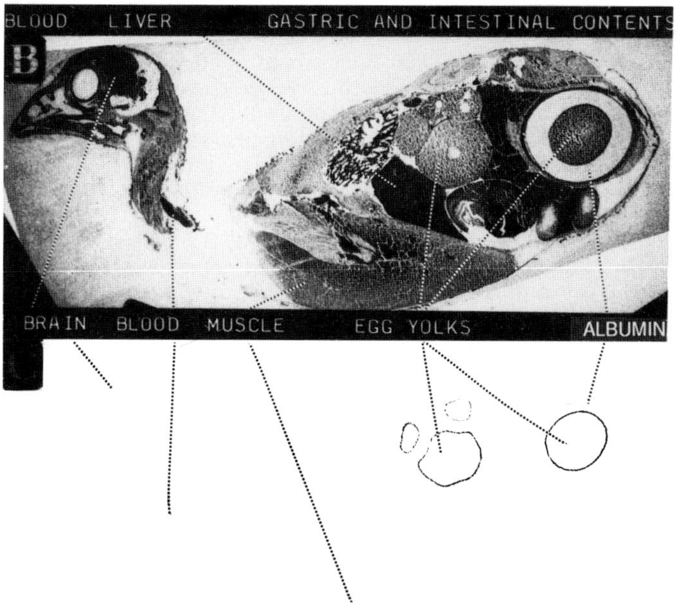

Figure 1. Whole-body autoradiogram (C) of a Japanese quail 24 h after a single intravenous injection of 70 ng/g body weight of ^{14}C-ochratoxin A, and a corresponding unstained 20-µm section (B). Exposure time, 600 days. Note the ring formed by accumulation of labelled toxin in egg yolks.

Although the toxic effects of ochratoxin A in humans have not been proven incontestably, data obtained on experimental and domestic animals indicate that this toxin may also be toxic to humans. As humans are exposed to ochratoxin A, food should be monitored to prevent human exposure.

Although the data obtained from regions of nephropathy in Bulgaria and Yugoslavia favour the hypothesis that ochratoxin A is an etiological agent for Balkan endemic nephropathy, the constitution of the disease itself makes this difficult to evaluate. The possibility that a number of other, still unidentified toxic metabolites of moulds are acting synergistically in these regions should also be considered. All of the currently available information, however, is in favour of the mycotoxic hypothesis.

References

Fuchs, R., Appelgren, L.E., Hagelberg, S. & Hult, K. (1988) Carbon-14-ochratoxin A distribution in the Japanese quail (Coturnix coturnix japonica) monitored by whole body autoradiography. *Poultry Sci.*, **67**, 707–714

Frye, C.E. & Chu, F.S. (1977) Distribution of ochratoxin A in chicken tissue and eggs. *J. Food Saf.*, **1**, 147–159

Hult, K., Hokby, E., Gatenbeck, S., Pleština, R. & Čeović, S. (1979a) Ochratoxin A and Balkan endemic nephropathy. IV. Occurrence of ochratoxin A in humans. *Chem. Rundschau*, **35**, 32

Hult, K., Hokby, E., Hagglund, U., Gatenbeck, S., Rutquist, L. & Sellyey, G. (1979b) Ochratoxin A in pig blood: method of analysis and use as a tool for feed studies. *Appl. Environ. Microbiol.*, **38**, 772–776

Hult, K., Pleština, R., Habazin-Novak, V., Radić, B. & Čeović, S. (1982) Ochratoxin A in human blood and Balkan endemic nephropathy. *Arch. Toxicol.*, **51**, 313–321

Hult, K., Fuchs, R., Peraica, M., Pleština, R. & Čeović, S. (1984) Screening for ochratoxin A in blood by flow injection analysis. *J. Appl. Toxicol.*, **4**, 326–329

Juszkiewicz, T., Piskorska-Pliszczynska, J. & Wisniewska, H. (1982) Ochratoxin A in laying hens: tissue deposition and passage into eggs. In: *Mycotoxins and Phycotoxins. Proceedings of the V International IUPAC Symposium, Vienna, Technical University, 1–3 September*, pp. 122–125

Krogh, P., Elling, F., Hald, B., Larsen, A.E., Lillehoj, E.B., Madsen, A & Mortensen, H.P. (1976a) The time-dependent disappearance of ochratoxin A residues in tissue of bacon pigs. *Toxicology*, **6**, 235–242

Krogh, P., Elling, F., Hald, B., Jyllimg, B., Petersen, V.E., Skadhauge, E. & Svendsen, C.K. (1976b) Experimental avian nephropathy. *Acta Pathol. Microbiol. Scand. A.*, **84**, 215–221

Mortensen, H.P., Hald, B. & Madsen, A. (1983) Feeding experiment with ochratoxin A contaminated barley for bacon pigs. 5. Ochratoxin A in pig blood. *Acta Agric. Scand.*, **33**, 135–239

Pavlović, M., Pleština, R. & Krogh, P. (1979) Ochratoxin A contamination of foodstuffs in an area with Balkan (endemic) nephropathy. *Acta Pathol. Microbiol. Scand. B*, **87**, 243–246

Prior, M.G., O'Neil, J.B. & Sisodia, C.S. (1980) Effects of ochratoxin A on growth response and residues in broilers. *Poultry Sci.*, **59**, 1254–1257

Reichman, K.G., Blaney, B.J., Connor, J.K. & Runge, B.M. (1982) The significance of aflatoxin and ochratoxin in the diet of Australian chickens. *Aust. Vet. J.*, **58**, 211–212

Steyn, P.S. (1984) Ochratoxins and related dihydroisocoumarins. In: Betina, V., ed., *Developments in Food Science 8*, Amsterdam, Elsevier Science Publishers, pp. 183–215

WHO (1979) *Mycotoxins* (Environmental Health Criteria 11), Geneva

Mycotoxins, Endemic Nephropathy and Urinary Tract Tumours
Ed. M. Castegnaro, R. Pleština, G. Dirheimer, I.N. Chernozemsky & H. Bartsch
Lyon, International Agency for Research on Cancer, pp. 135-137

OCHRATOXIN A IN HUMAN BLOOD IN RELATION TO BALKAN ENDEMIC NEPHROPATHY AND URINARY TRACT TUMOURS IN BULGARIA

T. Petkova-Bocharova[1] & M. Castegnaro[2]

[1]National Oncological Centre, Sofia, Bulgaria; and
[2]International Agency for Research on Cancer, Lyon, France

Summary

Ochratoxin A is suspected of being one of the etiological agents responsible for Balkan endemic nephropathy and the associated urinary tract tumours. Contamination of cereals by this mycotoxin has been found to be more frequent in areas of endemic nephropathy than in areas where the disease is absent. As ochratoxin A binds to serum albumin, it should be detectable in biological fluids from exposed populations. A survey was thus conducted to determine the occurrence of ochratoxin A in blood from people living in the endemic area who were either affected or unaffected by the two diseases and in blood from people living in control regions where these diseases do not occur. Blood samples were collected in 1984, 1986, 1989 and 1990. Ochratoxin A was found more frequently and at higher levels in blood from patients with Balkan endemic nephropathy and/or urinary tract tumours than in blood from unaffected people from endemic and control areas. These findings suggest further that ochratoxin A is involved in the etiology of the two diseases.

Introduction

The presence of ochratoxin A in food samples from areas of Bulgaria with Balkan endemic nephropathy and a high incidence of urinary tract tumours (Petkova-Bocharova & Castegnaro, 1985) shows that the populations of such areas are likely to consume this mycotoxin.

Ochratoxin A has also been found to be present in the blood of people living in areas endemic for nephropathy in Yugoslavia (Hult *et al.*, 1982) and Bulgaria (Petkova-Bocharova *et al.*, 1988).

We report here the occurrence of the mycotoxin in blood samples collected in 1984, 1986, 1989 and 1990 from people living in Vratza, a district of Bulgaria where there is a high incidence of the two diseases, and from people living in another area of Bulgaria where these diseases are not endemic.

Subjects and Methods

Blood samples were collected from 576 people in five groups: (I) patients with urinary tract tumours and/or Balkan endemic nephropathy; (II) healthy relatives of patients with urinary tract tumours and/or Balkan endemic nephropathy; (III) healthy people living in affected villages in the endemic area; (IV) healthy people living in unaffected villages in the endemic area; and (V) healthy people living in a nonendemic area of Bulgaria. All of the subjects had similar life styles, habits and diets, most of the food being produced at home.

Blood samples were collected from the same people in successive years, as far as was possible. The samples were analysed for ochratoxin A using the enzymatic method with carboxypeptidase A described by Hult *et al.* (1982). The detection limit of this method is 1–2 ng/g serum.

Results and Discussion

The results are presented in Table 1. Ochratoxin A was detected in all five groups; however, it was found more frequently and at higher levels in samples of blood from patients with urinary tract tumours and/or Balkan endemic nephropathy than in samples from unaffected people. The concentration of ochratoxin A ranged from 8 to 35 ng/g serum. Statistically significant differences ($p < 0.05$) in the proportion of samples with ochratoxin A were found only between group I and group V for 1986, 1989 and 1990. The absence of significance of this difference in 1984 is due to the small number of subjects in group V. The difference between group I and groups III and IV was also close to significance. A larger number of samples would probably have made them significant, as demonstrated when the results were pooled ($p < 0.01$); such pooling is justifiable because of the small annual variations in ochratoxin A-positive samples between these two groups.

Although there was little annual variation in the percentage of samples containing ochratoxin A in the different groups, the samples from patients with renal tumours and/or Balkan endemic nephropathy tended to contain the mycotoxin more frequently between 1984 and 1989. Those from unaffected people in endemic and control areas (groups IV and V) had virtually constant levels over the period of observation.

Table 1. Ochratoxin A in blood samples collected in different years from people living in areas with and without endemic nephropathy in Bulgaria

Year	Group[a]	No. of subjects	Samples containing ochratoxin A						
			Total		With 1–2 ng/g		With > 2 ng/g		
			No.	%	No.	%	No.	%	Mean ± SD (ng/g)
1984	I	31	9	29.0	5	16.1	4	12.9	21.0 ± 2.9
	II	33	6	18.2	3	9.1	3	9.1	15.0 ± 5.0
	III	26	3	11.5	2	7.6	1	3.8	12.0
	IV	26	3	11.5	2	7.6	1	3.8	18.0
	V	13	1	7.7	1	7.7	–		
1986	I	30	7	23.3	4	13.3	3	10.0	20.0 ± 2.0
	II	30	4	13.3	3	10.0	1	3.3	14.0
	III	37	4	10.8	3	8.1	1	2.7	13.0
	IV	34	4	11.7	3	8.8	1	2.9	12.0
	V	52	4	7.7	3	5.8	1	1.9	10.0
1989	I	24	7	29.2	4	16.7	3	12.5	27.2 ± 11.9
	II	28	5	17.8	2	7.1	3	10.7	15.7 ± 4.0
	III	28	4	14.3	3	10.7	1	3.6	10.0
	IV	30	3	10.0	2	6.7	1	3.3	15.0
	V	30	2	6.6	2	6.6	–		
1990	I	20	5	25.0	3	15.0	2	10.0	25.0 ±10.6
	II	20	3	15.0	1	5.0	2	10.0	12.5 ± 3.6
	III	25	3	12.0	2	8.0	1	4.0	12.0
	IV	29	3	10.3	2	6.9	1	3.4	10.0
	V	30	2	6.6	1	3.3	1	3.3	8.0

[a]Group I, patients with urinary tract tumours and/or Balkan endemic nephropathy; group II, healthy relatives of patients with urinary tract tumours and/or Balkan endemic nephropathy; group III, healthy people living in affected villages in the endemic area; group IV, healthy people living in unaffected villages in the endemic area; group V, healthy people living in a nonendemic area of Bulgaria

The prevalence of the presence of ochratoxin A in the blood of people with Balkan endemic nephropathy and/or urinary tract tumours adds further support to the hypothesis that this nephrotoxic and carcinogenic mycotoxin is involved in the etiopathogenesis of these two diseases.

References

Hult, K., Pleština, R., Habazin-Novak, V., Radić, B. & Čeović, S. (1982) Ochratoxin A in human blood and Balkan endemic nephropathy. *Arch. Toxicol.*, **51**, 313–321

Petkova-Bocharova, T., Chernozemsky, I.N. & Castegnaro, M. (1988) Ochratoxin A in human blood in relation to Balkan endemic nephropathy and urinary system tumours in Bulgaria. *Food Addit. Contam.*, **5**, 299–301

Petkova-Bocharova, T. & Castegnaro, M. (1988) Ochratoxin A contamination of cereals in an area of high incidence of Balkan endemic nephropathy in Bulgaria. *Food Addit. Contam.*, **2**, 267–270

Mycotoxins, Endemic Nephropathy and Urinary Tract Tumours
Ed. M. Castegnaro, R. Pleština, G. Dirheimer, I.N. Chernozemsky & H. Bartsch
Lyon, International Agency for Research on Cancer, pp. 139-143

OCHRATOXIN A AS A CONTAMINANT IN THE HUMAN FOOD CHAIN: A CANADIAN PERSPECTIVE

A.A. Frohlich, R.R. Marquardt & K.H. Ominski

Department of Animal Science, University of Manitoba, Winnipeg, Manitoba, Canada

Summary

Penetration of ochratoxin A into the human food chain in Canada was assessed by analysing stored grains, porcine blood and, finally, human blood. The potential for mycological growth and production of ochratoxin A was determined in 164 samples of stored grain (barley, wheat, maize and silage) collected from producers in Manitoba. A total of 34% were found to have the capacity to produce ochratoxin A; 14.5% had concentrations greater than 1.0 mg/kg. In 1988, 1200 blood samples were obtained from swine destined for slaughter in western Canada. High-performance liquid chromatography demonstrated that 3.6% of the 194 blood samples collected in February and March and 4.2% of the 1006 collected in May, June and July had concentrations of ochratoxin A that exceeded 20 ng/ml. In a subsequent survey of porcine blood carried out in 1989–90, 16–65% of the samples had detectable levels of ochratoxin A, at mean concentrations of 5.4–19.4 ng/ml. Subsequently, human blood samples were collected from 159 individuals, 69 of whom had some form of renal impairment. Of the latter, 40% had detectable levels of ochratoxin A, and 12% had concentrations greater than 0.5 ng/ml. Of the non-renal patients, 39% had detectable levels of ochratoxin A, and 11% had concentrations greater than 0.5 ng/ml. These studies demonstrate that ochratoxin A is present in the blood of people in Canada and that two possible points of entry of this toxin into the human food chain are contaminated grain and pork products.

Introduction

Contamination of agricultural commodities and animal products by ochratoxin A has been scrutinized extensively in Europe over the past several years. This mycotoxin occurs primarily in countries with temperate or continental climates, such as that found in Canada. Early reports by Prior (1976, 1981) and Abramson *et al.* (1983) confirmed the presence of ochratoxin A in cereal products in Canada. Examination of animal products and human body fluids in North America, however, is more limited, if not non-existent. In an attempt to assess the penetration of ochratoxin A into the human food chain in Canada, stored grains, swine blood and, finally, human blood were analysed for the presence of this mycotoxin.

Materials and Methods

In 1985–86, 164 grain samples were collected from various locations in Manitoba and subjected to mycological evaluation in an attempt to obtain a more comprehensive overview of the presence of ochratoxin A in grain, including information on the presence of indigenous fungi under various storage conditions, and the potential for mould growth and ochratoxin A production on moistened, incubated grain. The samples were initially analysed for the presence of different mycoflora using standard mycological methods (plating on potato dextrose agar and malt agar) and were then moistened to 24–25% and incubated at 28 °C for 14 days. Samples were divided into four classes depending on visual evaluation of the appearance of mycelia and sporulation. This was followed by analyses of mycoflora and screening for mycotoxins by thin-layer chromatography and quantitative high-performance liquid chromatography analyses (HPLC) (Frohlich *et al.*, 1988).

Results and Discussion

Figure 1 shows the mycological changes that resulted when samples were subjected to optimal conditions of moisture and temperature, which occur readily under natural conditions. *Penicillium* and *Aspergillus* species (storage fungi) predominated over *Alternaria* and *Fusarium* (field fungi) in all of the examined samples. HPLC analyses revealed that 44% of samples of high-moisture barley and 35% of moisturized, incubated samples of barley, 14% of moisturized, incubated samples of corn and 12% of moisturized, incubated samples of silage contained ochratoxin A. The percentages of samples with detectable levels of ochratoxin A were as follows: < 0.5 mg/kg, 56.4%; 0.5–2.0 mg/kg, 38.1%; and > 2.0 mg/kg, 5.4%.

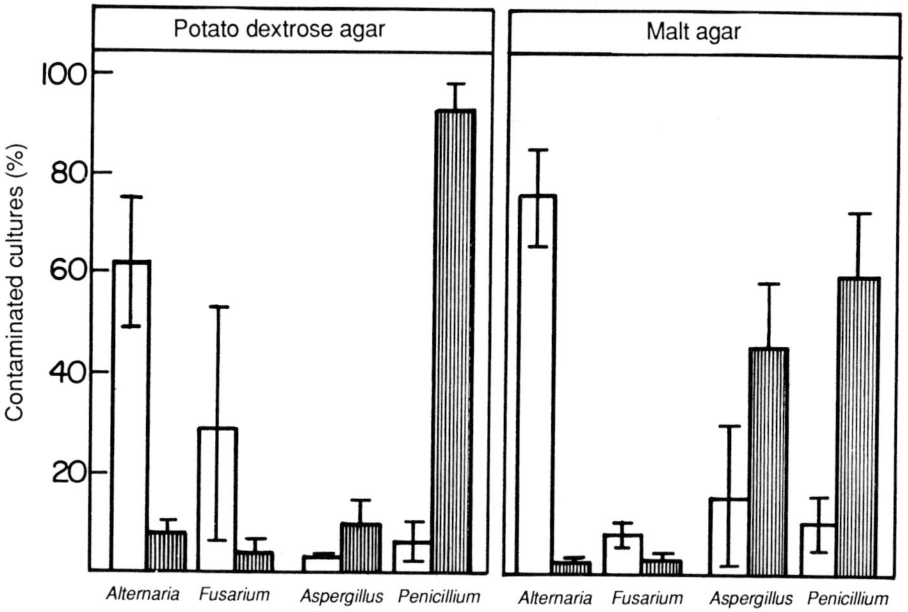

Figure 1. Fungi detected in barley before (open bars) and after (hatched bars) incubation (fermentation) of moisturized grain (25% moisture, 28 °C, 14 days) on two different media

The potential presence of ochratoxin A in cereals suggested that it might also be present in animal tissues, including pigs. As a result, porcine blood was collected and analysed for the presence of ochratoxin A to determine the incidence and extent of contamination with this toxin in pigs. In 1986, blood was collected from randomly selected pigs in February and March (n=194) and May and June (n=1006) at a slaughter house in western Canada (Marquardt *et al.*, 1988). Initially, samples were pooled in groups of two and extracted by the method of Hult *et al.* (1979), prior to HPLC analysis to determine the presence of ochratoxin A. Pooled samples found to contain > 10 ng/ml were analysed individually. Samples that contained > 20 ng/ml were confirmed using methylester derivatization (Nesheim *et al.*, 1973) and HPLC–mass spectroscopy (Abramson, 1987).

The overall distribution of ochratoxin A in the blood serum of slaughter pigs is presented in Table 1. The finding that a high proportion of samples collected in May, June and July contained detectable levels of ochratoxin A suggests that the risk of fungal growth in stored grains is greater in the spring and early summer owing to warmer temperatures and possibly higher moisture levels than in winter, which may lead to increased mould growth and corresponding ochratoxin A production.

Table 1. Concentration of ochratoxin A in the serum of slaughter pigs in a packing house in western Canada

Range of ochratoxin A detected (ng/ml)[a]	No. of positive samples	
	February–March	May, June, July
> 200	–	1
150–200	–	2
100–150	1	1
50–100	2	6
20–50	4	32
10–20	14	73
< 10[b]	76	698
Not detected	97	193
Total no. of samples	194	1006
Total no. confirmed at > 20 ng/ml	7 (3.6%)	42 (4.2%)

[a]Samples containing > 10 ng/ml ochratoxin A were reanalysed individually, and those containing > 20 ng/ml were confirmed. The highest value found was 229 ng/ml.
[b]In samples pooled from two pigs

In 1989–90, a second survey was conducted in which samples were collected at four intervals throughout the year and analysed individually by the methods described above. The incidence and mean concentration found at each collection period are presented in Table 2. Once again, a higher proportion of samples from the July or summer collection period contained detectable levels of ochratoxin A, confirming our previous observations.

Table 2. Prevalence of ochratoxin A in the serum of slaughter pigs

Month	No. of samples collected	% positive	Mean concentration (ng/ml)
April	368	38.0	8.32
July	429	65.0	16.25
October	389	21.3	19.36
January	402	16.7	5.85
Mean		36.0	12.33

Subsequently, human blood samples were collected from 159 individuals, 69 of whom had some form of renal impairment. Of these patients, 40% had detectable levels of ochratoxin A, and 12% had concentrations greater than 0.5 ng/ml; 39% of the non-renal patients had detectable levels of ochratoxin A, and 11% had concentrations greater than 0.5 ng/ml. Thus, renal and non-renal patients did not differ with regard to the distribution of ochratoxin A. The overall incidence and concentration of ochratoxin A in sera from nonrenal and renal patients are presented in Table 3. The distribution of detectable levels of ochratoxin A did not differ between males and females, nor was there any relationship between ochratoxin A distribution and age. The levels of ochratoxin A observed in this study are comparable to those observed in human blood in Europe (Bauer & Gareis, 1987).

Table 3. Prevalence of ochratoxin A in the serum of humans

Concentration range (ng/ml)	No. of positive samples	Mean concentration (ng/ml)
< 0.50	45	0.27
0.50–0.99	9	0.68
1.00–10.00	8	2.19
> 20.00	1	35.33

These studies demonstrate for the first time that ochratoxin A is present in the blood of Canadians. Two possible points of entry of this ubiquitous toxin into the human food chain are contaminated grain and pork products.

References

Abramson, D. (1987) Measurement of ochratoxin A in barley by liquid chromatography–mass spectrometry. *J. Chromatogr.*, **391**, 315–320

Abramson, D., Mills, J.T. & Boycott, B.R. (1983) Mycotoxins and mycoflora in animal feedstuffs in Western Canada. *Can. J. Comp. Med.*, **47**(1), 23–26

Bauer, J. & Gareis, M. (1987) Ochratoxin A in the food chain. *Z. Veterinarmed. B.*, **34**, 613–627 (in German)

Frohlich, A.A., Marquardt, R.R. & Bernatsky, A. (1988) Quantitation of ochratoxin A: use of reverse phase thin-layer chromatography for sample cleanup followed by liquid chromatography or direct measurement. *J. Assoc. Off. Anal. Chem.*, **71**(2), 949–953

Hult, K., Hokby, E., Hagglund, Y., Gotenbeck, S., Rutquist, L. & Sellyel, G. (1979) Ochratoxin A in pig blood: method of analysis and use as a tool for feed studies. *Appl. Environ. Microbiol.*, **38**, 772–776

Marquardt, R.R., Frohlich, A.A., Sreemannarayana, O. & Bernatsky, A. (1988) Ochratoxin A in blood from slaughter pigs in western Canada. *Can. J. Vet. Res.*, **52**, 186–190

Nesheim, S., Hardin, N.F., Francis, O.J. & Laughem, W.S. (1973) Analysis of ochratoxin A and B and their esters in barley using preparation and thin layer chromatography. *J. Assoc. Off. Anal. Chem.*, **56**, 817–821

Prior, M.G. (1976) Mycotoxin determinations on animal feedstuffs and tissues in western Canada. *Can. J. Comp. Med.*, **40**(1), 75–79

Prior, M.G. (1981) Mycotoxins in animal feedstuffs and tissues in western Canada 1975 to 1979. *Can. J. Comp. Med.*, **45**(2), 116–119

Mycotoxins, Endemic Nephropathy and Urinary Tract Tumours
Ed. M. Castegnaro, R. Pleština, G. Dirheimer, I.N. Chernozemsky & H. Bartsch
Lyon, International Agency for Research on Cancer, pp. 145-151

HUMAN OCHRATOXICOSIS IN FRANCE

E.E. Creppy[1], A.M. Betbeder[1], A. Gharbi[1], J. Counord[1], M. Castegnaro[2], H. Bartsch[2], P. Moncharmont[3], B. Fouillet[4], P. Chambon[4] & G. Dirheimer[5]

[1]Laboratoire de Toxicologie, Faculté de Pharmacie, Université de Bordeaux II, Bordeaux; [2]International Agency for Research on Cancer, Lyon; [3]Centre Régional de Transfusion Sanguine de Lyon, Beynost; [4]Unité de Formation et de Recherche des Sciences Pharmaceutiques, Lyon; and [5]Institut de Biologie Moléculaire et Cellulaire du Centre National de la Recherche Scientifique, Strasbourg, France

Summary

The prevalence of human ochratoxicosis in France is being determined using serum and plasma collected from apparently healthy people. The analytical method is based on the partition coefficient of ochratoxin A in aqueous and organic solvents, according to pH. High-performance liquid chromatography and spectrofluorimetry are used for detection and quantification (limit of detection, > 0.2 ng/ml). The presence of ochrotoxin A is confirmed by the action of carboxypeptidase to yield ochratoxin α or by derivitization of ochratoxin A with boron trifluoride. The significance of the interim values obtained and the number of positive samples is discussed. A comparison with the distribution of known values in Germany and Scandinavia could be helpful in risk assessment with a view to prevention.

Introduction

Ochratoxin A, a mycotoxin produced by a number of *Aspergillus* and *Penicillium* genera, is nephrotoxic to all animal species tested (Krogh, 1987) and is teratogenic (Arora & Fröelén, 1981) and carcinogenic to mice (Bendele *et al.*, 1985) and rats (Boorman, 1989). It is suspected to play a key role in the etiology of Balkan endemic nephropathy (Röschenthaler *et al.*,1984; Kuiper-Goodman & Scott, 1989) and/or the associated renal tumours (Castegnaro *et al.*, 1987).

Several studies have demonstrated that ochratoxin A can contaminate the food chain (Krogh, 1987). Human ochratoxicosis was first demonstrated in the Balkans (Hult *et al.*, 1982; Petkova *et al.*, 1988). Recent studies in Germany (Bauer and Gareis, 1987) and Scandinavia (Breitholtz *et al.*, 1991) have shown that human ochratoxicosis is not limited to the Balkans but is a more general European problem. All such investigations so far have been performed in north-western European countries, and we considered it worthwhile to extend them to countries such as France, Greece, Italy, Portugal and Spain, in which home-made cereal-based foods are consumed, in some cases on a large scale.

We therefore initiated a study in France of potential contamination of the food chain. Three areas were selected: (i) The Aquitaine region (Gironde, Dordogne, Landes, Lot-et-Garonne, Pyrénées Atlantiques) and a neighbouring department, the Gers, produce large amounts of corn which is used mainly for feeding poultry (ducks, geese, chicken), in order to obtain foie gras and other products. (ii) The Rhône and the neighbouring Ain were selected to provide a contrast between a purely urban population (Lyon and its surroundings) and a mixed urban and rural population (in the Ain). (iii) Alsace was selected as its population has a life style similar to that in Germany, where human ochratoxicosis was demonstrated recently (Bauer & Gareis, 1987). This region produces various cereals, which are used for pig and poultry feed and for beer production, both of which are potential candidates for contamination by ochratoxin A.

Materials and Methods

Blood and serum samples were obtained from blood banks and from health check-up centres. All blood samples were separated, frozen and stored at about –25 °C until analysis. All solvents and reagents were of analytical quality or high-performance liquid chromatography (HPLC) grade.

The method used for analysis of ochratoxin A is based on that described by Bauer and Gareis (1987) with modified HPLC conditions as described by Hietanen *et al.* (1986). In short, ochratoxin A was extracted by chloroform from serum acidified by a magnesium chloride-hydrochloric acid mixture adjusted to a pH of about 2.5. The ochratoxin A was then re-extracted from the chloroform phase with a sodium bicarbonate solution. After acidification of this solution to a pH of about 2.5, ochratoxin A was extracted into chloroform; this phase was evaporated to dryness under reduced pressure at 40 °C. The dried extract was taken up into methanol and separated into three aliquots: one was used for direct HPLC fluorimetric analysis, and the other two were stored at –25 °C for further confirmation, if necessary.

The HPLC conditions were as follows: column, 20 cm long, 6.4 mm o.d., 10 µm ODS; eluent, methanol:acetonitrile:sodium acetate 5 mM:acetic acid (300:300:400:14 v/v/v/v); fluorimetric parameters, excitation at 340 nm, emission at 465 nm. All positive samples were confirmed both by esterification, as described by the Association of Official Analytical Chemists (1980), and carboxypeptidation, as described by Hult *et al.* (1982).

The recovery of ochratoxin A was about 90%, so that results are expressed without further correction. Using this method with the new generation of spectrofluorimeters, a limit of detection below 0.1 ng/ml can be achieved.

Results and Discussion

Figure 1 represents typical chromatograms of the ochratoxin A standard (Fig. 1A) and of a sample derived by esterification (Fig. 1B); Figure 2 shows the same standard (Fig. 2A) and a sample contaminated at about 0.2 ng/ml (Fig. 2B). Table 1 summarizes the results of all the analyses so far performed in France.

Table 1. Prevalence of ochratoxicosis in humans in France

Samples	Rural population	Urban population
% positive at 2 ng/ml	1.60	0.20
% positive at 0.1 ng/ml	22.00	20.15
% confirmed positive at 2 ng/ml	1.40	0.18
% confirmed positive at 0.1 ng/ml	18.00	16.00
Range of levels (ng/ml)	0.1–6	0.1–1.3

Comparison of our results with those obtained in the Balkans, at the same limit of detection (2 ng/ml) used by Hult *et al* (1982) and Petkova-Bocharova *et al.* (1988), shows that both the numbers of positive samples in urban and rural populations in France and the concentration range are lower. For example, the percentage of ochratoxin A-containing samples with more than 2 ng/ml was only 1.6% in France and 10–20% in Bulgaria, the highest levels being 6 ng/ml in France and 20 ng/ml in Bulgaria. The higher values in Balkan countries may be due to accumulation of ochratoxin A in the blood because of renal impairment related to diagnosed or undiagnosed Balkan endemic nephropathy.

Comparison of our results with those from Germany and Scandinavia, with a limit of detection of 0.1 ng/ml (Bauer & Gareis,

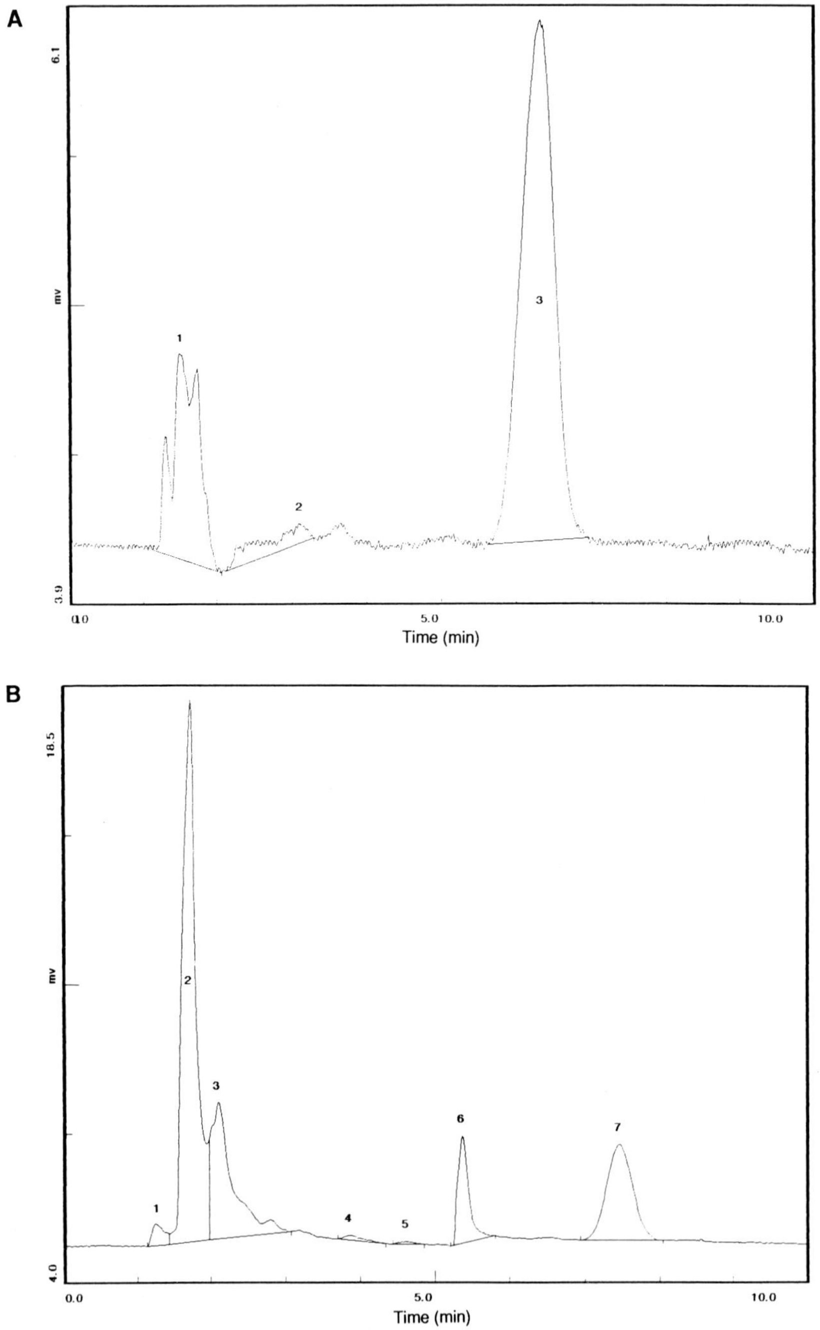

Figure 1. HPLC chromatogram of (A) ochratoxin A standard at 10 ng/ml by direct fluorimetric detection, with a limit of detection of 0.1 ng/ml, and (B) confirmation of the methylester derivative of same standard diluted at 4 ng/ml

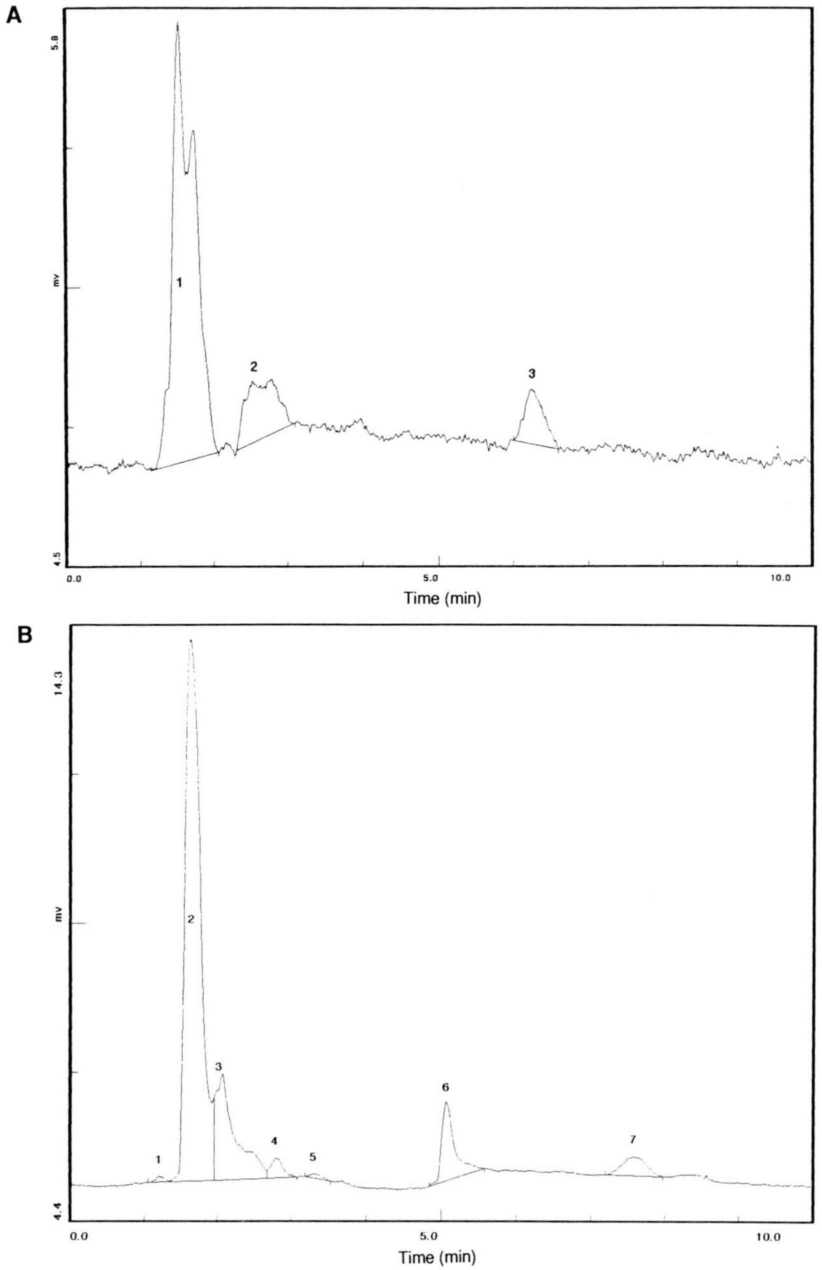

Figure 2. HPLC chromatogram of (A) an ochratoxin A-positive sample at the limit of detection of 0.1 ng/ml, and (B) confirmation of the same sample at a limit of detection of 0.2 ng/ml

1987; Breitholtz *et al.*, 1991) shows that the proportion of positive samples is far lower in France (22%) than in Denmark (50%) and Germany (50%) or Sweden (70%). This difference cannot be related to lack of confirmation in our study, since only 18–20% of the positive samples could not be confirmed by the two methods used.

The presence of ochratoxin A in serum is clearly related to consumption of cereals contaminated with this mycotoxin and perhaps to consumption of swine and poultry fed contaminated feed. The moulds that can produce ochratoxin A are of essentially two types: *Aspergillus* and *Penicillium*. Their growth is favoured by humidity, and a temperature of 28 °C is optimal for production of ochratoxin A by *A. ochraceus* (Trenk *et al.*, 1971). In contrast, *Penicillium* moulds that produce ochratoxin A grow well at 22% moisture over a wide range of temperatures (4–30 °C) (Mislivec & Tuite, 1970). The types of mould that produce ochratoxin A in northern and southern Europe should be determined precisely. It is possible, for example, that the ochratoxicosis in Germany and Scandinavia is related to *Penicillium* moulds and that in France to *Aspergillus*.

It is not known whether the low levels of ochratoxin A detected in western European countries affect human health. In most of the experimental studies carried out so far, high doses of ochratoxin A were used, the lowest being 2–4 mg/kg of feed to study nephrotoxicity and genotoxicity (Kane *et al.*, 1986). Studies should be carried out using lower doses of ochratoxin A in species that are more susceptible to this compound. For example, a 75% inhibition of the immunoglobulin M and G response to sheep red blood cells was found in Balb/c mice given ochratoxin A at 1 µg/kg intraperitoneally (Creppy *et al.*, 1983).

Acknowledgements

This research was funded by the Région d'Aquitaine, University of Bordeaux II, Direction de la Recherche, and the Ligue Nationale contre le Cancer, Comité Départementale de la Gironde. The authors are grateful to Mrs Guilcher for technical and secretarial assistance.

References

Arora, R.G. & Fröelén, H. (1981) Interference of mycotoxins with prenatal development of the mouse. II. Ochratoxin A induced teratogenic effects in relation to the dose and stage of gestation. *Acta Vet. Scand.*, **22**, 535–552

Association of Official Analytical Chemists (1980) AOAC Methods, Chapter 26, Cincinnati, Ohio, sections 26.096–26.102

Bauer, J. & Gareis, M. (1987) Ochratoxin A in the food chain. *Z. Veterinarmed. B*, **34**, 613–627 (in German)

Bendele, A.M., Carlton, W.W., Krogh, P. & Lillehoj, E.B. (1985) Ochratoxin A carcinogenesis in the (C57Bl/6JxC3H)F mouse. *J. Natl Cancer Inst.*, **75**, 733–742

Boorman, G., ed. (1989) *NTP Technical Report on the Toxicology and Carcinogenesis Studies of Ochratoxin A (CAS No. 303-47-9) in F344/N Rats (Gavage Studies)* (NIH Publication No. 89-2813), Research Triangle Park, North Carolina, National Toxicology Program, US Department of Health and Human Services

Breitholtz, A., Olsen, M., Dahlbäck, A. & Hult, K. (1991) Plasma ochratoxin A levels in three Swedish populations surveyed using an ion-pair HPLC technique. *Food Addit. Contam.* (in press)

Castegnaro, M., Bartsch, H. & Chernozemsky, I. (1987) Endemic nephropathy and urinary-tract tumors in the Balkans. *Cancer Res.*, **47**, 3608–3609

Creppy, E.E., Stormer, F.C., Röschenthaler, R. & Dirheimer, G. (1983) Effects of two metabolites of ochratoxin A, (4R)-4-hydroxy ochratoxin A and ochratoxin-alpha, on immune response in mice. *Infect. Immunol.*, **39**, 1015–1018

Hietanen, E., Malaveille, C., Camus, A.M., Béréziat, J.C., Brun, G., Castegnaro, M., Michelon, J., Idle, J.R. & Bartsch, H. (1986) Interstrain comparison of hepatic and renal microsomal carcinogen metabolism and liver S9-mediated mutagenicity in DA and Lewis rats phenotyped as poor and extensive metabolizers of debrisoquine. *Drug Metab. Disposition*, **14**, 118–126

Hult, K., Pleština, R., Habazin-Novak, V., Radić, B. & Čeović, S. (1982) Ochratoxin A in human blood and Balkan endemic nephropathy. *Arch. Toxicol.*, **51**, 313–321

Kane, A., Creppy, E.E., Röschenthaler, R. & Dirheimer, G. (1986) Changes in urinary and renal tubular enzymes caused by subchronic administration of ochratoxin A in rats. *Toxicology*, **42**, 233–243

Krogh, P. (1987) Ochratoxins in food. In: Krogh, P., ed., *Mycotoxins in Food*, New York, Academic Press, pp. 97–121

Kuiper-Goodman, T. & Scott, P.M. (1989) Risk assessment of the mycotoxin ochratoxin A. *Biomed. Environ. Sci.*, **2**, 179–248

Mislivec, P.B. & Tuite, J. (1970) Temperature and relative humidity requirements of species of penicillium isolated from yellow dentcorn kernels. *Mycologia*, **62**, 75–88

Petkova-Bocharova, T., Chernozemsky, I.N. & Castegnaro, M. (1988) Ochratoxin A in human blood in relation to Balkan endemic nephropathy and urinary system tumours in Bulgaria. *Food Addit. Contam.*, **5**, 299–301

Röschenthaler, R., Creppy, E.E. & Dirheimer, G. (1984) Ochratoxin A. On the mode of action of a ubiquitous mycotoxin. *J. Toxicol. Toxin Rev.*, **3**(1), 53–86

Trenk, H.L., Butz, M.E. & Chu, F.S. (1971) Production of ochratoxin in different cereal products by Aspergillus ochraceus. *Appl. Microbiol.*, **21**, 1032–1035

Mycotoxins, Endemic Nephropathy and Urinary Tract Tumours
Ed. M. Castegnaro, R. Pleština, G. Dirheimer, I.N. Chernozemsky & H. Bartsch
Lyon, International Agency for Research on Cancer, pp. 153-158

POSSIBLE SOURCES OF OCHRATOXIN A IN HUMAN BLOOD IN POLAND

P. Goliński[1], J. Grabarkiewicz-Szczęsna[1], J. Chełkowski[2], K. Hult[3] & M. Kostecki[1]

[1]Department of Chemistry, Agricultural University of Poznań, Poland; [2]Department of Plant Pathology, Agricultural University of Warsaw, Poland; and [3]Department of Biochemistry and Biotechnology, Royal Institute of Technology, Stockholm, Sweden

Summary

Samples of plant origin and human and porcine blood samples were screened over a long period for the presence of ochratoxin A. Of 1353 cereal samples, 11.7% contained the mycotoxin; of 1372 samples of feed, 1.5%; of 368 bread samples, 17.2%; of 215 flour samples, 22.3%; of 894 porcine serum samples, 37.4%; and of 1065 human serum samples, 7.2%. Seasonal variations in the natural occurrence of ochratoxin A were observed, with an increased percentage of positive samples in the spring. Individual daily intake of the mycotoxin, estimated on the basis of residues in human serum, was found to be 0.4 ng/g of food consumed.

Introduction

Ochratoxin A is a nephrotoxic metabolite of several fungal species frequently found in cereals and feeds. The spontaneous occurrence of ochratoxin A in cereals and in porcine tissues has been demonstrated in several countries. Consumption of grains contaminated with this toxin induces changes in renal function and structure in pigs (Krogh *et al.*, 1974), and similarities between mycotoxic porcine nephropathy and human endemic nephropathy have been noted (Krogh *et al.*, 1977).

Since ochratoxin A occurs in areas with a temperate climate, we decided to measure the natural contamination in Poland of cereals, feeds and food products of cereal origin as well as residues in human and porcine blood. Serum is an easy medium in which to measure ochratoxin A residues, as the metabolite has an affinity for plasma proteins (Hult *et al.*, 1982; Hagelberg *et al.*, 1989).

Materials and Methods

All samples were collected at random: cereals from various storage elevators, feed from different factories, flour from a number of mills and bread from various bakeries. Porcine blood was collected at slaughter houses, with no indication of the presence of nephropathy; and human blood samples were taken at hospital from individuals with no renal disease. The samples of cereals and feeds were collected between 1966 and 1987; those of flour, bread and other bakery products represent the grain crops of 1984 and 1985, and the human and porcine blood samples correspond to grain crops of 1983 and 1984.

Ochratoxin A was determined in cereals, food and feed according to the method of Nesheim (1973), with our modifications (Goliński & Grabarkiewicz-Szczęsna, 1984); the limit of detection was 5 ng/g. It was analysed in blood and serum samples using the method of Hult *et al.* (1979), with our modifications (Goliński *et al.*, 1984); the limit of detection was 1 ng/g.

Results and Discussion

Ochratoxin A was the most common secondary metabolite present in feeds and their cereal components in Poland. It was found more frequently in cereals than in feeds and oftener in rye than other cereals (Table 1). Contamination was greater in the spring than at other seasons, which may be due to the prolonged period of storage over the winter, the increases in temperature and humidity with thawing and rain, and the subsequent increase in activity of toxinogenic microflora.

The natural contamination of cereals results in the presence of ochratoxin A in flour and bakery products (Table 2). As among the cereals, the toxin was found more frequently and at higher concentration in rye flour and rye bread, suggesting that rye kernels are susceptible to fungal invasion and development and toxin biosynthesis. The number of flour and bread samples contaminated with ochratoxin A also increased in the spring.

The highest concentration of toxin found in feed (200 ng/g) was the lowest found by Krogh *et al.* (1974) to be able to induce porcine nephropathy. The finding of ochratoxin contamination of feeds suggested that residues might be found in porcine serum. The results of our two-year studies are shown in Table 3. About 55% of samples containing

Table 1. Contamination of mixed feeds and their cereal components with ochratoxin A[a]

Sample	Number of samples			Level of contamination (mg/kg)
	Analysed	Contaminated		
		No.	%	
Ground peanuts	609	0		
Mixed feed	1240	18	1.5	0.01–0.2
Protein concentrates	132	0		
All feeds	1372	20	1.5	0.010–0.2
Ground cereals	98	10	10.2	0.01–0.08
Wheat	239	28	11.7	0.005–2.4
Rye	228	62	27.2	0.005–0.8
Barley	616	54	8.8	0.005–1.2
Oats	49	2	4.1	0.080–0.8
Maize	123	2	1.6	0.025–0.4
All cereals	1353	158	11.7	0.005–2.4
Total	3334	178	5.3	0.005–2.4

[a]Based on the present and other investigations, from Goliński (1987)

Table 2. Contamination of bread and other cereal products with ochratoxin A

Sample	Number of samples			Average level of contamination (mg/kg)[a]
	Analysed	Contaminated		
		No.	%	
Bread	368	63	17.2	1.36
Wheat flour	137	27	19.7	3.7
Rye flour	78	21	26.9	5.41
All flour	215	48	22.3	4.37
Groats	35	4	11.4	1.14

[a]Calculated on the basis of all samples analysed, including those with no detectable concentration

ochratoxin A had residue levels lower than 2 ng/ml (Figure 1). The monthly frequency of ochratoxin A residues in porcine serum (Table 4) confirms our observations on seasonal variations in the natural occurrence of this toxin.

The logical sequel to the above results was to test for toxin residues in human blood. Ochratoxin A was found in 7.2% of all serum samples, at an average level of 0.27 ng/ml and a maximum of 40 ng/ml (Table 5).

Table 3. Occurrence of ochratoxin A residues in porcine blood and serum

Year	No. of samples analysed	Contaminated samples (%)	Average concentration (ng/ml)[a]	
			Serum	Blood
1983	388	40.7	7.80	2.89
1984	506	35.0	3.70	1.38
Total	894	37.4	5.48	2.03

[a]Calculated on the basis of all samples analysed, including those with no detectable concentration

Table 4. Monthly frequency and levels of contamination by ochratoxin A of porcine serum

Month	1983			1984		
	Frequency	Contamination (ng/ml)		Frequency	Contamination (ng/ml)	
		Maximum	Average[a]		Maximum	Average[a]
September	–	–	–	15/48	90	3.9
October	20/33	180	7.3	15/51	84	4.2
November	18/40	120	5.0	15/55	32	1.1
December	11/43	520	23.1	7/48	192	5.4
January	8/38	400	11.5	14/54	44	2.7
February	5/32	24	1.8	10/46	25	0.9
March	1/28	5	0.2	12/43	70	3.8
April	12/44	132	8.1	10/27	4	0.4
May	22/40	160	5.5	40/40	6	2.1
June	29/42	11	2.4	23/47	55	3.6
July	32/48	190	8.0	16/47	132	10.6

[a]Calculated on the basis of all samples analysed, including those with no detectable concentration

These values are lower than those found by Pleština (1981) but comparable with those of Hult *et al.* (1982) (Figure 2).

The individual daily intake of ochratoxin A was estimated on the basis of residues in human serum, using the coefficient of Mortensen *et al.* (1983), and found to be 0.4 ng/g of food consumed.

Acknowledgements

The authors are indebted to E. Rymaniak and I. Rissmann, Department of Chemistry, Agricultural University of Poznań for technical assistance and preparation of the manuscript.

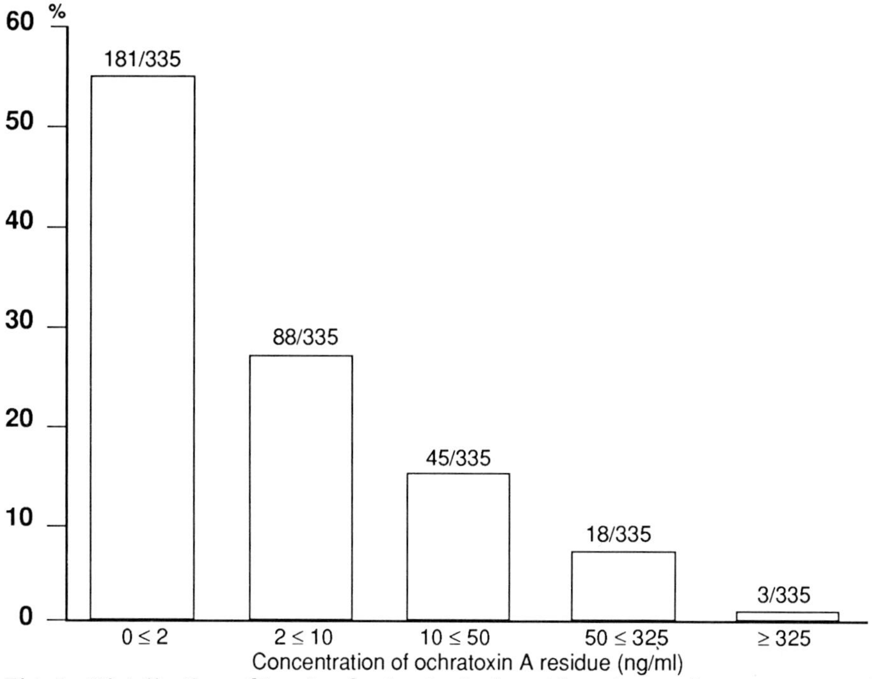

Fig. 1. Distribution of levels of ochratoxin A residues in porcine serum samples

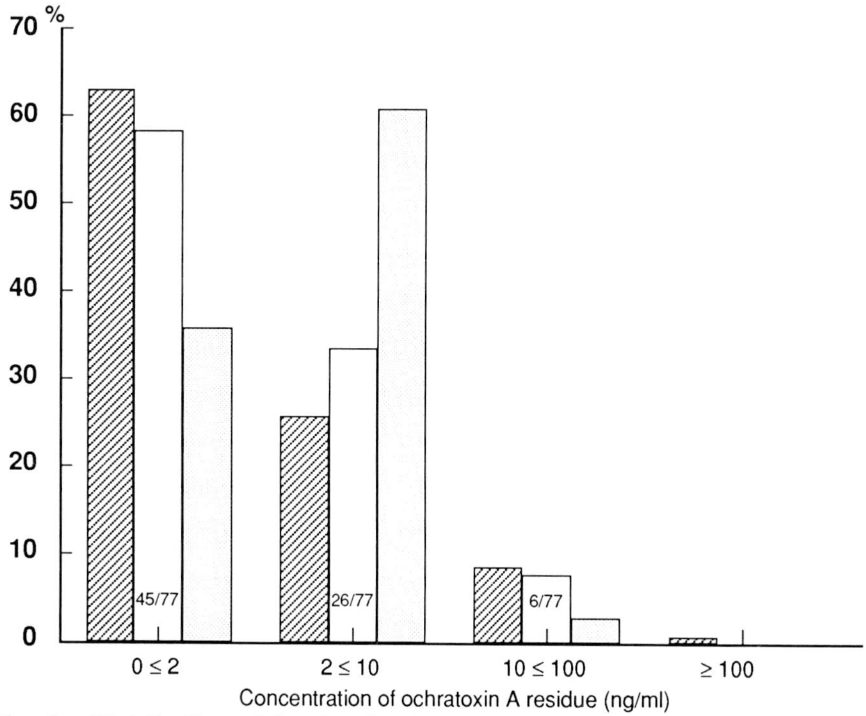

Fig. 2. Distribution of levels of ochratoxin A contamination in human serum samples in three studies: striped bars, Pleština (1981); open bars, our study; dotted bars, Hult *et al.* (1982)

Table 5. Occurrence of ochratoxin A residues in human blood and serum

Year	No. of samples analysed	Contaminated samples (%)	Average concentration (ng/ml)[a]	
			Serum	Blood
1983	397	6.3	0.21	0.08
1984	668	7.8	0.31	0.12
Total	1065	7.2	0.27	0.10

[a]Calculated on the basis of all samples analysed, including those with no detectable concentration

References

Goliński, P. (1987) Ochratoxin A in the human organism as a result of food and feed contamination. *Rocz. AR Poznaniu*, **168** (in Polish)

Goliński, P. & Grabarkiewicz-Szczęsna, J. (1984) Chemical confirmatory test for ochratoxin A, citrinin, penicillic acid, sterigmatocystin and zearalenone, performed directly on thin layer chromatographic plates. *J. Assoc. Off. Anal. Chem.*, **67**, 1108–1110

Goliński, P., Hult, K., Grabarkiewicz-Szczęsna, J., Chełkowski, J., Kneblewski, J. & Szebiotko, K. (1984) Mycotoxic porcine nephropathy and spontaneous occurrence of ochratoxin A residues in kidney and blood of Polish swine. *Appl. Environ. Microbiol.*, **47**, 1210–1212

Hagelberg, S., Hult, K. & Fuchs, R. (1989) Toxicokinetics of ochratoxin A in several species and its plasma-binding properties. *J. Appl. Toxicol.*, **9**, 91–96

Hult, K., Hökby, E., Hägglund, U., Gatenbeck, S., Rutqvist, L. & Sellyey, G. (1979) Ochratoxin A in pig blood: method of analysis and use as a tool for feed studies. *Appl. Environ. Microbiol.*, **38**, 772–776

Hult, K., Pleština, R., Habazin-Novac, V., Radić, B. & Čeović, S. (1982) Ochratoxin A in human blood and Balkan endemic nephropathy. *Arch. Toxicol.*, **51**, 313–321

Krogh, P., Axelsen, N.H., Elling, F., Gyrd-Hansen, N., Hald, B., Hyldegaard-Jensen, J., Larsen, A.E., Madsen, A., Mortensen, H.P., Moller, T., Petersen, P.K., Ravnskov, U., Rostgaard, M. & Aalund, O. (1974) Experimental porcine nephropathy. *Acta Pathol. Microbiol. Scand. Sect. A*, **246** (Suppl.), 1–21

Krogh, P., Hald, B., Pleština, R. & Čeović, S. (1977) Baklan (endemic) nephropathy and foodborn ochratoxin A: preliminary results of survey of foodstuffs. *Acta Pathol. Microbiol. Scand. Sect. B*, **85**, 238–240

Mortensen, H.P., Hald, B. & Madsen, A. (1983) Feeding experiment with ochratoxin A contaminated barley for bacon pigs. 5. Ochratoxin A in pig blood. *Acta Agric. Scand.*, **33**, 238–239

Nesheim, S. (1973) Analysis of ochratoxin A and B in barley using partition and thin layer chromatography. I. Development of the method. *J. Assoc. Off. Anal. Chem.*, **56**, 817–821

Pleština, R. (1981) Human exposure to ochratoxin A in areas of Yugoslavia with endemic nephropathy. In: Krogh, P., ed., *Mycotoxins and Phycotoxins*, New York, Pathotox Publishing

Mycotoxins, Endemic Nephropathy and Urinary Tract Tumours
Ed. M. Castegnaro, R. Pleština, G. Dirheimer, I.N. Chernozemsky & H. Bartsch
Lyon, International Agency for Research on Cancer, pp. 159-164

OCHRATOXIN A IN HUMAN BLOOD IN EUROPEAN COUNTRIES

B. Hald

Department of Veterinary Microbiology,
Royal Veterinary and Agricultural University, Frederiksberg, Denmark

Summary

Ochratoxin A is a mycotoxin which contaminates cereals in particular and occurs all over the world. Humans are undoubtedly exposed to this toxin through foods of vegetable and animal origin and through airborne dust. The presence of ochratoxin A in human blood has been suggested as an indicator for indirect assessment of exposure to this nephrotoxic agent. In several countries, therefore, human blood has been collected with the purpose of obtaining more information on the intake of ochratoxin A. Analyses of serum samples in European countries revealed that blood from healthy humans was contaminated with ochratoxin A at concentrations of 0.1–14.4 µg/l. The frequency of contamination of human sera seems to indicate continuous, widespread exposure of humans to ochratoxin A.

Introduction

Ochratoxin A is found predominantly in foods of cereal origin. Several species of *Aspergillus* and *Penicillium* are capable of producing this toxin, which is nephrotoxic in animals, affecting primarily the proximal tubule (Krogh, 1987).

Exposure to Ochratoxin A

Humans are exposed to ochratoxin A mainly through contaminated foods, although the existence of other routes, such as airborne exposure to ochratoxin A-containing spores and mycelial fragments, cannot be excluded. Usually, fungi invade only a minor fraction of food particles, and this is the main reason for the pronounced variation in results of monitoring foodstuffs for ochratoxin A contamination.

The formation of ochratoxin A in foodstuffs is strongly influenced by a number of environmental factors, the most important of which are temperature and water activity, the latter reflecting in particular precipitation during crop harvest (Lacey, 1989). The combined effect of these factors results in a pronounced variation in ochratoxin A contamination of foodstuffs, even in the same crop from the same area, when compared on an annual basis. Thus, it is difficult if not impossible to assess ochratoxin A contamination of foodstuffs, retrospectively or prospectively. Furthermore, because of variation from batch to batch and even within the same crop year, data on ochratoxin A contamination of foodstuffs cannot easily be translated into exposure data based on human intake.

Furthermore, human susceptibility to ochratoxin A is difficult to assess, since toxicological experiments are not carried out on humans. The most useful estimates of human susceptibility and exposure would be made by analysing ochratoxin A levels in food prepared for consumption, but no such data are available. Indirect assessment by measurements of blood residues of ochratoxin A is not always successful, as ochratoxin A has a serum half-time ranging from 4 to more than 500 h, depending on the species (Krogh 1987).

Occurrence of Ochratoxin A in Humans

When ochratoxin A is given orally to animals, most absorption takes place in the small intestine and primarily in the proximal jejunum and stomach (Kumagai *et al.*, 1982). Absorption rates vary from 40 to 60%. Ochratoxin A is present in blood bound to both serum albumin, with an association constant in the order of 10^5/M (Chu, 1971; Galtier, 1974), and to yet unidentified low-molecular-weight plasma constituents (Stojković *et al.*, 1984), and as free ochratoxin A. The low-molecular-weight species can cross the glomerular membrane and accumulate in the kidney (Stojković *et al.*, 1984), where the highest tissue concentration of ochratoxin A is found. Binding of ochratoxin A to serum albumin is particularly strong in humans, as determined *in vitro* (Galtier, 1979).

Denmark

Human serum was collected in Denmark over a three-year period to obtain information on intake of ochratoxin A. Samples were taken randomly from a blood bank every week and extracted with chloroform after acidification. The chloroform layers were purified over sodium hydrogen carbonate, and the ochratoxin A content was determined by high-performance liquid chromatography. The detection limit was 0.1 μg/l, and recovery at 0.5 μg/l was 87.5 ± 3.2. Confirmation was carried out by reaction with boron trifluoride methanol (Hunt *et al.*, 1980).

Overall, 54.2% of the samples contained ochratoxin A (Table 1), with means of 1.5 μg/l in 1986, 2.3 μg/l in 1987 and 1.6 μg/l in 1988. These high contamination levels are surprising, as the quality of cereals for human consumption should be better than that of animal feed (Pedersen, 1984). Surveys of ochratoxin A in cereals in Denmark from 1986 to 1989, however, showed that only in dry and fairly normal years can contamination of cereals be kept below 10 μg/kg (Tholstrup & Rasmussen, 1990), the limit aimed at in pork by setting the limit of ochratoxin A in pig kidneys to 25 μg/kg. The annual variations in ochratoxin A levels in cereals are clearly reflected by the ochratoxin A levels in the pigs (see Hald, this volume, Table 2).

Table 1. Ochratoxin A (μg/l) in 144 Danish blood bank samples, 1986–88

Month	1986		1987		1988	
	Range	Mean	Range	Mean	Range	Mean
January	ND–0.7	0.2	ND–9.4	4.5	ND–2.1	0.6
February	ND–7.3	2.6	ND–2.9	1.2	ND–2.3	2.2
March	ND–5.1	2.5	ND–0.2	0.1	ND–13.2	4.4
April		ND	ND–4.3	1.1	ND–0.9	2.1
May		ND	ND–7.4	3.3	ND–4.7	1.9
June	ND–3.1	1.5	ND–2.9	0.8	ND–3.8	2.1
July	ND–0.5	0.1	ND–4.9	1.2	ND–0.6	0.8
August	ND–3.3	1.6	ND–4.3	3.9	ND–0.8	0.4
September	ND–9.7	4.5	ND–6.9	3.6	ND–4.1	2.4
October	ND–5.3	2.9	ND–9.2	5.6	ND–1.9	1.1
November	ND–5.3	2.2	ND–4.9	1.5		ND
December		ND	ND–2.0	0.9	ND–1.7	0.8
Overall mean	1.5		2.3		1.6	

ND, not detected

Other European countries

Surveys on the occurrence of ochratoxin A in human blood serum have also been conducted in other European countries during the last nine years (Table 2).

In Germany, three studies were conducted, in Bayern, Niedersachsen and Hessen, and 56.5–68.1% of 691 serum samples contained ochratoxin A in a range of 0.1–14.4 μg/l. The Deutschesforschungs-

Table 2. Ochratoxin A in human blood in some European countries

Country	% positive	Concentration (µg/l)		Reference
		Mean	Range	
Czechoslovakia	24.5		0.1–1.2	Fukal & Reisnerova (1990)
Germany:				
Bayern	56.5	0.6	0.1–14.4	Bauer *et al.* (1986)
Niedersachsen	45.5[a]		0.1–0.2	Scheuer & Leistner (1986)
Hessen	68.1	1.1	0.1–8.4	Hadlok (1989)
Poland	4.2	0.3	1.3–4.8	Goliński & Grabarkiewicz-Szezęsna (1985, 1989)
	7.2	0.27	1.0–40.0	
Sweden:				
Visby	31.0	0.3	0.3–6.0	Breitholz *et al.* (1991)
Uppsala	4.0	0.02	0.3–0.8	
Östersund	8.0	0.03	0.3–0.8	
Balkan peninsula:				
Bulgaria, endemic area	26.0	20.0	1–35	Petkova-Bocharova *et al.* (1988)
Bulgaria, non-endemic area	7.7	10.0	1–?	
Yugoslavia, endemic area	6.0	7.2	1–40	Hult *et al.* (1982a,b)
Yugoslavia, non-endemic area	7.7	5.4	1–10	

[a] Whole blood

gemeinschaft Mitteilung XII of der Senatskommission zur Prüfung von Lebensmittelzusatz- und -inhaltsstoffen produced estimates in 1990 of human exposure to ochratoxin A based on data from German surveys of ochratoxin A levels in meat and cereals, and on consumption figures for these two food sources. The exposure varied from 81.6 to 158 ng/day, depending on the daily intake of corn products.

In Sweden, 297 serum samples were collected from blood donors in three districts (Uppsala, Östersund and Visby) during the first two weeks of October 1989. Samples from Visby were found to have a more frequent occurrence and higher levels of ochratoxin A than those from the two other districts (Breitholz *et al.*, 1991). This finding reflects local levels of ochratoxin A in foodstuffs, since Visby is situated on the island of Gotland and its population is thus more heavily dependent on local products than in most other parts of the country. The detection limit in this study was 0.3 µg/l. In random sampling and analysis of pig blood for ochratoxin A carried out in Sweden (Hult *et al.*, 1984; Holmberg *et al.*, 1990a), levels of ochratoxin A were higher in pig blood from the district of Visby than the mean values for other parts of the country.

The daily intake of ochratoxin A in Sweden was calculated by Breitholz *et al.* (1991) from the levels found in serum samples from the three districts and from analyses of food and food consumption. The calculations based on plasma levels gave levels of 0.03–0.35 ng/kg bodyweight per day, and those based on food analyses gave 0.36 ng/kg bodyweight per day.

Surveys of human blood serum have also been carried out in Poland and Czechoslovakia. In Czechoslovakia, 24.5% of 143 samples analysed by radioimmunoassay contained ochratoxin A. Published surveys of ochratoxin A in human blood in the Balkan peninsula (endemic and nonendemic areas) are also summarized in Table 2; more recent information is provided by Fuchs *et al.* and Petkova-Bocharova *et al.* (this volume).

The high incidence of ochratoxin A in Danish and German blood samples may reflect the use of very sensitive analytical methods (detection limit, 0.1 μg/l); nevertheless, it also reflects continuous, widespread exposure of humans to ochratoxin A.

Conclusion

It is reasonable to assume that humans may respond to acute or chronic effects of ochratoxin A whenever they are exposed through contamination of dietary components. It also seems reasonable to assume that the character and intensity of the human response might vary, depending on factors such as age, sex, nutritional status, concurrent exposure to other agents and the level and duration of exposure. Furthermore, it is apparent that there are fluctuations in the levels of ochratoxin A in human serum, which may reflect local, seasonal or annual fluctuations in the levels of ochratoxin A in food. Although data on the occurrence of ochratoxin A indicate the geographically widespread nature of the problem, little information has been obtained because samples were collected randomly.

As the toxicokinetic profile of ochratoxin A in humans is still unknown, it is difficult to interpret the relevance of the concentrations found in humans. Results from experiments with animals seem to indicate that ochratoxin A remains in human blood for a long time. Daily intake of small doses over a long period, however, could give the same effect.

Ochratoxin A contamination of food and feed constitutes a public health problem of unknown dimensions. It is therefore still relevant to investigate to what extent ochratoxin A is hazardous to humans.

References

Bauer, J., Gareis, M. & Gedek, B. (1986) Incidence of ochratoxin A in blood serum and kidneys of man and animals. In: *Proceedings of the 2nd World Congress on Foodborne Infections and Intoxications, Berlin*, p. 907 (unpublished)

Breitholz, A., Olsen, M., Dahlbäck, Å. & Hult, K. (1991) Plasma ochratoxin A levels in three Swedish populations surveyed using an ion-pair HPLC technique. *Food Addit. Contam.* , **8**, 183–192

Chu, F.S. (1971) Interaction of ochratoxin A with bovine serum albumin. *Arch. Biochem. Biophys.*, **147**, 359–366

Fukal, L. & Reisnerova, H. (1990) Monitoring of aflatoxin and ochratoxin A in Czechoslovak human serum by immunoassay. *Bull. Environ. Contam. Toxicol.*, **44**, 345–349

Galtier, P. (1974) Fate of ochratoxin A in the animal organism. I. Blood transport of the toxin in the rat. *Ann. Rech. Vet.*, **5**, 311-318 (in French)

Galtier, P. (1979) Toxicological and pharmacokinetic study of a mycotoxin, ochratoxin A. Thesis, University of Toulouse, France, pp. 1–50

Goliński, P. & Grabarkiewicz-Szezęsna, J. (1985) The first Polish cases of the detection of ochratoxin A residues in human blood. *Rocz. Panstw. Zakl. Hig.*, **36**, 378–381 (in Polish)

Goliński, P. & Grabarkiewicz-Szezęsna, J. (1989) Trial of determination of the daily dose of ochratoxin A consumed by humans in Poland. *Rocz. Panstw. Zakl. Hig.*, **40**, 50–52 (in Polish)

Hadlok, R.M. (1989) Ochratoxin. Occurrence in blood from slaughter pigs and humans. In: *Vortrag 11, Mycotoxin Workshop, Berlin* (unpublished, in German)

Hult, K., Pleština, R., Čeović, S., Habazin-Novak, V. & Radić, B. (1982a) Ochratoxin A in human blood: analytical results and confirmational tests from a study in connection with Balkan endemic nephropathy. In: *Proceedings of the Vth International IUPAC Symposium on Mycotoxins and Phycotoxins*, Vienna, Austrian Chemical Society, pp. 334–341

Hult, K., Pleština, R., Habazin-Novak, V., Radić, B. & Čeović, S. (1982b) Ochratoxin A in human blood and Balkan endemic nephropathy. *Arch. Toxicol.*, **51**, 313–321

Hult, K., Rutquist, L., Holmberg, T., Thafbelin, B. & Gatenbeck, S. (1984) Ochratoxin A in blood of slaughter pigs. *Nord. Vet. Med.*, **36**, 314–316

Holmberg, T., Breitholz, A., Bengtsson, A. & Hult, K. (1990) Ochratoxin A in swine blood in relation to moisture content in feeding barley at harvest. *Acta Agric. Scand.*, **40**, 201–204

Hunt, D.C., McConnie, B.R. & Crosby, N.T. (1980) Confirmation of ochratoxin A by chemical derivatisation and high-performance liquid chromatography. *Analyst (London)*, **105**, 89–90

Krogh, P. (1987) Ochratoxins in food. In: Krogh, P., ed., *Mycotoxins in Food, Science and Technology*, New York, Academic Press, pp. 97–121

Kumagai, S. & Aibara, K. (1982) Intestinal absorption and secretion of ochratoxin A in the rat. *Toxicol. Appl. Pharmacol.*, **64**, 94–102

Lacey, J. (1989) Pre- and post-harvest ecology of fungi causing spoilage of foods and other stored products. *J. Appl. Bacteriol., Symposium Supplement*, 11S–25S

Pedersen, A., ed. (1984) *Cereals and Cereal Products, Trace Elements, Pesticides, Nutrients, Food Additives, Nitrate and Nitrite, Ochratoxin A and Ergots* (Publication No. 93), Copenhagen, Statens Levnedsmiddelinstitut, p. 68 (in Danish)

Petkova-Bocharova, T., Chernozemsky, I.N. & Castegnaro, M. (1988) Ochratoxin A in human blood in relation to Balkan endemic nephropathy and urinary system tumours in Bulgaria. *Food Addit. Contam.*, **5**, 299–301

Scheuer, R. & Leistner, L. (1986) Occurrence of ochratoxin A in pork and pork products. *Proc. Eur. Meet. Meat Res. Work.*, **32**(1, 4:2), 191

Stojković, R., Hult, K., Gamulin, S. & Pleština, R. (1984) High affinity binding of ochratoxin A to plasma constituents. *Biochem. Int.*, **9**, 33–38

Tholstrup, B. & Rasmussen, G. (1990) *Ochratoxin A in Cereals 1986–1989* (Publication No. 199), Copenhagen, Statens Levnedsmiddelinstitut, p. 50 (in Danish)

Mycotoxins, Endemic Nephropathy and Urinary Tract Tumours
Ed. M. Castegnaro, R. Pleština, G. Dirheimer, I.N. Chernozemsky & H. Bartsch
Lyon, International Agency for Research on Cancer, pp. 165-169

CONCENTRATIONS OF OCHRATOXIN A IN THE URINE OF ENDEMIC NEPHROPATHY PATIENTS AND CONTROLS IN BULGARIA: LACK OF DETECTION OF 4-HYDROXYOCHRATOXIN A

M. Castegnaro[1], V. Maru[2], T. Petkova-Bocharova[3], I. Nikolov[3] & H. Bartsch[1]

[1]*International Agency for Research on Cancer, Lyon, France;* [2]*Biochemistry Section, Pathology Department, Haslok Hospital & Research Centre, Bombay, India; and* [3]*National Oncological Centre, Sofia, Bulgaria*

Summary

Ochratoxin A has been detected more frequently and at higher levels as a contaminant in staple food consumed by subjects affected by Balkan endemic nephropathy or urinary tract tumours in the Vratza district (Bulgaria) than in samples from control populations in and outside the endemic area. Serum from patients with Balkan endemic nephropathy also contained ochratoxin A more frequently and at higher levels than serum from controls. Metabolic phenotyping of subjects in the Vratza district with debrisoquine revealed a preponderance of extensive metabolizers among subjects at high risk for Balkan endemic nephropathy. In rats, ochratoxin A is metabolized to 4-hydroxy-ochratoxin A, and rat strains shown to be poor or extensive metabolizers of debrisoquine were also poor or extensive metabolizers of ochratoxin A. In order to determine whether the metabolic phenotype for debrisoquine also parallels that of ochratoxin A in humans, a sensitive method was developed for quantifying ochratoxin A and its 4-hydroxy metabolite in human urine. This method was subsequently used to analyse urine from subjects who had previously been phenotyped for debrisoquine. Ochratoxin A was detected more frequently and at higher levels in urine from members of families affected by Balkan endemic nephropathy than in samples taken from subjects in control areas. No 4-hydroxyochratoxin A was found in any of these samples (detection limit, 15 ng/l urine). On the basis of results from human studies and animal models, the role of genetic polymorphism in drug oxidation and disease susceptibility is discussed briefly.

Introduction

Ochratoxin A has been found in samples of beans and maize from endemic and nonendemic areas for Balkan endemic nephropathy and the associated urinary tract tumours in Bulgaria (Petkova-Bocharova & Castegnaro, 1985). Although there was no significant difference between the mean levels of ochratoxin A in contaminated samples from endemic and control areas, a much higher proportion ($p < 0.05$) of samples collected in 1981 from the endemic area were contaminated. Follow-up studies on crops from 1984, 1985, 1986, 1989 and 1990 confirmed these results (Petkova-Bocharova *et al.*, this volume). Despite annual variations (in some years, ochratoxin A levels were in the milligrams per kilogram range), cereal and bean samples consumed by affected families were more heavily contaminated than those consumed by families from unaffected households in both endemic and nonendemic areas. As ochratoxin A binds to serum albumin, it can also be detected in biological fluids from exposed populations. In Bulgaria, it was found more frequently and at higher levels in samples of blood from affected families in endemic and control areas (Petkova-Bocharova *et al.*, 1988). Subsequent investigations have confirmed this finding (Petkova-Bocharova & Castegnaro, this volume). In rats, ochratoxin A is metabolized to ochratoxin α and 4-hydroxyochratoxin A (Støren *et al.*, 1982), and these compounds are found in the urine at levels of 25–27% and 1–1.5%, respectively, of an initial ochratoxin A dose of 6.6 mg/kg body weight, together with unchanged ochratoxin A (6%). Ochratoxin A and its 4-hydroxy metabolites thus could occur in the urine of people exposed to this mycotoxin.

Metabolic phenotyping with debrisoquine of people in Vratza district revealed a preponderance of fast metabolizers (metabolic ratio, < 0.8; % dose as unchanged debrisoquine:% dose as 4-hydroxydebrisoquine in urine) among people at high risk for Balkan endemic nephropathy and urinary tract tumours (Ritchie *et al.*, 1983; Nikolov *et al.*, this volume). Using rat strains which show genetic polymorphism for debrisoquine-4-hydroxylation (DA strain as the slow debrisoquine metabolizer and Lewis strain as the fast metabolizer), we demonstrated that the metabolic ratio of ochratoxin A:4-hydroxyochratoxin A was 2–5 times greater in DA than in Lewis rats, as was the metabolic ratio of debrisoquine:4-hydroxydebrisoquine (Castegnaro *et al.*, 1989). These results suggest that fast metabolizers of ochratoxin A and debrisoquine are more prone to ochratoxin A-induced carcinogenicity. A higher tumour yield in response to ochratoxin A was found in rats than in mice and in males than in females in both species (Bendele *et al.*, 1985; Boorman, 1988). Rats are known to be faster metabolizers of debrisoquine than mice, and in both species males are faster metabolizers than females (Castegnaro *et al.*, 1990a).

We have investigated the excretion of ochratoxin A and its 4-hydroxy metabolite in urine from people living in high- and low-incidence areas for Balkan endemic nephropathy and who had previously been phenotyped for debrisoquine.

Materials and Methods

Urine samples were collected from persons living in areas of high and low incidence of Balkan endemic nephropathy and urinary tract tumours and stored in plastic vials at –30 °C until analysis.

Ochratoxin A and its 4-hydroxy metabolite were analysed using the method described by Castegnaro *et al.* (1990b). Briefly, after chloroform extraction from urine acidified to pH 2.4–2.5 and two purification steps using column chromatography and high-performance liquid chromatography (HPLC), ochratoxin A was analysed by HPLC/spectrofluorimetry. For samples suspected to contain ochratoxin A, a confirmatory step was performed using derivatization of ochratoxin A to its methylester and HPLC/spectrofluorimetry analysis. The recoveries of ochratoxin A were 65–75% for samples spiked at 10–500 ng/l, and the limit of detection for ochratoxin A was 5 ng/l. Recoveries of 4-hydroxyochratoxin A were 60–75% for samples spiked at 30–500 ng/l, but they decreased to 40–50% for samples spiked at 20 ng/l. The limit of detection for 4-hydroxyochratoxin A was about 15 ng/l.

Results and Discussion

A total of 152 urine samples collected from patients with Balkan endemic nephropathy or urinary tract tumours and from control families were analysed, and ochratoxin A was detected in about 33% (Table 1), more often in endemic villages than in nonendemic ones. The highest levels were detected in patients with Balkan endemic nephropathy or urinary tract tumours, with a maximum of 600 ng/l, which probably reflected recent consumption of highly contaminated food, although the pharmacokinetics of ochratoxin A excretion in humans is unknown. In no case was free 4-hydroxyochratoxin A detected.

Several explanations can be put forward to explain the lack of detection of the hydroxy metabolite in human urine samples: (i) Ochratoxin A and its metabolite are found as such in rat urine, but it is possible that 4-hydroxyochratoxin A is present as the glucuronide or

Table 1. Urinary excretion of ochratoxin A by patients with endemic nephropathy or urinary tract tumours and by healthy people

Group	No. of samples	Samples containing ochratoxin A		
		No.	%	Level (range; ng/l)
Patients with endemic nephropathy or urinary tract tumour	36	14	38.9	5–604
People with suspected endemic nephropathy	25	9	36	5–32
Family members of patients with endemic nephropathy	25	12	48	5–33
Healthy members of healthy families in endemic villages	32	11	44	5–43
Healthy people in unaffected villages in endemic area	31	4	12.9	17–41
Healthy people in villages in nonendemic area	3	0	–	–

sulfate conjugate in human urine. (ii) While rats and humans may have similar activities of debrisoquine-4-hydroxylase, the P450 isozymes that catalyse 4-hydroxylation of ochratoxin A may be different from those that catalyse 4-hydroxylation of debrisoquine, and the form may be absent or less active in humans. Indeed, it has been demonstrated that hepatic and renal ochratoxin A 4-hydroxylase activity is significantly lower in DA than in Lewis rats and is inducible by 3-methylcholanthrene and phenobarbital to a greater extent in DA than in Lewis rats; however, debrisoquine-4-hydroxylation was not inducible. This suggests that ochratoxin A and debrisoquine-4-hydroxylase are not identical (Hietanen *et al.*, 1985, 1986). Further work to identify the P450 isozymes (CYP genes) responsible for the 4-hydroxylation of ochratoxin A is required. (iii) 4-Hydroxyochratoxin A may be formed as a metabolite in humans *in vivo* but not excreted or further metabolized. (iv) Chakor *et al.* (1988), using inhibitors and enhancers of liver microsomal mono-oxygenases, demonstrated that ochratoxin A metabolism leads to a reduction of its acute toxicity, although 4-hydroxyochratoxin A was as toxic as ochratoxin A itself. They proposed either that 4-hydroxyochratoxin A is further metabolized to non-toxic or less toxic compounds (which can be conjugated), or that the formation of non-toxic derivatives like ochratoxin α or ochratoxin B (ochratoxin A without chlorine) is also inducible. These two hypothesis remain to be explored.

The reason for the apparent parallelism in rodent strains between their sensitivity to the tumorigenicity of ochratoxin A and their capacity to hydroxylate debrisoquine and ochratoxin A remains to be elucidated.

References

Bendele, A.M., Carlton, W.W., Krogh, P. & Lillehoj, E.B. (1985) Ochratoxin A carcinogenesis in the (C57BL/6J X C3H)F_1 mouse. *J. Natl Cancer Inst.*, **75**, 733–742

Boorman, G. (1988) *NTP Technical Report on the Toxicology and Carcinogenesis Studies of Ochratoxin A* (NTP TR358, NIH publication No. 88-2813), Research Triangle Park, NC

Castegnaro, M., Bartsch, H., Béréziat, J.-C., Arvela, P., Michelon, J. & Broussolle, L. (1989) Polymorphic ochratoxin A hydroxylation in rat strains phenotyped as poor and extensive metabolizers of debrisoquine. *Xenobiotica*, **19**, 225–230

Castegnaro, M., Chernozemsky, I.N., Hietanen, E. & Bartsch, H. (1990a) Are mycotoxins risk factors for endemic nephropathy and associated urothelial cancers? *Arch. Geschwulstforsch.*, **60**, 295–303

Castegnaro, M., Maru, V., Maru, G.B. & Ruiz-Lopez, M.D. (1990b) High-performance liquid chromatographic determination of ochratoxin A and its 4R-4-hydroxy metabolite in human urine. *Analyst*, **115**, 129–131

Chakor, K., Creppy, E.E. & Dirheimer, G. (1988) *In vivo* studies on the relationship between hepatic metabolism and toxicity of ochratoxin A. *Arch. Toxicol.*, **Suppl. 12**, 201–204

Hietanen, E., Bartsch, H., Castegnaro, M., Malaveille, C., Michelon, J. & Broussolle, L. (1985) Use of antibodies (Ab) against cytochrome P-450 isozymes to study genetic polymorphism in drug oxidation. *J. Pharm. Clin.*, **4**, 71–78

Hietanen, E., Malaveille, C., Camus, A.-M., Béréziat, J.-C., Brun, G., Castegnaro, M., Michelon, J., Idle, J.R. & Bartsch, H. (1986) Interstrain comparison of hepatic and renal microsomal carcinogen metabolism and liver S9-mediated mutagenicity in DA and Lewis rats phenotyped as poor and extensive metabolizers of debrisoquine. *Drug Metab. Disposition*, **14**, 118–126

Petkova-Bocharova, T. & Castegnaro, M. (1985) Ochratoxin A contamination of cereals in an area of high incidence of Balkan endemic nephropathy in Bulgaria. *Food Addit. Contam.*, **2**, 267–270

Petkova-Bocharova, T., Chernozemsky, I.N. & Castegnaro, M. (1988) Ochratoxin A in human blood in relation to Balkan endemic nephropathy and urinary system tumours in Bulgaria. *Food Addit. Contam.*, **5**, 299–301

Ritchie, J.C., Crothers, M.J., Idle, J.R., Greig, J.B., Connors, T.A., Nikolov, I.G. & Chernozemsky, I.N. (1983) Evidence for an inherited metabolic susceptibility to endemic (Balkan) nephropathy. In: Strahinjić, S. & Stefanović, V., eds, *Current Research in Endemic (Balkan) Nephropathy*, Niš, University of Niš, pp. 23–27

Støren, O., Holm, H. & Størmer, F.C. (1982) Metabolism of ochratoxin A by rats. *Appl. Environ. Microbiol.*, **44**, 785–789

Mycotoxins, Endemic Nephropathy and Urinary Tract Tumours
Ed. M. Castegnaro, R. Pleština, G. Dirheimer, I.N. Chernozemsky & H. Bartsch
Lyon, International Agency for Research on Cancer, pp. 171-186

MECHANISM OF ACTION OF OCHRATOXIN A[1]

G. Dirheimer[2] & E.E. Creppy[3]

[2]Institut de Biologie Moléculaire et Cellulaire du Centre National de la Recherche Scientifique, Strasbourg; and [3]Laboratoire de Toxicologie, Faculté de Pharmacie, Université de Bordeaux II, Bordeaux, France

Summary

Ochratoxin A has a number of toxic effects in mammals, the most notable of which is nephrotoxicity. It is also immunosuppressive, teratogenic and carcinogenic. The biochemical and molecular aspects of its action were first studied in bacteria. The appearance of 'magic spots' (ppGpp and pppGpp) pointed to inhibition of the charging of transfer ribonucleic acids (tRNA) with amino acids. This suggestion was confirmed by the demonstration that ochratoxin A inhibits bacterial, yeast and liver phenylalanyl-tRNA synthetases. The inhibition is competitive to phenylalanine and is reversed by an excess of this amino acid. As a consequence, protein synthesis is inhibited, as shown with hepatoma cells in culture, with Madin Darby canine kidney cells (which are much more sensitive) and *in vivo* in mouse liver, kidney and spleen, the inhibition being more effective in the latter two organs. An excess of phenylalanine also prevents inhibition of protein synthesis in cell cultures and *in vivo*. Analogues of ochratoxin A in which phenylalanine has been replaced by other amino acids have similar inhibitory effects on the respective amino acid-specific aminoacyl tRNA synthetases. 4R-Hydroxyochratoxin A, a metabolite of ochratoxin A, has a similar action, whereas ochratoxin α (the dihydroisocoumarin moiety) and ochratoxin B (ochratoxin A without chlorine) have no effect. Ochratoxin A might act on other enzymes that use phenylalanine as a substrate. We showed

[1]This paper is dedicated to the memory of Professor R. Röschenthaler.

recently that it inhibits phenylalanine hydroxylase. In addition, the phenylalanine moiety of ochratoxin A is partially hydroxylated to tyrosine by incubation with hepatocytes and *in vivo*. This competitive action with phenylalanine might explain why this amino acid prevents the immunosuppressive effect of ochratoxin A and partially prevents its teratogenic and nephrotoxic actions. The effect of ochratoxin A on protein synthesis is followed by an inhibition of RNA synthesis, which might affect proteins with a high turnover. Ochratoxin A also lowers the level of phosphoenolpyruvate carboxykinase, a key enzyme in gluconeogenesis; this inhibition is reported to be due to a specific degradation of mRNA that codes for this enzyme. Recently, ochratoxin A was also found to enhance lipid peroxidation both *in vitro* and *in vivo*. This inhibition might have an important effect on cell or mitochondrial membranes and be responsible for the effects on mitochondria that have been shown by several authors. Finally, the recent results of Pfohl-Leszkowicz *et al.* (this volume), who showed the formation of DNA adducts mainly in kidney but also in liver and spleen, explain the DNA single-strand breaks observed previously in mice and rats after acute and chronic treatment.

Introduction

A discussion of the mode of action of a compound requires some general considerations. The aim of research in this field is to track down the possible primary target of biological activity and to gain some insight into interactions between the reactive components. This objective is, however, usually not easy to attain. Many factors must be taken into account, and particularly the concentration. A well-established example in this respect is the antibiotic tetracycline (Gale *et al.*, 1981), which has a specific action on prokaryotic ribosomes at limiting bactericidal concentrations but is an uncoupler of the respiratory chain at 100–1000 times this concentration. Similar results were obtained with ochratoxin A.

Furthermore, the effect of a mycotoxin at its primary site of action may be masked by regulatory mechanisms that are triggered by the affected cell. Such control mechanisms may be particularly complex in higher animals. They are generally better understood in prokaryotes and sometimes even indicate certain pharmaceutical activities; they can often be measured more easily, avoided or recognized more readily in prokaryotic cells.

Bacteria

The mechanism of action of ochratoxin A was first investigated by Röschenthaler and coworkers (Singer & Röschenthaler, 1977) in bacteria, because, although these are prokaryotic cells, results obtained with these organisms may indicate where to search for effects in eukaryotes. At a pH lower than 7, ochratoxin A is a potent inhibitor of growth in Gram-positive bacteria like *Bacillus subtilis*, *Bacillus stearothermophilus* and *Streptococcus faecalis*. In the minimal medium of Anagnostopoulos and Spizizen (1961), the growth of *B. subtilis* was inhibited by 50% with ochratoxin A at 6 µg/ml (15 µM). With high doses of ochratoxin A (more than 30 µM), an uncoupling of the respiratory chain could be demonstrated. Growth of *S. faecalis*, which lacks a complete respiratory chain, is also inhibited: at 15 µM or less, strong inhibition of protein synthesis followed by inhibition of RNA synthesis was observed (Figure 1). DNA synthesis was not affected at this concentration. When protein synthesis was blocked by chloramphenicol, however, RNA synthesis, which then continues for about half an hour, was not sensitive to ochratoxin A. This suggests that the inhibition of RNA synthesis by ochratoxin A observed in the absence of chloramphenicol is due only to a regulation phenomenon.

In fact, accumulation of the regulatory nucleotides ppGpp and pppGpp to about four fold was demonstrated after addition of ochratoxin A in *B. subtilis* (Singer & Röschenthaler, 1977) (Figure 2), and stabilization of the polysome profiles was shown in *S. faecalis* (Heller & Röschenthaler, 1977, 1978). These results rule out direct inhibition by ochratoxin A of RNA synthesis, which appears to be inhibited by a secondary regulation mechanism involving ppGpp and pppGpp nucleotides ('magic spot' nucleotides). The accumulation of these nucleotides is generaly interpreted as a sign that a tRNA species is incompletely aminoacylated in the cells (Cashel & Gallant, 1974; Goldman & Jakubowski, 1990). Konrad and Röschenthaler (1977) were able to show a lower degree of aminoacylation of tRNA for phenylalanine in bacteria.

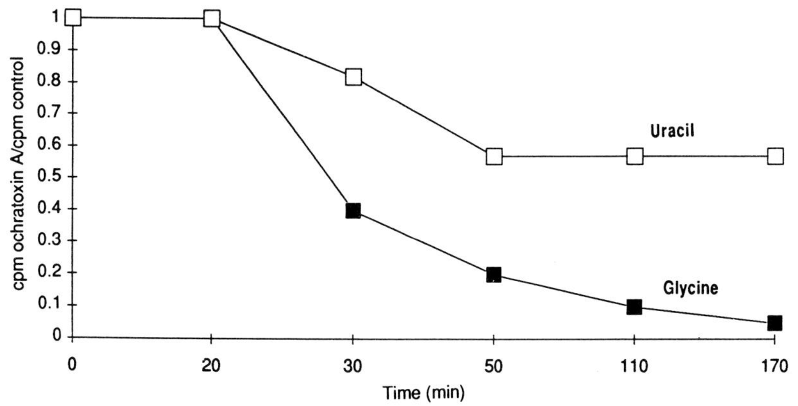

Figure 1. Protein and RNA synthesis, as measured by incorporation of ^{14}C-glycine and ^{14}C-uracil into acid-precipitable material, after addition of ochratoxin A at 10 µg/ml

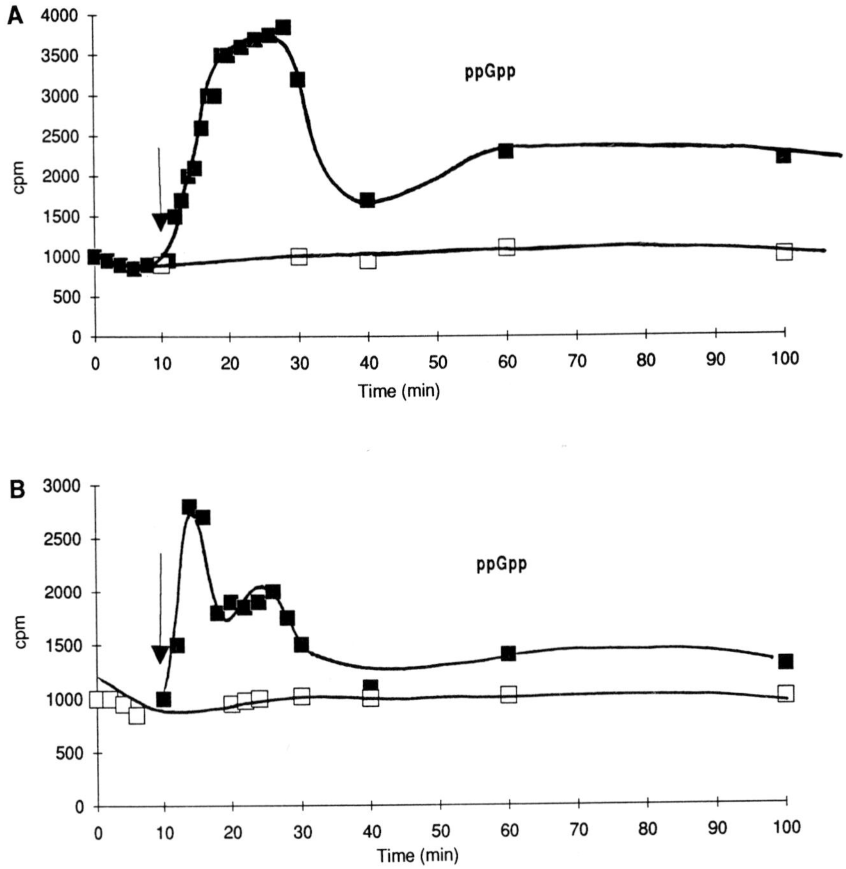

Figure 2. Increased levels of ppGpp (A) and pppGpp (B) nucleotides after addition of ochratoxin A at 10 µg/ml to a culture medium (arrow) of *Bacillus subtilis*

The susceptibility to ochratoxin A of the phenylalanyl-tRNA synthetase (PheRS) of *B. subtilis* and of *B. stearothermophilus* was therefore investigated. The PheRS of these bacteria is inhibited competitively and specifically by ochratoxin A *versus* phenylalanine (Konrad & Röschenthaler, 1977), indicating that ochratoxin A acts as an analogue to phenylalanine. We showed subsequently (Bunge *et al.*, 1978) that in a polyuridylic acid-dependent peptide synthesis system of *B. stearothermophilus*, synthesis of polyphenylalanine is also competitively inhibited. When aminoacylated phenylalanine-tRNA was included in the system, peptide synthesis was not inhibted. This result demonstrates that the only reaction of protein synthesis involved in this inhibition was aminoacylation.

Eukaryotic Systems in vitro

Ochratoxin A did not inhibit the growth of *Saccharomyces cerevisiae* at concentrations as high as 1 mM at pH 6.0. *In vitro*, however, a specific, competitive inhibition of the aminoacylation of phenylalanine-tRNA was shown by Creppy *et al.* (1979a). Both components, the PheRS and the $tRNA^{Phe}$, were pure. The two reactions catalysed by this enzyme—the activation of the amino acid and the aminoacylation of the tRNA—are inhibited by ochratoxin A. The inhibition constant, K_i is 1.5 mM for ochratoxin A in both reactions, whereas the K_m of the enzyme for phenylalanine in the activation reaction (measured by pyrophosphate-ATP exchange) is 30 µM and that in the aminoacylation reaction is 5 µM. These results prove that the inhibition by ochratoxin A of PheRS is a general phenomenon found in both prokaryotic and eukaryotic systems.

The [4R]-4-hydroxyochratoxin A epimer, a metabolite of ochratoxin A, had an effect similar to that of ochratoxin A on protein synthesis in this yeast system, but ochratoxin α lacking the phenylalanine moiety (Creppy *et al.*, 1983a) and ochratoxin B lacking the C5 chlorine had no effect.

In order to confirm that the mode of action of ochratoxin A is competition with phenylalanine in the PheRS-catalysed reaction, we checked the effect of synthetic analogues of ochratoxin A, in which the phenylalanine moiety had been replaced by other amino acids, on the corresponding aminoacyl-tRNA synthetases; for example, the effect of valine-ochratoxin A was assayed on valyl-tRNA synthetase.

All the experiments were performed using the conditions of initial velocities, which were established using the partially purified yeast aminoacyl-tRNA synthetases. The inhibition constants were determined by comparing the rates of tRNA charging for increasing concentrations of the amino acid in the absence and in the presence of two different concentrations of the ochratoxin A analogue, using Lineweaver–Burk plots. The concentrations of the analogues are indicated in Table 1 for the eight analogues of ochratoxin A and for ochratoxin B and ochratoxin α. The results indicate (i) that the ochratoxin A-PheRS system exhibits the highest affinity of all the systems tested (K_i approximately 1.3 mM); and (ii) that only ochratoxin A inhibits PheRS: neither ochratoxin B nor ochratoxin α is active as an inhibitor. The alanine, tyrosine and tryptophan analogues all have similar K_i values (between 1.5 and 2.3 mM), whereas serine-ochratoxin A has a K_i five times higher and valine-, methionine-, glutamic acid- and proline-ochratoxin A have K_i about 10 times higher.

Experiments on partially purified PheRS from hepatoma cell cultures and from rat and guinea-pig livers *in vitro* showed an inhibition constant of 0.3 mM. The affinities of the hepatoma cell cultures and animal enzymes for ochratoxin A were five times higher than in yeast. Thus, the liver system is more sensitive to ochratoxin A than the yeast system.

Table 1. Inhibition constants for homologues of ochratoxin A

Product	Concentration (mM)	Inhibition constant (mM)	K_m of amino acid (mM)	K_i/K_m
Ochratoxin A[a]	0.72, 1.44	1.3	0.0033[a]	433
Ochratoxin B[a]	1.2	None	0.0033[a]	–
Ochratoxin α	1.1	None	0.0033[a]	–
Ala-ochratoxin A	0.7, 1.4	1.6	0.18	8.9
Glu-ochratoxin A	2.5, 5.0	14.5	0.37	39.2
Met-ochratoxin A	4.3	13.0	0.006	2166.6
Pro-ochratoxin A	5.0, 10.0	19.0	0.19	99.9
Ser-ochratoxin A	5.3, 14.6	6.0	0.48	12.5
Trp-ochratoxin A	1.5, 3.0	2.3	0.017	135.3
Tyr-ochratoxin A	1.5, 3.0	1.5	0.032	46.9
Val-ochratoxin A	0.5	11.0	0.034	323.5

[a]The amino acid is phenylalanine.

Hepatoma Cell Cultures

In a concentration range between 40 and 45 μM, ochratoxin A had a cytostatic effect on hepatoma cell cultures derived from a malignant tumour of rat liver parenchymal cells (Morris hepatoma 7288c) adapted for growth in suspension. Higher concentrations were cytotoxic. At a concentration of 90 μM, a decreased rate of protein synthesis was observed 30 min after addition of ochratoxin A (Figure 3). After 150 min, protein synthesis was blocked almost completely, and a decrease in the rate of RNA synthesis became measurable which came to a halt after a further 150 min. The effect of ochratoxin A on RNA synthesis is therefore most probably a consequence of an effect on protein synthesis. DNA synthesis is not inhibited for at least 5 h after addition of ochratoxin A (Creppy *et al.*, 1979b).

When hepatoma cell cultures received 150 μM phenylalanine in addition to 90 μM ochratoxin A, no inhibition of cell growth occurred. The reversion of inhibition by phenylalanine cannot be due to competition with cell transport receptors, as postulated by Delacruz and Bach (1991), because when phenylalanine was given 2 h after ochratoxin A, when the toxin was already in the cell and had inhibited growth and protein synthesis, it still reversed the inhibition. In addition, we have shown recently that the presence of phenylalanine does not impair the entry of ochratoxin A into hepatoma cell cultures (Fig. 4) or into hepatocytes (data not shown). Reciprocally, the entry of phenylalanine is not impaired by ochratoxin A (data not shown). Thus, inhibition of growth and protein synthesis in these cells can be traced back to the inhibition by ochratoxin A of the aminoacylation reaction of $tRNA^{Phe}$.

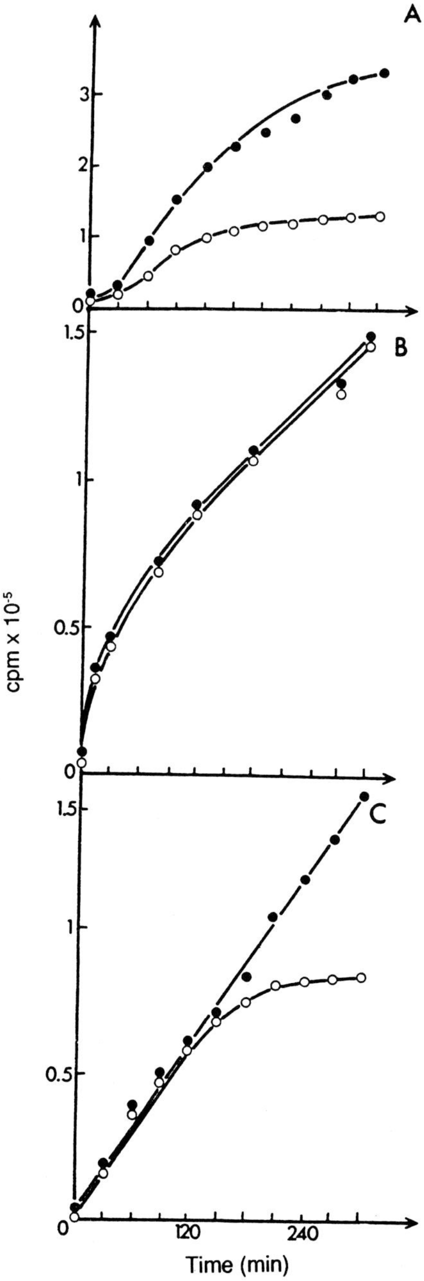

Figure 3. Inhibition of protein (A), DNA (B) and RNA (C) synthesis by ochratoxin A in hepatoma cell cultures; control, filled circles; ochratoxin A at 90 µM, open circles

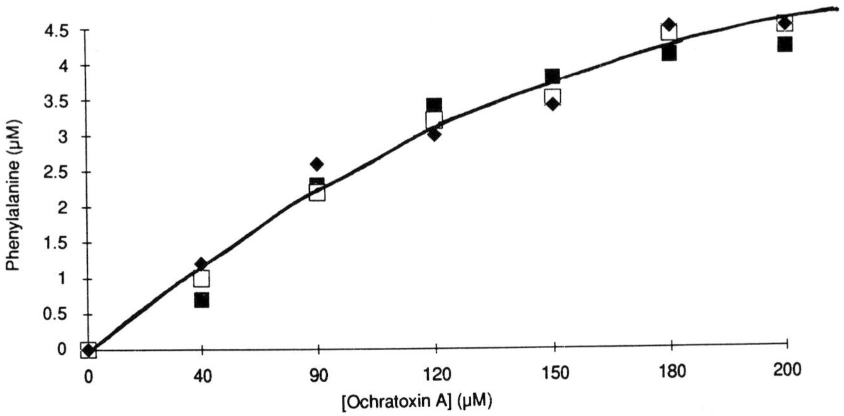

Figure 4. Entry of ochratoxin A into hepatoma cell cultures as a function of phenylalanine added to the culture medium, which already contains 400 μM phenylalanine; no ochratoxin A, open squares; 90 μM ochratoxin A, diamonds; 180 μM ochratoxin A, filled squares

Protein synthesis was also measured in hepatoma cell cultures after preincubation of the cells for various times with 90 μM of the synthetic ochratoxin A analogues (Creppy *et al.*, 1983a). At the indicated times (Figure 5), the cell culture was placed in a flask containing 2 μCi of ^{3}H-leucine, and incorporation of this amino acid into proteins was measured after 15 min. As these analogues decrease cell multiplication to different degrees, protein synthesis was determined on the basis of the number of cells remaining in each experiment. Figure 5 shows that in

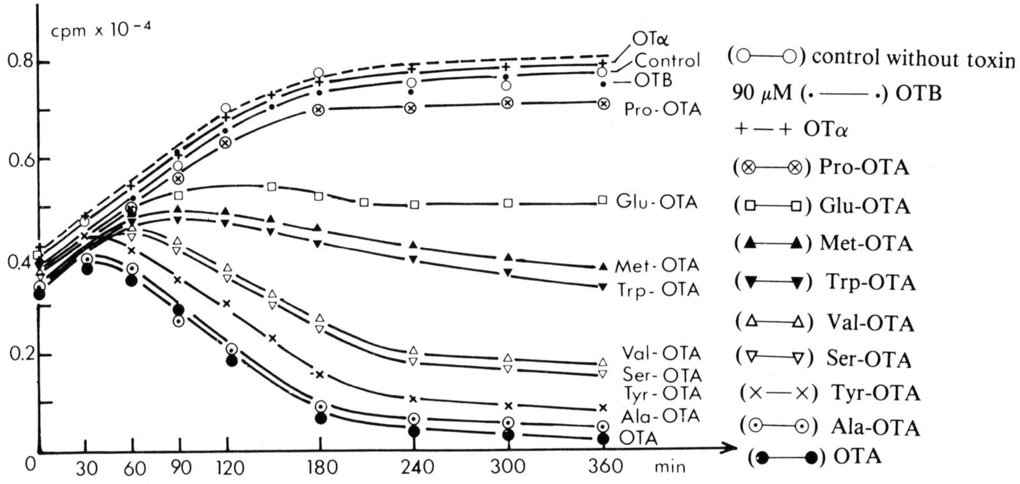

Figure 5. Protein synthesis in hepatoma cell cultures followed by incorporation of ^{3}H-leucine

the control protein synthesis increases for 180 min and then parallels the rate of cell multiplication. Neither ochratoxin α nor ochratoxin B seems to have an effect at the 90-μM level. Pro-ochratoxin A has very little effect. Glu-ochratoxin A has an inhibitory effect which can be seen after 60 min, but protein synthesis reaches a stable level after 240 min and does not decrease even after 6 h. Met-ochratoxin A and Trp-ochratoxin A have similar effects, with an inhibitory action which starts at 90 min and becomes more and more pronounced. The most marked effects were obtained (in order of increasing effect) with Val-ochratoxin A and Ser-ochratoxin A (which have similar effects), Tyr-ochratoxin A, and then Ala-ochratoxin A, which has an effect after only 30 min that is closely similar to that of ochratoxin A. These results are also interesting in connection with the recent finding of ochratoxin A analogues in *A. ochraceus* cultures (R. Hadidane, personal communication).

We also tested the effect of ochratoxin A on the Madin Darby canine kidney (MDCK) cell line, which we used as a model system to study kidney damage (Creppy *et al.*, 1986). MDCK cells were cultured in a monolayer, whereas hepatoma cell cultures were cultured in suspension. With 12.5 μM ochratoxin A, protein synthesis was inhibited to 50%, and inhibition was almost complete with 25 μM of the toxin. Figure 6 shows the effects of ochratoxin A on protein synthesis in hepatoma cell cultures and in MDCK cells. It is clear that the renal cells were more sensitive than the hepatoma cells to the effect of ochratoxin A. Recent studies (Delacruz & Bach, 1990) showed that the proximal tubules and glomeruli have an exceptionally high rate of incorporation of aromatic amino acids into macromolecules, which is reduced by 50% at 5–10 μM ochratoxin A.

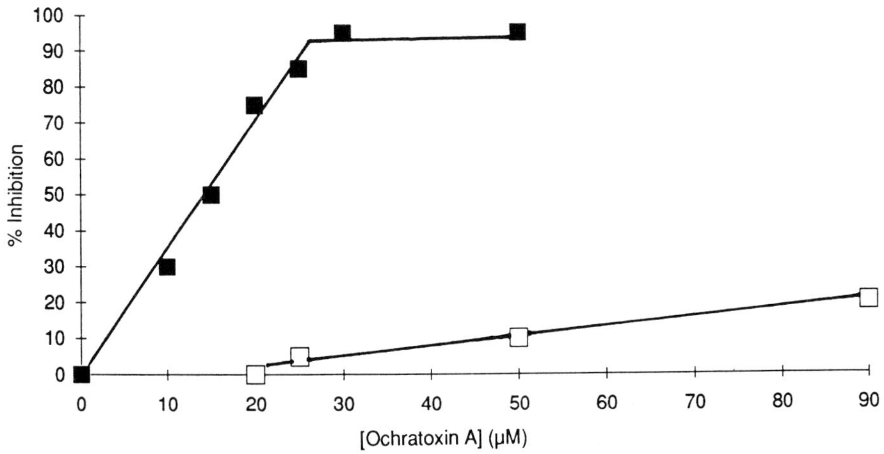

Figure 6. Effects of ochratoxin A on protein synthesis in hepatoma cell cultures (open squares) and in Madin Darby canine kidney (MDCK) cells (filled squares)

Finally, the effect of ochratoxin A on protein synthesis was also assayed on Balb/c mouse spleen lymphocytes (Creppy *et al.*, 1982) by preincubating them with ochratoxin A at 0.005–10 µM for 72 h (Figure 7). Ochratoxin A very effectively inhibited protein synthesis in this system: 2 µM ochratoxin A resulted in 50% inhibition.

Studies Using Whole Animals

We finally tested the effect of ochratoxin A on protein synthesis *in vivo* by measuring the incorporation of ^{14}C-amino acids into the proteins of liver, kidney and spleen of Balb/c mice treated with single doses of ochratoxin A at 1, 2.5, 5 or 10 mg/kg (Creppy *et al.*, 1984). Ochratoxin A was injected either simultaneously or 5 h before intraperitoneal injection of the amino acids. In the latter case, protein synthesis was measured 1 h after administration of the amino acids. Protein synthesis was inhibited even with the lowest dose, to 26% in liver, 68% in kidney and 75% in spleen, as compared with the control; similar degrees of inhibition in the different organs were also found with the higher doses, independently of the mode of application of ochratoxin A. Protein synthesis thus appears to be affected even more in the spleen than that in kidney but to a significantly lesser extent in liver. This effect was particularly striking with doses of less than 2.5 mg/kg ochratoxin A. It should be noted that, as was observed *in vitro*, a dose of phenylalanine 10 times that of ochratoxin A prevented the inhibition of protein synthesis caused by 10 mg/kg of ochratoxin A in all organs. Thus, inhibition of protein synthesis by ochratoxin A is observed in all systems tested, from bacteria to mammals, *in vitro* and *in vivo*.

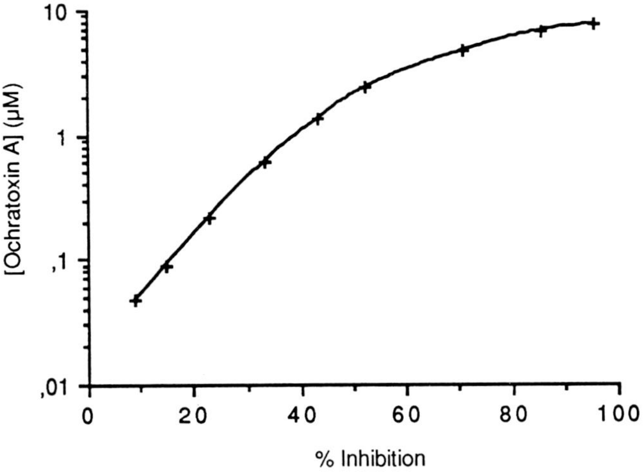

Figure 7. Effect of ochratoxin A on protein synthesis in Balb/c mouse spleen lymphocytes

The strong suppression in Balb/c mice of the immunoglobulin M and G responses to sheep red blood cells by single intraperitoneal doses of ochratoxin A as low as 1 μg/kg may be due to an effect on protein synthesis. Figure 8 shows that a dose of 0.5 μg/kg induces 75% immunosuppression, as measured by the standard plaque-counting technique for estimation of antibody-producing spleen lymphocytes. When phenylalanine was given at a concentration approximately twice that of ochratoxin A (wt/wt), no immunosuppressive effect was observed, or the effect was strongly attenuated (Haubeck *et al.*, 1981). [4R]-4-Hydroxyochratoxin A was almost as effective as ochratoxin A, but ochratoxin α was ineffective (Creppy *et al.*, 1983b). The high sensitivity of the immune cells may be due to selective uptake and concentration of the toxin by lymphocytes.

Both the lethal effect of ochratoxin A and its nephrotoxicity could also be due to an effect on protein synthesis and should thus be prevented by phenylalanine. We therefore tested whether phenylalanine would impair the lethal effect of ochratoxin A in mice. All animals that received an intraperitoneal dose of 0.8 mg/mouse died within 24 h, but when this lethal dose was injected together with 0.8 mg phenylalanine, 97% of the animals survived; 100% survived when 1 mg phenylalanine was injected. The mice survived the observation period of 28 days (Creppy *et al.*, 1979c,d).

Thus, doses of phenylalanine only slightly higher than the lethal doses of ochratoxin A efficiently counteracted the toxic effect of ochratoxin A when given simultaneously. These relatively low doses may be effective because phenylalanine has a greater affinity for PheRS than ochratoxin A, as discussed above. In cell cultures as well, doses only slightly higher than those of ochratoxin A could reverse the toxic action.

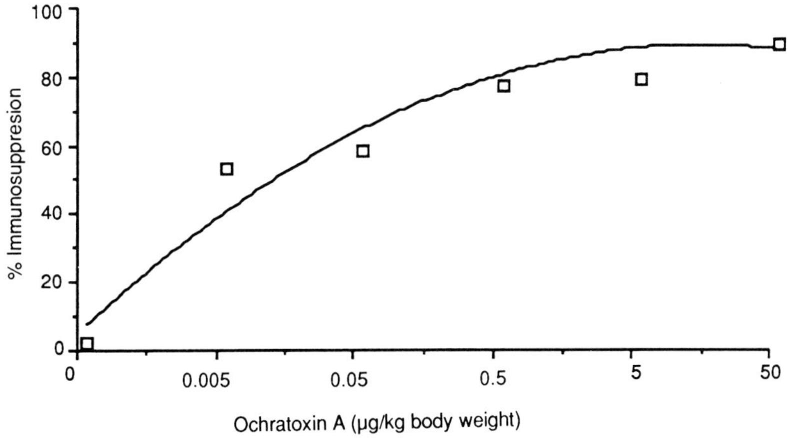

Figure 8. Immunosuppressive effect of ochratoxin A in Balb/c mice

Our results were confirmed by Moroi *et al.* (1985), who also showed a reduction in the toxicity of intraperitoneally injected ochratoxin A in male ddY mice when phenylalanine was injected simultaneously; the oral toxicity of ochratoxin A, however, was not affected by simultaneous treatment with phenylalanine.

Ochratoxin A is also teratogenic (for references, see Mayura *et al.*, 1984), but the incidence of ochratoxin A-induced malformations in rats was significantly diminished in the presence of added phenylalanine (20 mg/kg phenylalanine for 1.7 mg/kg ochratoxin A).

Other Biochemical Effects of Ochratoxin A

Action on phenylalanine hydroxylase

Since ochratoxin A can potentially act on all metabolic systems involving phenylalanine, we assayed its action on phenylalanine hydroxylase in isolated rat hepatocytes and *in vivo* (Creppy *et al.*, 1990). When ochratoxin A (at 0.12–1.4 mM) was incubated with freshly isolated rat hepatocytes, it inhibited both the hydroxylation of phenylalanine (at 0.05 mM) to tyrosine, catalysed by phenylalanine hydroxylase, and the subsequent metabolism of tyrosine, as measured by homogentisate oxidation. The IC_{50} of ochratoxin A for phenylalanine hydroxylation was 0.43 mM. Ochratoxin α (at 0.5–1.0 mM) did not inhibit phenylalanine hydroxylase activity under these conditions. Incubation of hepatocytes with uniformly labelled ^{3}H-ochratoxin A and unlabelled phenylalanine resulted in formation of Tyr-ochratoxin A (up to 6% of the total mycotoxin added, after 30 min), indicating that ochratoxin can act as a substrate for phenylalanine hydroxylase. Tyr-ochratoxin A was also found *in vivo* in the livers of poisoned animals. It could be metabolized further, to give rise to [4R]- or [4S]-4-hydroxytyrosine-ochratoxin A and other metabolites, which may be responsible for the appearance of the small peaks of radioactivity on silica gel columns (Creppy *et al.*, 1990).

Action on phosphoenolpyruvate carboxykinase

Oral administration of ochratoxin A (at 2 mg/kg body weight for two days) led to a decrease in cytosolic phosphoenolpyruvate carboxykinase (PEPCK) in rat kidneys (Meisner & Selanik, 1979; Meisner & Meisner, 1981; for a review, see Frank *et al.*, 1990). This decrease correlated well with the decrease in renal gluconeogenesis. Meisner *et al.* (1983) propose that ochratoxin A could act by favouring degradation of PEPCK mRNA. This could, however, be an indirect effect of protein synthesis inhibition, in that the mRNAs with a high turnover are no longer protected by ribosomes when protein synthesis stops. Recently, Krogh *et al.* (1988) showed that the level of PEPCK was also decreased in renal cytosol of pigs fed low levels of ochratoxin A (0.2 and 1.0 mg/kg) for 1–5 weeks.

Lipid peroxidation

Rahimtula *et al.* (1988) found that addition of ochratoxin A to rat liver and kidney microsomes greatly enhanced the rate of NADPH-dependent lipid peroxidation. Peroxidation of polyunsaturated fatty acids present in membrane lipids has been proposed as a mechanism by which a number of foreign compounds produce structural tissue injury. It remains to be demonstrated, however, whether the toxic effects observed in rodents following administration of ochratoxin A are due to the onset of lipid peroxidation.

Action on DNA

The genotoxicity of ochratoxin A was recently reviewed (Kuiper-Goodman & Scott, 1989). Ochratoxin A has been shown to be nonmutagenic in various microbial and mammalian gene mutation assay systems. Tests for DNA damage and repair in bacteria also gave negative results (Table 2). A weakly positive response for induction of unscheduled DNA synthesis in primary hepatocytes from ACI C3H strain mice and ACI strain rats was found by Mori *et al.* (1984) but not by Bendele *et al.* (1985). Contradictory results were also found for sister chromatid exchange in CHO cells (for a review, see Kuiper-Goodman & Scott, 1989).

Table 2. Results of genotoxicity tests with ochratoxin A

Endpoint	Test system	Activation[a]	Dose	Result
Prokaryotes				
Gene mutation	*Salmonella typhimurium* TA98, 100, 1535, 1537, 1538, G46, C3076, D3052	+/−	0.1–1000 mg/plate	−/−
DNA repair	*Escherichia coli* WP2 (SOS)		1–2 mg/ 100 µl	−/−
DNA damage	*Bacillus subtilis rec*	+/−	20–100 µg/disc	−
Eukaryotes in vitro				
Gene mutation	*Saccharomyces cerevisiae* D3	+/−	200 µg/ phase	−/−
Gene mutation	Mouse lymphoma TK, C3H mouse mammary carcinoma 8-AG	+/−	0.1–12.5	−/−
Sister chromatid exchange	Human HPBL cells	+/−	5–10 µg/ml	−/−
Chromosomal aberration	Chinese hamster ovary cells	+/−	30–300 µg/ml	−/−

[a]With (+) and without (−) activation with microsomal preparation of mouse, rat or hamster liver

Creppy *et al.* (1985a,b) and Kane *et al.* (1986) have detected single-stranded breaks in DNA both in cell cuture and *in vivo*. The *in-vitro* experiments were done on Balb/c mouse spleen cells stimulated for 48 h with phytohaemagglutinin and exposed to ochratoxin A at doses of 10 μg/ml for 48 h. The cells were loaded on polyvinylchloride filters and lysed with an alkaline (pH 10) solution. Single-stranded DNA was eluted from the filters by washing with a solution at pH 12.6. The eluted DNA and the DNA remaining on the filters was measured by a fluorimetric technique which made it possible to visualize single-strand breaks in the DNA (Creppy *et al.*, 1985a,b).

In vivo, male Balb/c mice were injected intraperitoneally with 2.5 mg/kg ochratoxin A. At 6, 12, 24, 48 and 72 h after injection, liver, spleen and kidneys were excised and minced. The cell suspensions obtained were filtered and centrifuged, and the cells were treated by the alkaline elution technique, as above. Single-stranded breaks were found in the DNA of all three tissues. The largest number of breaks was observed after 24 h; these lesions were no longer present in the kidney 48 h after treatment, whereas complete recovery was observed at 72 h in the liver. Unfortunately, the experiments in the spleen were not pursued long enough to see if breaks observed after 24 h were repaired (Creppy *et al.*, 1985a,b).

The most interesting results were obtained with Wistar rats that were given ochratoxin A at 290 μg/kg body weight by intubation every 48 h for 12 weeks. This dose of ochratoxin A to a contamination of feed with 2 mg/kg, and this concentration can be encountered in naturally contaminated feeds. A very significant number of single-strand breaks was obtained in DNA from both liver and kidney (Kane *et al.*, 1986). Stetina and Votava (1986) confirmed that ochratoxin A can induce single-strand breaks in the DNA of CHO cells after 3 h, but they used a high concentration of ochratoxin A (500 μM). They obtained negative results in a line of rat fibroblasts.

Overall, these findings suggest that ochratoxin A is genotoxic to mammalian cells. This conclusion is confirmed by the results of Pfohl-Leszkowicz *et al.* (this volume), who found DNA adducts in liver, kidney and spleen, 24, 48 or 72 h after intraperitoneal treatment of mice with ochratoxin A at 2.5 mg/kg using the method of ^{32}P-postlabelling. The genotoxic effects observed in liver and kidney cells of ochratoxin A-exposed animals correlate with the carcinogenic effects seen in those organs (for review, see Kuiper-Goodman & Scott, 1989).

Conclusion

The toxic effects of a compound are rarely due to a unique action on a target reaction. This is probably true for ochratoxin A, because its complex structure comprises a phenylalanine moiety, which makes it a possible inhibitor of all reactions in which phenylalanine intervenes, and a dihydroisocoumarin moiety, which might have as yet undiscovered long-term toxic properties. Thus, much research remains to be done to understand completely the mode of action of ochratoxin A.

Acknowledgements

This work was supported by grants from the Institut National de la Santé et de la Recherche Médicale (external research contract 89/2007), the Ligue Nationale contre le Cancer (Comités départementaux du Haut-Rhin et de la Gironde), the Association de la Recherche contre le Cancer and the Ministère de la Recherche et de la Technologie (Action Génotoxicité).

References

Anagnostopoulos, C. & Spizizen, J. (1961) Requirements for transformation of *Bacillus subtilis*. *J. Bacteriol.*, **80**, 741–746

Bendele, A.M., Neal, S.B., Oberly, T.J., Thompson, C.Z., Bewsey, B.J., Hill, L.E., Rexroat, M.A., Carlton, W.W. & Probst, G.S. (1985) Evaluation of ochratoxin A for mutagenicity in a battery of bacterial and mammalian cell assays. *Food Chem. Toxicol.*, **23**, 911–918

Bunge, I., Dirheimer, G. & Röschenthaler R. (1978) *In vivo* and *in vitro* inhibition of protein synthesis in *Bacillus stearothermophilus* by ochratoxin A. *Biochem. Biophys. Res. Commun.*, **83**, 398–405

Cashel, M. & Gallant, J. (1974) Cellular regulation of guanosine tetraphosphate and guanosine pentaphosphate. In: Nomura, M., Tissières, A. & Lengyel, P., eds, *Ribosomes*, Cold Spring Harbor, NY, CSH Press, pp. 733–746

Creppy, E.E., Lugnier, A.A.J., Fasiolo, F., Heller, K., Röschenthaler, R. & Dirheimer, G. (1979a) *In vitro* inhibition of yeast PheRS by ochratoxin A. *Chem.-Biol. Interactions*, **24**, 257–262

Creppy, E.E., Lugnier, A.A.J., Beck, G., Röschenthaler, R. & Dirheimer, G. (1979b) Action of ochratoxin A on cultured hepatoma cells. Reversion of inhibition by ochratoxin A. *FEBS Lett.*, **104**, 287–290

Creppy, E.E., Schlegel, M., Röschenthaler, R. & Dirheimer, G. (1979c) Preventive action of phenylalanine on acute poisoning by ochratoxin A. *C.R. Acad. Sci. Paris*, **289**, 915–918 (in French)

Creppy, E.E., Schlegel, M., Röschenthaler, R. & Dirheimer, G. (1979d) Phenylalanine prevents acute poisoning by ochratoxin A in mice. *Toxicol. Lett.*, **6**, 77–80

Creppy, E.E., Lafarge-Frayssinet, C., Röschenthaler, R. & Dirheimer, G. (1982) Splenic lymphocytes and immunosuppressor effect of ochratoxin A. In: *3rd International School of Molecular Biology, Alger*, Vol. 1, Alger, Office des Publications Universitaires, pp. 91-94 (in French)

Creppy, E.E., Kern, D., Steyn, P.S., Vleggaar, R., Röschenthaler, R. & Dirheimer, G. (1983a) Comparative study of the effect of ochratoxin A analogues on yeast aminoacyl-tRNA synthetases and on growth and protein synthesis in hepatoma cells. *Toxicol. Lett.*, **19**, 217–224

Creppy, E.E., Stormer, F.C., Röschenthaler, R. & Dirheimer, G. (1983b) Effects of two metabolites of ochratoxin A, [4R]-4-hydroxyochratoxin A and ochratoxin α, on the immune response in mice. *Infect. Immunol.*, **39**, 1015–1018

Creppy, E.E., Röschenthaler, R. & Dirheimer, G. (1984) Inhibition of protein synthesis in mice by ochratoxin A and its prevention by phenylalanine. *Food Chem. Toxicol.*, **22**, 883–886

Creppy, E.E., Kane, A., Dirheimer, G., Lafarge-Frayssinet, C., Mousset, S. & Frayssinet, C. (1985a) Genotoxicity of ochratoxin A in mice: DNA single-strand break evaluation in spleen, liver and kidney. *Toxicol. Lett.*, **28**, 29–35

Creppy, E.E., Kane, A., Dirheimer, G., Lafarge-Frayssinet, C., Mousset, S. & Frayssinet, C. (1985b) DNA breaks in the spleen, liver and kidneys of mice treated with ochratoxin A. *C.R. Soc. Biol.*, **179**, 688–695 (in French)

Creppy, E.E., Kane, A., Giessen-Crouse, E., Roth, A., Röschenthaler, R. & Dirheimer, G. (1986) Effect of ochratoxin A on enzyme activities and macromolecules synthesis in MDCK cells. *Arch. Toxicol.*, **Suppl. 9**, 310–314

Creppy, E.E., Chakor, K., Fischer, M.J. & Dirheimer, G. (1990) The mycotoxin ochratoxin A is a substrate for phenylalanine hydroxylase in isolated rat hepatocytes and *in vivo*. *Arch. Toxicol.*, **64**, 279-284

Delacruz, L. & Bach, P.H. (1990) The role of ochratoxin A metabolism and biochemistry in animal and human nephrotoxicity. *J. Biopharm. Sci.*, **1**, 277–304

Frank, H.K., Dirheimer, G., Grunow, W., Netter, K.J., Osswald, H. & Schlatter, J. (1990) *Ochratoxin A, Vorkommen und toxikologische Bewertung*, Weinheim, VCH Verlagsgesellschaft, pp. 1–68 (in German)

Gale, E.F., Cundliffe, E., Reynolds, P.E., Richmond, M.H. & Waring, M.J. (1981) *The Molecular Basis of Antibiotic Action*, London, John Wiley & Sons, p. 23

Goldman, E. & Jakubowski, H. (1990) Uncharged tRNA, protein synthesis, and the bacterial stringent response. *Mol. Microbiol.*, **4**, 2035–2040

Haubeck, H.D., Lorkowski, G., Kölsch, E. & Röschenthaler, R. (1981) Immunosuppression by ochratoxin A and its prevention by phenylalanine. *Appl. Environ. Microbiol.*, **41**, 1040–1042

Heller, K. & Röschenthaler, R. (1977) Some results on the inhibition of cellular function by ochratoxin A on *Streptococcus faecalis*. In: Larcan, A. & Lambert, H., eds, *Mycotoxins*, Paris, Masson, pp. 35–42 (in French)

Heller, K. and Röschenthaler, R. (1978) Inhibition of protein synthesis in *Streptococcus faecalis* by ochratoxin A. *Can. J. Microbiol.*, **24**, 466–472

Kane, A., Creppy, E.E., Roth, A., Röschenthaler, R. & Dirheimer, G. (1986) Distribution of the [^{3}H]-label from low doses of radioactive ochratoxin A ingested by rats, and evidence for DNA single-strand breaks caused in liver and kidneys. *Arch. Toxicol.*, **58**, 219–224

Konrad, I. & Röschenthaler, R. (1977) Inhibition of phenylalanyl-tRNA synthetase from *Bacillus subtilis* by ochratoxin A. *FEBS Lett.*, **83**, 341

Krogh, P., Gyrd-Hansen, N., Hald, B., Larsens, S., Nielsen, J.P., Smith, M., Ivanov, C. & Meisner, H. (1988) Renal enzyme activities in experimental ochratoxin A-induced porcine nephropathy: diagnostic potential of phosphoenolpyruvate carboxykinase and gamma-glutamyl transpeptidase activity. *J. Toxicol. Environ. Health*, **23**, 1–14

Kuiper-Goodman, T. & Scott, P.M. (1989) Risk assessment of the mycotoxin ochratoxin A. *Biomed. Environ. Sci.*, **2**, 179–248

Mayura, K., Parker, R., Berndt, W.O. & Phillips, T.D. (1984) Ochratoxin A-induced teratogenesis in rats: partial protection by phenylalanine. *Appl. environ. Microbiol.*, **48**, 1186–1188

Meisner, H. & Meisner, P. (1981) Ochratoxin A, *in vivo* inhibitor of renal phosphoenolpyruvate carboxykinase. *Arch. Biochem. Biophys.*, **208**, 146–153

Meisner, H. & Selanik, P. (1979) Inhibition of renal gluconeogenesis in rats by ochratoxin. *Biochem. J.*, **180**, 681–684

Meisner, H., Cimbala, M.A. & Hanson, R.W. (1983) Decrease of renal phosphoenolpyruvate carboxykinase RNA and poly(A)$^+$ RNA level by ochratoxin A. *Arch. Biochem. Biophys.*, **223**, 264–270

Mori, H., Kawai, K., Ohbayashi, F., Kuniyasu, T., Yamasaki, M., Hamasaki, T. & Williams, G.M. (1984) Genotoxicity of a variety of mycotoxins in the hepatocyte primary culture/DNA repair test and mouse hepatocytes. *Cancer Res.*, **44**, 2918–2923

Moroi, K., Suzuki, S., Kuga, I., Yamazaki, M. & Kanisawa, M. (1985) Reduction of ochratoxin A toxicity in mice treated with phenylalanine and phenobarbital. *Toxicol. Lett.*, **25**, 1–5

Rahimtula, A.D., Béréziat, J.-C., Bussacchini-Griot, V. & Bartsch, H. (1988) Lipid peroxidation as a possible cause of ochratoxin A toxicity. *Biochem. Pharmacol.*, **37**, 4469–4477

Singer, U. & Röschenthaler, R. (1977) Inhibition of protein synthesis by ochratoxin A on Bacillus subtilis. In: Larcan, H. & Lambert, D., eds, *Mycotoxins*, Paris, Masson, pp. 43–49 (in French)

Stetina, R. & Votava, M. (1986) Induction of DNA single-strand breaks and DNA synthesis inhibition by patulin, ochratoxin A, citrinin, and aflatoxin B1 in cell lines CHO and AWRF. *Folia Biol.*, **32**, 128–144

Mycotoxins, Endemic Nephropathy and Urinary Tract Tumours
Ed. M. Castegnaro, R. Pleština, G. Dirheimer, I.N. Chernozemsky & H. Bartsch
Lyon, International Agency for Research on Cancer, pp. 187-200

PHARMACOKINETICS OF OCHRATOXIN A IN ANIMALS

P. Galtier

Laboratoire de Pharmacologie-Toxicologie, Institut National de la Recherche Agronomique, Toulouse, France

Summary

The fate of ochratoxin A has been studied in laboratory rodents and in breeding animals. In rats, orally administered ochratoxin A is readily absorbed, and considerable amounts of the toxin are detected in plasma, where maximal concentrations occur 2–4 h after administration. Pharmacokinetic analysis of curves of plasma level *versus* time suggests its distribution in two distinct body compartments. The half-time of the toxin depends on both the dose and the animal species, varying from 0.7 h in fish to 840 h in monkeys. In plasma, the toxin is bound to albumin, like many acidic compounds. This interaction is competitively inhibited by phenylbutazone, ethylbiscoumacetate and sulfamethoxypyridazine and is decreased in albumin-deficient rats. The hydrolysis of ochratoxin A to an isocoumarin derivative (ochratoxin α) is the major metabolic pathway. This detoxication is brought about by animal and bacterial carboxypeptidases and takes place in the rumen and large intestine. 4-Hydroxyochratoxin A is the main hepatic metabolite, and its formation appears to be polymorphic, like debrisoquine 4-hydroxylation. The ratio of 4-hydroxyochratoxin A to ochratoxin A excreted in urine may be linked to the carcinogenic potential of the toxin, as the metabolite is almost as effective an immunosuppressor as ochratoxin A. After undergoing enterohepatic circulation, the toxin and ochratoxin α are excreted in faeces and urine as various unidentified metabolites. Transport of the mycotoxin in the kidney is mediated by the renal organic anion transport system, and renal metabolism may contribute to detoxification. Although dose-dependent placental transfer of ochratoxin A has been described in rodents, the toxin does not cross the placenta into fetuses of sows administered a low dose (0.38 mg/kg) orally. Its diffusion into the milk of female rabbits is seen after intravenous administration, but in cows given 50 mg of the mycotoxin, barely detectable amounts of ochratoxin α were recovered in milk. Ochratoxin A is preferentially distributed in liver, kidney, muscle and fat. The experimental data are in close accordance with several reports on the spontaneous occurrence of unchanged toxin residues in blood and kidneys of slaughter pigs.

Introduction

Ochratoxin A (7-carboxy-5-chloro-8-hydroxy-3,4-dihydro-3-methyl isocoumarin linked through the 7-carboxy group by an amide bond to L-phenylalanine) is a nephrotoxic mycotoxin, which is produced by *Aspergillus ochraceus* Wilh. (Van der Merwe *et al.*, 1965) and by *Penicillium viridicatum* West. (Van Walbeek *et al.*, 1969). Because of its natural occurrence in foods, feeds and also in tissues of slaughter pigs, numerous experimental investigations have been carried out on the fate of ochratoxin A. The aim of this paper is to review contributions on the characteristics of the absorption, distribution, metabolism and elimination of this mycotoxin in laboratory rodents and in breeding animal species.

Absorption of Ochratoxin A and its Fate in the Bloodstream

Gastrointestinal absorption

The phenolic hydroxy group in the dihydroisocoumarin ring and the carboxylic group in phenylalanine are responsible for the weak acidic properties of ochratoxin A. Pitout (1968) determined a global pK of 7.1; the existence of two distinct dissociation constants, pK_1 = 6.75 and pK_2 = 10.25 was also demonstrated (Galtier *et al.*, 1977). The pH values indicate that the toxin may exist in both nonionized and ionized forms in aqueous media. Diffusion of the nonionized form across the lipid membrane is generally considered to be the main mechanism of gastrointestinal transfer for many weak electrolytes.

On the basis of the rapid decline in the gastric level in rats administered the toxin orally (Galtier, 1974a; Suzuki *et al.*, 1977), absorption was assumed to take place in the upper parts of the gastrointestinal tract. Subsequently, Kumagai and Aibara (1982) reported that the capacity for ochratoxin A absorption was maximal in the proximal jejunum, intermediate in the duodenum and midjejunum to midileum and least in the distal ileum after the toxin was introduced into the lumen of a closed loop at various sites in the rat intestine. Lee *et al.* (1984), however, suggested that absorption of orally administered ochratoxin A in mice would take place maximally in the duodenum and to a lesser degree in the jejunum. The absence of ochratoxin A from the ileum was probably the result of a very low luminal concentration of the toxin at that point in the gastrointestinal tract, rather than any lack of uptake by the ileal mucous membrane. Some evidence of ochratoxin A absorption was also seen in the stomach, which is consistent with the findings of a previous report by Kumagai and Aibara (1982). In a more recent investigation, Kumagai (1988) showed that ochratoxin A was absorbed from the jejunum of anaesthetized rats, even when its level was higher in the plasma than in the jejunal lumen. Uptake of ochratoxin A by the everted jejunum of rats was enhanced by a decrease in the pH of the medium, a decrease in uptake coinciding with an increase in the proportion of toxin present in the nonionized form. These results suggested that absorption of ochratoxin is attributable mainly to transfer of the nonionized form and that a concentration gradient of the nonionized form from the jejunal lumen to the blood and lymph may be achieved even when the total ochratoxin A level is higher in the plasma than in the lumen.

Blood distribution and plasma protein binding

Ochratoxin A administered orally to male adult rats was found only in traces in leukocytes and red blood cells, while considerable amounts were detected in plasma. In these animals, 95% of the toxin was bound to plasma proteins, and saturation of binding sites occurred at about 70 μg/ml. Plasma electrophoresis with fluorodensitometric assay of the nonstained electrophoregram demonstrated binding to serum albumin (Galtier, 1974b). This preliminary observation was confirmed subsequently in rats given ^{14}C-ochratoxin A intraperitoneally (Chang & Chu, 1977). More recently, binding was studied in plasma from carp, quail, mouse, rat, monkey, pig and humans: 99–100% of the toxin was bound to plasma proteins in all species except fish (78%) (Hagelberg *et al.*, 1989). Further studies were designed to determine the binding constant of ochratoxin A to serum albumin. In the case of bovine albumin, this parameter was found to be dependent on temperature (Chu, 1971); at 37 °C, significant interspecies differences were demonstrated, with maximal affinity in cattle, pigs and humans and lowest affinity in chicken, rats and sheep (Galtier, 1979). Phenylbutazone, ethylbiscoumacetate and sulfamethoxypyridazine each competitively inhibited the binding of ochratoxin A to porcine serum albumin *in vitro*, indicating that these acidic drugs and the toxin probably bind at the same site on the protein and that toxicologically significant interactions could occur between the toxin and drugs widely used in human and veterinary therapy (Galtier *et al.*, 1980).

Pharmacokinetic studies in animals

When ochratoxin A is administered orally to animals, the profile of plasma levels corresponds to a rapid phase of absorption followed by a slow elimination. Fitting these data and comparing them with data obtained following intravenous injection allows determination of pharmacokinetic parameters, such as peak concentration of toxin in plasma (Cp_{max}) and its corresponding time (t_{max}), the biological half-time ($t_{1/2}$) and the relative bioavailability or fraction of dose absorbed (*F*). Table 1 shows these values in laboratory rodents and in breeding animals following oral administration of the mycotoxin at different doses. The results clearly demonstrate the possible dose-dependence of these parameters and the large differences in the pharmacokinetics of ochratoxin A among these species. The longer values for the biological half-time (57–510 h) were obtained exclusively in monogastric species, such as monkey, pig and rat, whereas short half-times characterized birds (4–6 h) and fish (1.6 h in carp). A similar variability was seen in the calculated clearance by renal filtration (Hagelberg *et al.*, 1989). The relative bioavailability of ochratoxin A was generally around 60%, except in chicken (40%) and especially fish (only 1.6%).

Table 1. Pharmacokinetic parameters of ochratoxin A in animals following oral administration

Animal species	Body weight (kg)	Dose (mg/kg)	Cp_{max} (μg/ml)	t_{max} (h)	$t_{1/2}$ (h)	F (h)	Reference
Rat	0.25	10	87	8	68	–	Galtier (1974b)
	0.20–0.25	15	34	4	–	–	Suzuki *et al.* (1977)
	0.25	2.5	25.2	1	57.7	67	Galtier *et al.* (1979)
	0.25–0.30	0.05	0.39	–	120	44	Hagelberg *et al.* (1989)
Mouse	0.02	0.05	0.37	–	39	57	Hagelberg *et al.* (1989)
Rabbit	2–3	2	2.13	1	8.2	56	Galtier *et al.* (1981)
Monkey	–	0.05	0.50	–	510	57	Hagelberg *et al.* (1989)
Pig	35	0.5	1.74	10	88.8	66	Galtier *et al.* (1981)
Chicken	1.8	2	0.78	0.33	4.1	40	Galtier *et al.* (1981)
Quail	0.16	0.05	0.26	–	6.7	6.2	Hagelberg *et al.* (1989)
Carp	1	0.05	0.014	–	0.68	1.6	Hagelberg *et al.* (1989)

Cp_{max}, peak concentration of toxin in plasma; t_{max}, corresponding time; $t_{1/2}$, biological half-time of toxin; F, relative bioavailability; –, not determined

Tissue Distribution of Ochratoxin A in Animals

Experiments in laboratory rodents

In an early experiment on the fate of ochratoxin A in rats, Nel and Purchase (1968) found that liver and kidneys of intraperitoneally treated rats contained a green fluorescent spot, which was identified as the parent toxin. Wide distribution of orally administered toxin was described in the same rodent species by Galtier (1974a), Suzuki *et al.* (1977) and Lillehoj *et al.* (1979). Whatever the dose and form of toxin (unlabelled or ^{14}C-ochratoxin A), native ochratoxin A was found to be present in all organs investigated. During the first 24 h after oral administration, maximal accumulation was demonstrated in the gastrointestinal tract, but significant levels were observed in the kidneys, liver, brain, muscle, skin and fat at 24 h after exposure and thereafter. Distribution in the kidneys was closely associated with the tissue-specificity of ochrotoxin A-induced nephropathy (Suzuki *et al.*, 1977). In a pharmacokinetic study (Galtier *et al.*, 1979), ^{14}C-ochratoxin A was distributed in two kinetically distinct body compartments in rats. Cluster analysis showed that the central compartment corresponded to all well-perfused organs, such as kidneys, liver, spleen, brain, lungs, heart and testes; the peripheral compartment, in which the possibility of residues was assumed to be greater, included skin, muscle, fat, eyes and certain glands.

The immunohistochemical fate of ochratoxin A in kidney and liver was investigated in mice by Lee *et al.* (1984). In the kidney, the vast majority of the toxin was localized within the cytoplasm of proximal convoluted tubular cells. The mycotoxin was also found within the epithelium of distal tubules, the loop of Henle and the epithelium of Bowman's capsule and the glomerulus. In liver, ochratoxin A was concentrated predominantly in hepatocytes of the periportal area rather than around the central vein; stained hepatocytes exhibited strong cytoplasmic but weak and infrequent nuclear staining.

Studies in breeding animals

The results of these basic studies were confirmed in breeding animals. Muscle of rainbow trout injected intravenously with ^{14}C-ochratoxin A contained almost no radioactivity during the whole experiment; it was therefore concluded that contamination of fish feed does not play an important role in human exposure to ochratoxin A (Fuchs *et al.*, 1986). The fate of the toxin in pig and poultry is well documented because of the spontaneous occurrence of ochratoxin A residues in the tissues of these two species. The kinetics of the toxin has been studied particularly in edible tissues such as liver, kidneys, muscle and fat (Table 2). Generally, the toxin levels found corresponded to concentrations that could occur naturally in animal feed (0.3–4 mg/kg).

Table 2. Levels of ochratoxin A residues in liver, kidney, muscle and adipose tissues of pigs and hens fed contaminated diets

Animal species	Feed contamination (mg/kg)	Period of exposure (days)	Withdrawal time (h)	Ochratoxin concentration (µg/kg)				Reference
				Liver	Kidney	Muscle	Fat	
Pig	4	120	I[a]	38	54	5	37	Krogh *et al.* (1974)
	1	120	I	8	14	3	8	
	0.2	120	I	2	4	ND[b]	ND	
Pig	1	30	1	18	26	12	6	Krogh *et al.* (1976a)
	1	30	8	5	8	2	3	
	1	30	15	0.6	3	ND	ND	
	1	30	29	ND	0.5	ND	ND	
Pig	1.38	14	I	21	71	38	29	Madsen *et al.* (1982)
	1.38	28	I	22	62	37	30	
	1.38	42	I	20	49	44	59	
	1.38	56	I	20	45	33	51	
Pullet	0.3	341	I	2	12	2	ND	Krogh *et al.* (1976b)
	1	341	I	5	19	2	ND	
	1	15	I	9	32	5	ND	
Hen	0.5	42	I	26	37	0–8	0	Prior & Sisodia (1978)
	1	42	I	58	77	5–13	0	
	4	42	I	73	107	16–21	0	
	4	42	24	26	31	0	0	
	4	42	48	9	13	0	0	

[a]I, immediate or unstated time of slaughter of animals
[b]ND, not detected

In pigs exposed to ochratoxin A for two weeks to four months, the only observable lesion was kidney damage identical to that seen in spontaneous porcine nephropathy, and at slaughter most carcasses contained sizable amounts of toxin. Residue levels were highest in the kidneys and then, in decreasing order, in muscle, liver and fat. The content of ochratoxin A increased with increasing levels in the feed and was high when the toxin was fed just prior to slaughter. Very low residue levels were present even when the feed contained as little as 25–200 µg/kg. When ochratoxin A-containing feed was replaced by uncontaminated feed some time before slaughter, the residue levels decreased rapidly (Madsen *et al.*, 1982). When chickens and laying hens were fed contaminated diets, ochratoxin A residues were recovered in kidneys, liver and muscle but not in fat or skin (Krogh *et al.*, 1976b). As in the case of pigs, these birds could have passed a meat inspection because no macroscopic lesion was apparent. This could represent a possible health problem (Krogh *et al.*, 1974).

Biliary excretion and enterohepatic circulation of ochratoxin A

Biliary excretion of ochratoxin A was demonstrated and reached maximum levels between 1 and 2 h after administration to rats (Kumagai & Aibara, 1982). Ochratoxin A was found to be located principally in the lumen of biliary ducts and not within biliary cells (Lee *et al.*, 1984). Following intubation of a single low dose of ^{3}H-ochratoxin A into mice, the oscillating pattern of plasma radioactivity led to the assumption that the toxin underwent enterohepatic circulation (Roth *et al.*, 1988). This idea was confirmed in another experiment in which bile from rats treated orally with 2 mg/kg ochratoxin A was given by gavage into the stomach or duodenum of other rats. The finding that the plasma ochratoxin A level was higher when the bile was administered into the duodenum confirmed that enterohepatic recirculation of ochratoxin A occurs in these animals (Fuchs *et al.*, 1988a).

Metabolism of Ochratoxin A

Hydrolysis and esterification of the mycotoxin

Ochratoxin A undergoes rapid biotransformation to unidentified green fluorescent metabolites and to a blue fluorescent spot, which was designated ochratoxin α (Nel & Purchase, 1968). This compound was subsequently identified as the dihydro-isocoumarin derivative produced by hydrolysis of the peptide bond of ochratoxin A and has since been recovered in all metabolic studies of ochratoxin A in rats, regardless of the route of administration (Van Walbeek *et al.*, 1971; Galtier, 1974a; Chang & Chu, 1977; Suzuki *et al.*, 1977; Galtier *et al.*, 1979).

Because of the importance of this derivative in all animals tested, several studies were designed to determine the origin of the biotransformation. Pitout (1969) concluded that the mycotoxin could be hydrolysed by carboxypeptidase A and possibly by trypsin, α-chymotrypsin and cathepsin C. When ochratoxin A was incubated with rat tissue homogenates (Doster & Sinnhuber, 1972; Suzuki *et al.*, 1977), it was converted to ochratoxin α in pancreas and small intestine, but only traces or no hydrolysate were detected in liver or kidney. The main metabolite was found only in the large intestine of orally treated rats (Galtier, 1974a); and hydrolysis by the bacterial flora of rat caecum was demonstrated by incubation of the native toxin in rat caecal contents. Similar results were obtained with rumenal fluid from cows and sheep, which in some cases esterified ochratoxin A to ochratoxin C (Galtier & Alvinerie, 1976). After centrifugation of rumenal fluid, only a fraction containing protozoae could hydrolyse the mycotoxin efficiently, whereas the sterile supernatant—like that found in caecal contents—neither transformed nor destroyed ochratoxin A. The results of this study and that of Hult *et al.* (1976) confirmed that the toxin undergoes hydrolytic detoxification by microbial flora, particularly in rumenal fluid. Subsequently, Kiessling *et al.* (1984) demonstrated a decreased capacity of rumenal fluid to cleave ochratoxin A to ochratoxin α and phenylalanine after feeding, although this activity was gradually restored by the next feeding time.

Hydroxylation of ochratoxin A

Intraperitoneal injection of ochratoxin A to rats resulted in urinary excretion of a compound with the same chromatographic properties as 4-hydroxyochratoxin A (Hutchinson & Steyn, 1971). This identity was confirmed later (Storen *et al.*, 1982), and hydroxylation of the mycotoxin was studied extensively in *in-vitro* models. After incubating ochratoxin A with rat liver microsomes and NADPH, 90% of metabolites corresponded to 4-hydroxyochratoxin A. This metabolism was inhibited by carbon monoxide and metyrapone. The rate of formation increased in microsomal preparations from phenobarbital-treated animals, and a type-I spectrum appeared after binding of the toxin to microsomes. These findings suggest the involvement of cytochrome P450 in the 4–hydroxylation of ochratoxin A by hepatic microsomes (Stormer & Pedersen, 1980). When ochratoxin A was incubated with pig liver microsomes, two epimeric hydroxylated metabolites, (4R)- and (4S)-hydroxyochratoxin A, were formed in approximately equal amounts. These metabolites were also formed, but in a different ratio, in experiments with microsomes from human and rat liver (Störmer *et al.*, 1981) and with primary cultures of rat hepatocytes (Hansen *et al.*, 1982). The formation of another metabolite, identified as 10-hydroxyochratoxin A, has been described after incubation of ochratoxin A with rabbit liver microsomes, although this compound has not yet been described *in vivo* (Störmer *et al.*, 1983). When these three hydroxylated metabolites were incubated with alcohol dehydrogenase in the presence of NAD, only (4R)- and 10-hydroxyochratoxin A acted as substrates for this enzyme (Syvertsen & Störmer, 1983).

Possible polymorphism in 4-hydroxylation of the mycotoxin was investigated by Hietanen *et al.* (1986) in rats. Their findings gave further support to a possible cosegregation of the genes that regulate ochratoxin A and debrisoquine 4-hydroxylation (Castegnaro *et al.*, 1989).

Toxicity of metabolites

It is not currently known whether ochratoxin A requires metabolic conversion by enzymes in the liver or kidney to exert its nephrotoxic and carcinogenic properties. It was established early on, however, that the toxicity of the hydrolysate, ochratoxin α, was very low (Yamazaki *et al.*, 1971; Chu *et al.*, 1972). In contrast, ochratoxin C was as toxic as the parent toxin for one-day-old ducks (Steyn & Holzapfel, 1967) and chicks

(Chu *et al.*, 1972). A similar order of potency was observed with respect to enhancement of lipid peroxidation in rat liver microsomes (Rahimtula *et al.*, 1988). When (4R)-hydroxyochratoxin A and ochratoxin α were investigated for their immunosuppressive properties in mice, only the hydroxy derivative was almost as effective an immunosuppressor as ochratoxin A (Creppy *et al.*, 1983). Hydroxyochratoxin A was previously described as ineffective in comparison to the parent toxin in inducing lethality in rats receiving six oral doses daily (Hutchinson & Steyn, 1971).

Routes of Elimination of Ochratoxin A and its Metabolites

Excretion in urine and faeces

Studies describing elimination of ochratoxin A after oral administration to rats indicated that the native toxin and ochratoxin α are eliminated slowly in urine and faeces, representing about 10 and 25% of the administered dose, respectively. Faecal elimination of these two compounds was significantly lower in animals that received a similar intravenous dose (Galtier, 1979). When the mycotoxin was injected into albumin-deficient rats, plasma concentrations were much decreased, whereas the rate of excretion in bile and urine was 20–70 times higher than in normal rats (Kumagai, 1985).

Information about the elimination of ochratoxin A in breeding animals is limited. In pregnant pigs, a daily oral dose of 0.38 mg/kg given for eight days was recovered as ochratoxin α in faeces and urine (Patterson *et al.*, 1976). Ribelin (1978) showed that intravenously administered ochratoxin A was excreted in the urine of cows; however, when doses up to 100 mg/kg were given orally, only ochratoxin α appeared in urine. Excretion of the toxin was also studied in goats given a single 0.5 mg/kg oral dose of ^{3}H-ochratoxin A (Nip & Chu, 1979); within seven days, 53 and 38% of the radiolabel was found to have been excreted in faeces and urine, respectively.

The mechanism of renal excretion of ochratoxin A has been well documented because of its nephrotoxic properties. *In-vitro* studies have suggested that the toxin interacts with the transport of organic anions, by entering the proximal tubular cells and competing with *para*-aminohippurate transport in the renal brush border and basolateral membrane vesicles (Stein *et al.*, 1985; Friis *et al.*, 1988; Sokol *et al.*, 1988). Renal tubular secretion and further reabsorption of the mycotoxin could facilitate its residual persistence in the kidneys.

Placental transfer of ochratoxin A

As teratogenic effects of ochratoxin A have been reported in chicken and in laboratory rodents, studies were designed to examine transplacental transfer of the toxin. In an experiment on the effects of feeding mixed ochratoxins to pigs in early pregnancy, although ochratoxin A accumulated in maternal tissues, including the placenta, it did not reach the fetus (Shreeve *et al.*, 1977). Intrauterine contamination of piglets by the toxin has, however, been described by Rajic *et al.* (1986) when sows were given feeds containing 2.2 mg/kg of the toxin four days before parturition. Residues were recovered in muscles, lung and liver of piglets slaughtered immediately after birth. When pregnant ewes were given ochratoxin A at 1 mg/kg intravenously, no toxin was detected in amniotic fluid, but fetal tissue levels were 1/400 to 1/1000 of those in maternal blood (Ribelin, 1978). In mice given ^{14}C-ochratoxin A intravenously, the toxin crossed the placental barrier on day 9 of pregnancy, the time at which it is most effective in producing fetal malformations (Appelgren & Arora, 1983). Similarly, when pregnant mice were given a single intraperitoneal dose of 5 mg/kg on day 11 or 13 of pregnancy, the concentrations of ochratoxin A in embryos increased up to 30–48 h after injection; pharmacokinetically, the embryos were found to correspond to a 'deep' compartment (Fukui *et al.*, 1987).

Excretion in milk and eggs

Excretion of ochratoxin A in milk was demonstrated after intravenous administration of the toxin to female rabbits (Galtier *et al.*, 1977). At the highest dose (4 mg/kg), the level in milk reached 1 mg/l. Mammary excretion was also studied with a constant plasma concentration of the toxin, to determine the percentages of protein-bound toxin in plasma and milk. Because of the similarity between the theoretical and the experimental ratio of mycotoxin levels in milk and plasma ultrafiltrates, transfer through the blood–milk barrier by diffusion of the nonionized form of the free toxin was proposed.

In dairy cows receiving pure ochratoxin A at an oral dose of 50 or 1000 mg for four days, ochratoxin α was recovered in the milk; the parent toxin was present in milk only after the highest dose (Ribelin, 1978). Ochratoxin A was also excreted in milk of goats given 0.5 mg/kg of ^{14}C-labelled toxin (Nip & Chu, 1979) and of sows receiving feeds containing 2.3 mg/kg of the toxin for eight days (Rajić *et al.*, 1978).

In pullets fed an ochratoxin A-contaminated diet at 0.3 or 1 mg/kg for 341 days, no mycotoxin was found in any of the eggs analysed (Krogh *et al.*, 1976b). In an experimental study on tissue distribution of ^{14}C labelled ochratoxin A in laying Japanese quail, however, Fuchs *et al.* (1988b) demonstrated specific retention of unidentified radioactivity as a ring-shaped deposition in eggs, indicating that the toxin could deposit over a short time period and would not diffuse into the yolk.

Conclusion

Pharmacokinetic data obtained in laboratory rodents and in breeding animals indicate that ochratoxin A is a weakly acidic compound that can bind to circulating and intracellular proteins. Because of differences in the physiology of animal species (monogastric and polygastric mammals, birds, fish), wide variations are seen in the absorption, distribution, metabolism and elimination of the mycotoxin. For instance, microbial detoxification into ochratoxin α might represent a protection against the toxic effects of ochratoxin A in ruminants and in species such as rabbits and birds that have a rapid digestive flow rate and metabolically active microflora in the gastrointestinal tract.

The excretion of native ochratoxin A into milk and eggs would appear limited in view of its metabolism and its binding to plasma proteins. A more real hazard is presented by the limited biotransformation of ochratoxin A in pigs, as confirmed by several reports of the spontaneous occurrence of ochratoxin A residues in blood and carcasses of slaughter pigs.

References

Appelgren, L.E. & Arora, R.G. (1983) Distribution of ^{14}C-labelled ochratoxin A in pregnant mice. *Food Chem. Toxicol.*, **21**, 563–568

Castegnaro, M., Bartsch, H., Béréziat, J.C., Arvela, P., Michelon, J. & Broussolle, L. (1989) Polymorphic ochratoxin A hydroxylation in rat strains phenotyped as poor and extensive metabolizers of debrisoquine. *Xenobiotica*, **19**, 225–230

Chang, F.C. & Chu, F.S. (1977) The fate of ochratoxin A in rats. *Food Cosmet. Toxicol.*, **15**, 199–204

Chu, F.S. (1971) Interaction of ochratoxin A with bovine serum albumin. *Arch. Biochem. Biophys.*, **147**, 359–366

Chu, F.S., Noh, I. & Chang, C.C. (1972) Structural requirements for ochratoxin intoxication. *Life Sci.*, **11**(1), 503–508

Creppy, E.E., Störmer, F.C, Roschenthaler, R. & Dirheimer, G. (1983) Effects of two metabolites of ochratoxin A, (4R)-4-hydroxyochratoxin A and ochratoxin α, on immune response in mice. *Inf. Immun.*, **39**, 1015–1018

Doster, R.C. & Sinnhuber, R.O. (1972) Comparative rates of hydrolysis of ochratoxins A and B in vitro. *Food Cosmet. Toxicol.*, **10**, 389–394

Friis, C., Brinn, R. & Hald, B. (1988) Uptake of ochratoxin A by slices of pig kidney cortex. *Toxicology*, **52**, 209–217

Fuchs, R., Appelgren, L.E. & Hult, K. (1986) Distribution of ^{14}C-ochratoxin A in the rainbow trout. *Acta Pharmacol. Toxicol.*, **59**, 220–227

Fuchs, R., Radić, B., Peraica, M., Hult, K. & Pleština, R. (1988a) Enterohepatic circulation of ochratoxin A in rats. *Period. Biol.*, **90**, 39–42

Fuchs, R., Appelgren, L.E., Hagelberg, S. & Hult, K. (1988b) ^{14}C-Ochratoxin A distribution in the Japanese quail monitored by whole body autoradiography. *Poult. Sci.*, **67**, 707–714

Fukui, Y., Hoshino, K., Kameyama, Y., Yasui, T., Toda, C. & Nagano, H. (1987) Placental transfer of ochratoxin A and its cytotoxic effect on the mouse embryonic brain. *Food Cosmet. Toxicol.*, **25**, 17–24

Galtier, P. (1974a) Fate of ochratoxin A in the animal organism. II. Tissue distribution and elimination in the rat. *Ann. Rech. Vét.*, **5**, 319–328 (in French)

Galtier, P. (1974b) Fate of ochratoxin A in the animal organism. I. Blood transport of the toxin in the rat. *Ann. Rech. Vét.*, **5**, 311–318 (in French)

Galtier, P. (1979) Toxicological and pharmacokinetic study of a mycotoxin, ochratoxin A. Thesis, Université Paul Sabatier, Toulouse, France (in French)

Galtier, P. & Alvinerie, M. (1976) In vitro transformation of ochratoxin A by animal microbial floras. *Ann. rech. Vét.*, **7**, 91–98

Galtier, P., Baradat, C. & Alvinerie, M. (1977) Study on the elimination of ochratoxin A through the milk in rabbits. *Ann. Nutr. Aliment.*, **31**, 911–918 (in French)

Galtier, P., Charpenteau, J.L., Alvinerie, M. & Labouche, C. (1979) The pharmacokinetic profile of ochratoxin A in the rat after oral and intravenous administration. *Drug Metab. Disposition*, **7**, 429–434

Galtier, P., Camguilhem, R. & Bodin, G. (1980) Evidence for in vitro and in vivo interaction between ochratoxin A and three acidic drugs. *Food Cosmet. Toxicol.*, **18**, 493–496

Galtier, P., Alvinerie, M. & Charpenteau, J.L. (1981) The pharmacokinetic profiles of ochratoxin A in pigs, rabbit and chickens. *Food Cosmet. Toxicol.*, **19**, 735–738

Hagelberg, S., Hult, K. & Fuchs, R. (1989) Toxicokinetics of ochratoxin A in several species and its plasma-binding properties. *J. Appl. Toxicol.*, **9**, 91–96

Hansen, C.E., Dueland, S., Drevon, C.A. & Störmer, F. (1982) Metabolism of ochratoxin A by primary cultures of rat hepatocytes. *Appl. Environ. Microbiol.*, **43**, 1267–1271

Hietanen, E., Malavieille, C., Camus, A.M., Béréziat, J.C., Brun, G., Castegnaro, M., Michelon, J., Idle, J.R. & Bartsch, H. (1986) Interstrain comparison of hepatic and renal microsomal carcinogen metabolism and liver S9-mediated mutagenicity in DA and Lewis rats phenotyped as poor and extensive metabolizers of debrisoquine. *Drug Metab. Disposition*, **14**, 118–126

Hult, K., Teiling, A. & Gatenbeck, S. (1976) Degradation of ochratoxin A by a ruminant. *Appl. Environ. Microbiol.*, **32**, 443–444

Hutchinson, R.D. & Steyn, P.S. (1971). The isolation and structure of 4-hydroxyochratoxin A and 7-carboxy-3,4-dihydro-8-hydroxy-3-methylisocoumarin from Penicillium viridicatum. *Tetrahedron Lett.*, **43**, 4033–4036

Kiessling, K.H., Petterson, H., Sandholm, K. & Olsen, M. (1984) Metabolism of aflatoxin, ochratoxin, zearalenone, and three trichothecenes by intact rumen fluid, rumen protozoa, and rumen bacteria. *Appl. Environ. Microbiol.*, **47**, 1070–1073

Krogh, P., Axelsen, N.H., Elling, F., Gyrd-Hansen, N., Hald., B., Hyldgaard-Jensen, J., Larsen, A.E., Madsen, A., Mortensen, H.P., Möller, T., Petersen, O.K., Ravnskov, V., Rostgaard, M. & Aalund, O. (1974) Experimental porcine nephropathy. *Acta Pathol. Microbiol. Scand. A*, **246**, 1–21

Krogh, P., Elling, F., Hald, B., Larsen, A.E., Lillehoj, E.B., Madsen, A. & Mortensen, H.P. (1976a) Time-dependent disappearance of ochratoxin A residues in tissues of bacon pigs. *Toxicology*, **6**, 235–242

Krogh, P., Elling, F., Hald, B., Jvilling, B., Petersen, V.E., Skadhauge, E. & Svenden, C.K. (1976b) Experimental avian nephropathy. *Acta Pathol. Microbiol. Scand.*, **84**, 215–221

Kumagai, S. (1985) Ochratoxin A: plasma concentration and excretion into bile and urine in albumin-deficient rats. *Food Chem. Toxicol.*, **23**, 941–943

Kumagai, S. (1988) Effects of plasma ochratoxin A and luminal pH on the jejunal absorption of ochratoxin A in rats. *Food Chem. Toxicol.*, **26**, 753–758

Kumagai, S. & Aibara, K. (1982) Intestinal absorption and secretion of ochratoxin A in the rat. *Toxicol. Appl. Pharmacol.*, **64**, 94–102

Lee, S., Beery, J.T. & Chu, F.S. (1984) Immunohistochemical fate of ochratoxin A in mice. *Toxicol. Appl. Pharmacol.*, **72**, 218–227

Lillehoj, E.B., Kwolek, W.F., Elling, F. & Krogh, P. (1979) Tissue distribution of radioactivity from ochratoxin-^{14}C in rats. *Mycopathologia*, **68**, 175–177

Madsen, A., Mortensen, H.P. & Hald, B. (1982) Feeding experiments with ochratoxin A contaminated barley for bacon pigs. I. Influence on pig performances and residues. *Acta Agric. Scand.*, **32**, 225–239

Nel, W. & Purchase, I.F.H. (1968) The fate of ochratoxin A in rats. *J. S. Afr. Chem. Inst.*, **21**, 87–88

Nip, W.K. & Chu, F.S. (1979) The fate of ochratoxin A in goats. *J. Environ. Sci. Health.*, **B14**, 319–333

Patterson, D.S.P., Roberts, B.A. & Small, B.J. (1976) Metabolism of ochratoxins A and B in the pig during early pregnancy and the accumulation in body tissues of ochratoxin A only. *Food Cosmet. Toxicol.*, **14,** 439–442

Pitout, M.J. (1968) The effect of ochratoxin A on glycogen storage in the rat liver. *Toxicol. Appl. Pharmacol.*, **13**, 299–306

Pitout, M.J. (1969) The hydrolysis of ochratoxin A by some proteolytic enzymes. *Biochem. Pharmacol.*, **18**, 485–491

Prior, M.G. & Sisodia, C.S. (1978) Ochratoxicosis in white Leghorn hens. *Poult. Sci.*, **57**, 619–623

Rahimtula, A.D., Béréziat, J.C., Busacchini-Griot, V. & Bartsch, H. (1988) Lipid peroxidation as a possible cause of ochratoxin A toxicity. *Biochem. Pharmacol.*, **37**, 4469–4477

Rajić, I., Masić, Z., Knežević, N., Obradović, V., Jezdić, R. & Matić, G. (1986) Intrauterine contamination of piglets by ochratoxin A and its excretion through colostrum and milk of sows. In: Ozegović, L., ed., *Proceedings of Second Symposium on Mycotoxins, Sarajevo, Yugoslavia*, Sarajevo, Academy of Sciences and Arts of Bosnia and Hercegovina, pp. 83–88

Ribelin, W.E. (1978) Ochratoxicosis in cattle. In: Willie, T.D. & Morehouse, L.G., eds, *Mycotoxic Fungi, Mycotoxins, Mycotoxicoses: An Encyclopedic Handbook*, Vol. 2., New York, Marcel Dekker, pp. 28–35

Roth, A., Chakor, K., Creppy, E.E., Kane, A., Roschenthaler, R. & Dirheimer, G. (1988) Evidence for an enterohepatic circulation of ochratoxin A in mice. *Toxicology*, **48**, 293–308

Shreeve, B.J., Patterson, D.S.P., Pepin, G.A., Roberts, B.A. & Wrathall, A.E. (1977) Effect of feeding ochratoxin to pigs during early pregnancy. *Br. Vet. J.*, **133**, 412–417

Sokol, P.P., Ripich, G., Holohan, P.D. & Ross, C.R. (1988) Mechanism of ochratoxin A transport in kidney. *J. Pharmacol. Exp. Ther.*, **246**, 460–465

Stein, A.F., Phillips, T.D., Kubena, L.F. & Harvey, B.B. (1985) Renal secretion and reabsorption as factors in ochratoxicosis: effects of probenecid on nephrotoxicity. *J. Toxicol. Environ. Health*, **16**, 593–605

Steyn, P.S. & Holzapfel, C.W. (1967) The isolation of the methyl and ethyl esters of ochratoxins A and B, metabolites of Aspergillus ochraceus Wilh. *J. S. Afr. Chem. Inst.*, **20**, 186–189

Stören, O., Holm, H. & Störmer, F.C. (1982) Metabolism of ochratoxin A by rats. *Appl. Environ. Microbiol.*, **44**, 785–789

Störmer, F.C. & Pedersen, J.I. (1980) Formation of 4-hydroxyochratoxin A from ochratoxin A by rat liver microsomes. *Appl. Environ. Microbiol.*, **39**, 971–975

Störmer, F.C., Hansen, C.E., Pedersen, J.I., Hvistendahl, G. & Aasen, A.J. (1981) Formation of (4R)- and (4S)-4-hydroxyochratoxin A from ochratoxin A by liver microsomes from various species. *Appl. Environ. Microbiol.*, **42**, 1051–1056

Störmer, F.C., Stören, O., Hansen, C.E., Pedersen, J.I. & Aasen, A.J. (1983) Formation of (4R)- and (4S)-4-hydroxyochratoxin A and 10-hydroxyochratoxin A from ochratoxin A by rabbit liver microsomes. *Appl. Environ. Microbiol.*, **45**, 1183–1187

Suzuki, S., Satoh, T. & Yamazaki, M. (1977) The phamacokinetics of ochratoxin A in rats. *Jpn. J. Pharmacol.*, **27**, 735–744

Syvertsen, C. & Störmer, F.C. (1983) Oxidation of two hydroxylated ochratoxin A metabolites by alcohol dehydrogenase. *Appl. Environ. Microbiol.*, **45**, 1701–1703

Van der Merwe, K.J., Steyn, P.S., Fourie, L., Scott, B. & Theron, J.J. (1965) Ochratoxin A, a toxic metabolite produced by Aspergillus ochraceus Wilh. *Nature*, **205**, 1112–1113

Van Walbeek, W., Scott, P.M., Harwig, J. & Lawrence, J.W. (1969) Penicillium viridicatum Westling: a new source of ochratoxin A. *Can. J. Microbiol.*, **15**, 1281–1285

Van Walbeek, W., Moodie, C.A., Scott, P.M., Harwig, J. & Grice, H.C. (1971) Toxicity and excretion of ochratoxin A in rats intubated with pure ochratoxin A or fed cultures of Penicillium viridicatum. *Toxicol. Appl. Pharmacol.*, **20**, 439–441

Yamazaki, M., Suzuki, S., Sakakibara, Y. & Miyaki, K. (1971) The toxicity of 5-chloro-8-hydroxy-3,4-dyhydro-3-methylisocoumarin-7-carboxylic acid, a hydrolyzate of ochratoxin A. *Jpn. J. Med. Sci. Biol.*, **24**, 245–250

Mycotoxins, Endemic Nephropathy and Urinary Tract Tumours
Ed. M. Castegnaro, R. Pleština, G. Dirheimer, I.N. Chernozemsky & H. Bartsch
Lyon, International Agency for Research on Cancer, pp. 201-203

DISTRIBUTION OF ^{14}C-OCHRATOXIN A AND ^{14}C-OCHRATOXIN B IN RATS: A COMPARISON BASED ON WHOLE-BODY AUTORADIOGRAPHY

A. Breitholtz Emanuelsson[1], R. Fuchs[2], K. Hult[1,4] & L.-E. Appelgren[3]

[1]*Department of Biochemistry and Biotechnology, Royal Institute of Technology, Stockholm, Sweden;*
[2]*Department of Toxicology, Institute for Medical Research and Occupational Health, University of Zagreb, Zagreb, Yugoslavia; and*
[3]*Department of Pharmacology and Toxicology, Faculty of Veterinary Medicine, Swedish University of Agricultural Sciences, Biomedicum, Uppsala, Sweden*

Summary

The distribution patterns of ochratoxin A and its nontoxic dechloro-analogue ochratoxin B were studied in rats using whole-body autoradiography. No prominent difference in distribution patterns was found that could explain why the rat is the animal most susceptible to ochratoxin A-induced renal cancer or why ochratoxin B is less toxic than ochratoxin A.

Text

In order to determine why ochratoxin B is less toxic than ochratoxin A, a whole-body autoradiography study was performed. ^{14}C-Ochratoxin A and ^{14}C-ochratoxin B labelled in the phenylalanine moiety were prepared by a new method (A. Breitholtz Emanuelsson *et al.*, in preparation), to give specific activities for each of 513 Ci/mol. The toxins were administered to Wistar rats by injection into the tail vein at a dose of 2 μCi per animal, corresponding to about 75 ng/g body weight. The experimental design was similar to that of our earlier studies on mouse (Fuchs *et al.*, 1988a), Japanese quail (Fuchs *et al.*, 1988b) and rainbow trout (Fuchs *et al.*, 1986).

---

[4]To whom correspondence should be addressed

A summary of the results is given in Table 1, presented as relative degrees of blackening on autoradiograms from animals killed at different times after injection. The distribution pattern of the two toxins was dominated by their binding to serum albumin, and no prominent difference in the patterns could be seen. The large difference in clearance between ochratoxin A and ochratoxin B reported in toxicokinetic studies (Hagelberg *et al.*, 1989) was confirmed here, in that ochratoxin B was excreted much faster than ochratoxin A. This was best seen in blood, myocardium and salivary glands after survival times of 24 h and 4 days (Table 1). The greater clearance of ochratoxin B can be attributed at least partly to more efficient elimination through glomerular filtration, as more free ochratoxin B occurs in serum compared to ochratoxin A (Hagelberg *et al.*, 1989). Increased filtration was seen on the autoradiograms as a much higher concentration of radiolabel in urine after administration of ^{14}C-ochratoxin B than after administration of ^{14}C-ochratoxin A at 5 min, 60 min and 4 h after injection (Table 1).

The intestinal contents had higher concentrations of radiolabel after administration of ^{14}C-ochratoxin B than after administration of ^{14}C-ochratoxin A 4 h and 24 h after injection (Table 1). The larger amount of radiolabel in the intestinal contents of ^{14}C-ochratoxin B-treated rats may be due to more efficient biliary excretion or to a lower level of enterohepatic circulation of ochratoxin B compared to ochratoxin A. A reduction in enterohepatic circulation could be the result of rapid hydrolysis of the amide bond in ochratoxin B by microorganisms and enzymes in the intestines. It is known that the amide bond of ochratoxin B is hydrolysed by carboxypeptidase A much faster than that of ochratoxin A (Doster & Sinnhuber, 1972).

Radiolabel was analysed qualitatively in the different organs by thin-layer chromatography. At long survival times after injection of ^{14}C-ochratoxin B, a substantial part of the radiolabel in the intestinal contents consisted of polar compounds that did not move from the origin. One interpretation of this finding is that this radiolabel represents phenylalanine cleaved from ochratoxin B and subsequently incorporated into proteins by intestinal microorganisms. The same phenomenon was seen with ^{14}C-ochratoxin A, but to a lower extent.

Thin-layer chromatography of other tissue samples showed that only ochratoxin A or B was present in blood, muscle and testis. In urine, several metabolites of ochratoxin B but none of ochratoxin A were detected. Elimination of ochratoxin B by metabolism and biliary excretion is 10 times faster than that of ochratoxin A (Hagelberg *et al.*, 1989). This large difference may explain why metabolites of ochratoxin B, but not of ochratoxin A, accumulate and can be detected.

In conclusion, ochratoxin B was metabolized and cleared more rapidly than ochratoxin A, but there was no prominent difference in the

Table 1. Tissue distribution of radioactivity from ^{14}C-ochratoxin A (OA) and ^{14}C-ochratoxin B (OB) in rats at different time intervals after intravenous injection of a single dose corresponding to about 75 ng/g body weight. The relative degree of blackening in different organs after 26 days' exposure is compared with the blackening of a ^{14}C-staircase represented by 2^{12} (4096) and then forming a geometrical series, with the next step half the strongest one, 2^{11} (2048), and the next 2^{10} and so on.

Tissue	Toxin	Time after injection				
		5 min	60 min	4 h	24 h	4 days
Blood	OA	1024	512	256	256	128
	OB	4096	512	256	128	≤ 8
Myocardium	OA	256	128	32	64	32
	OB	512	128	64	4	≤ 8
Salivary gland	OA	32	128	64	16	8
	OB	≤ 2	512	256	4	2
Small intestinal contents	OA	≤ 2	512	256	4	2
	OB	≤ 2	32	4096	64	16
Large intestinal contents	OA	≤ 2	≤ 2	128	32	128
	OB	≤ 2	≤ 2	≤ 2	256	32
Liver	OA	512	256	128	128	32
	OB	512	256	128	32	16
Kidney/cortex	OA	128	128	64	64	32
	OB	512	128	64	16	8
Kidney/medulla	OA	512	256	128	64	32
	OB	512	256	≤ 128	16	8
Urine	OA	512	1024	256	256	256
	OB	2048	2048	2048	64	–
Bone marrow	OA	128	32	32	4	4
	OB	256	64	32	8	≤ 2
Muscle	OA	16	16	8	8	8
	OB	16	8	16	≤ 2	≤ 2
Brown fat	OA	128	32	32	64	–
	OB	128	32	64	≤ 2	≤ 2

–, not investigated

distribution that could explain the difference in the toxicities of the two compounds. Furthermore, no substantial difference was observed in the behaviour of ochratoxin A in rats and mice that could explain the high susceptibility of rats to ochratoxin A-induced renal cancer.

References

Doster, R.C. & Sinnhuber, R.O. (1972) Comparative rates of hydrolysis of ochratoxin A and B in vitro. *Food Cosmet. Toxicol.*, **10**, 389–394

Fuchs, R., Appelgren, L.-E. & Hult, K. (1986) Distribution of ^{14}C-ochratoxin A in the rainbow trout (*Salmon gairdneri*). *Acta Pharmacol. Toxicol.*, **59**, 220–227

Fuchs, R., Appelgren, L.-E. & Hult, K. (1988a) Distribution of ^{14}C-ochratoxin A in the mouse monitored by whole body autoradiography. *Acta Pharmacol. Toxicol.*, **63**, 355–360

Fuchs, R., Appelgren, L.-E. & Hult, K. (1988b) Carbon-14-ochratoxin A distribution in the Japanese quail (*Coturnix coturnix japonica*) monitored by whole body autoradiography. *Poultry Sci.*, **67**, 707–714

Hagelberg, S., Fuchs, R. & Hult, K. (1989) Toxicokinetics of ochratoxin A in several species and its plasma-binding properties. *J. Appl. Toxicol.*, **9**, 91–96

ADVERSE BIOLOGICAL EFFECTS OF OCHRATOXIN A AND STUDIES OF MECHANISM OF ACTION

Mycotoxins, Endemic Nephropathy and Urinary Tract Tumours
Ed. M. Castegnaro, R. Pleština, G. Dirheimer, I.N. Chernozemsky & H. Bartsch
Lyon, International Agency for Research on Cancer, pp. 207–214

ALTERATIONS IN CALCIUM HOMEOSTASIS AS A POSSIBLE CAUSE OF OCHRATOXIN A NEPHROTOXICITY

A.D. Rahimtula & X. Chong

Biochemistry Department, Memorial University, St John's, Newfoundland, Canada

Summary

Disruption of calcium homeostasis, leading to a sustained increase in cytosolic calcium level, has been associated with cytotoxicity in response to a variety of agents in different cell types. We have observed that a single high dose or multiple lower doses of ochratoxin A administered to rats resulted in an increase in renal endoplasmic reticulum calcium pump activity. The increase was very rapid, being evident within 10 min of ochratoxin A administration and remained elevated for at least 6 h thereafter. Ochratoxin A also decreased renal mitochondrial state-3 respiration and calcium uptake. The latter may lead to an increase in cytosolic calcium level, and the increase in microsomal calcium uptake activity may be an attempt to restore calcium homeostasis. Repeated moderate doses of ochratoxin A led to an eventual decrease in microsomal calcium pump activity, and this could lead to even higher cytosolic calcium levels. Changes in the rate of microsomal calcium uptake correlated with changes in the steady-state levels of the phosphorylated Mg^{2+}/Ca^{2+}-ATPase intermediate, indicating that this enzyme is responsible for the calcium pump activity.

Introduction

Disruption of calcium homeostasis, leading to a sustained increase in cytosolic calcium levels, has been associated with cytotoxicity in response to a variety of agents in different cell types (Orrenius *et al.*, 1989). The purpose of this study was to determine if ochratoxin A, a nephrotoxic and carcinogenic mycotoxin, alters calcium uptake by renal mitochondria and microsomes either subsequent to administration *in vivo* or after addition *in vitro*.

Materials and Methods

Male Sprague-Dawley rats (200–225 g) were obtained from Charles River Canada. Ochratoxin A was purchased from Sigma Chemical Co. (St Louis, USA). All other chemicals were of the highest grade commercially available. Rats received ochratoxin A intraperitoneally in sodium bicarbonate and were sacrificed at timed intervals. Controls received vehicle only. Mitochondria and microsomes were prepared from the renal cortex by established methods (Rahimtula *et al.*, 1979; Cain & Skilleter, 1987). Microsomal calcium pump activity was determined by measuring the ATP-dependent sequestration of ^{45}Ca (Moore, 1982). Phosphoenzyme levels were measured by labelling microsomes with [γ-^{32}P]ATP followed by sodium dodecyl sulfate-polyacrylamide gel electrophoresis and determination of radiolabel in the appropriate band (Heilmann *et al.*, 1985). Mitochondrial calcium uptake was measured as described by Gunter *et al.* (1978). Lipid peroxide levels were estimated by measuring the amount of malondialdehyde formed (Uchiyama & Mihara, 1978).

Results

Administration of a single high dose of ochratoxin A (10 mg/kg body weight intraperitoneally) resulted in a significant elevation of renal microsomal calcium pump activity. Figure 1 shows the rate of calcium uptake by renal microsomes isolated from rats treated 2 h earlier with ochratoxin A or vehicle. The rate of calcium uptake was linear for about 20 min but continued to increase throughout the 40-min incubation. The increase in the rate of calcium uptake was apparent within 10 min of ochratoxin A administration and remained elevated for at least 6 h, before returning to control levels (Figure 2). Multiple lower doses of ochratoxin A (0.5–2.0 mg/kg body weight daily for four days) also caused an elevation of calcium pump activity, although multiple higher doses (4 mg/kg body weight daily for four days, Figure 3, or 2 mg/kg body weight daily for eight days, data not shown) lowered the rate of calcium uptake.

The presence of ochratoxin A (1 mM) in incubations *in vitro* inhibited microsomal calcium uptake by about 40%. Preincubation of microsomes with NADPH also strongly suppressed their ability to sequester calcium (about 85%). Inclusion of ochratoxin A (1 mM) during preincubation with NADPH, however, largely overcame the inhibitory

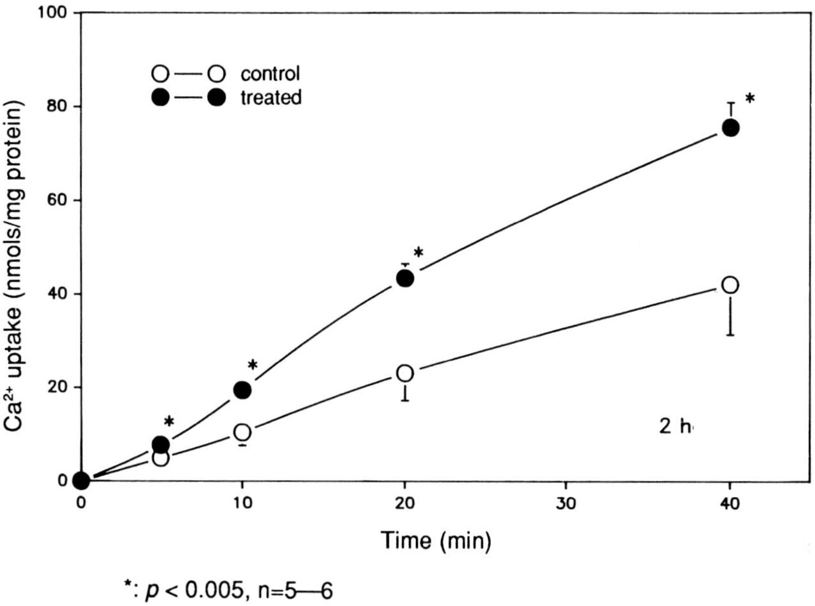

Figure 1. Comparison of the rates of calcium uptake by renal cortical microsomes isolated from control rats and rats treated 2 h previously with ochratoxin A (10 mg/kg body weight intraperitoneally)

Results are means ± SD of triplicate determinations from 5–6 individual rats per treatment group.

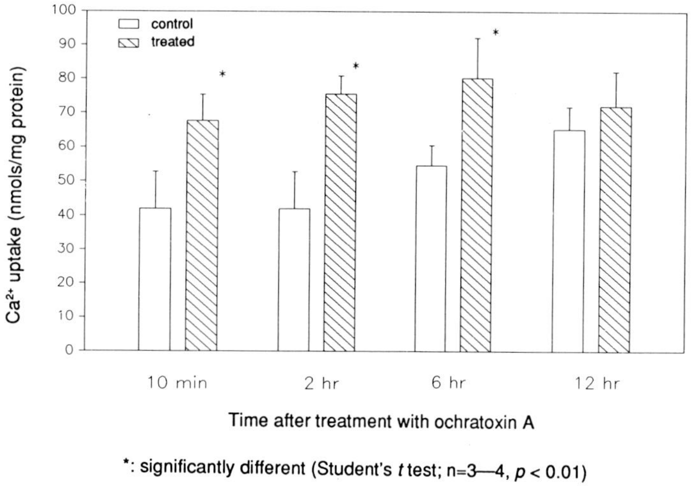

Figure 2. Time course of increase in microsomal calcium pump activity subsequent to administration of ochratoxin A (10 mg/kg body weight intraperitoneally)

Results are means ± SD of triplicate determinations from 3–4 individual rats per treatment group.

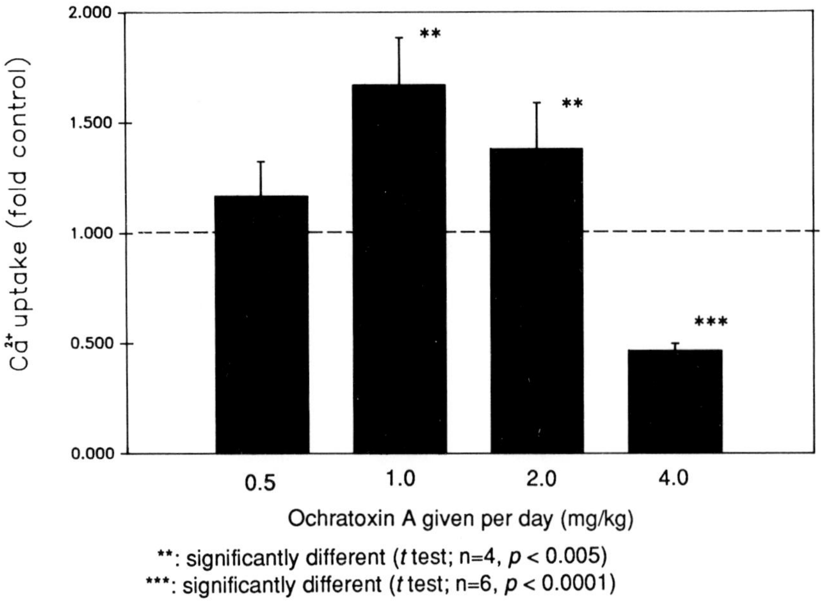

Figure 3. Dose-response of ochratoxin A-induced alterations in calcium uptake activity in renal microsomes

Results are means ± SD of triplicate determinations from 4–6 individual rats per treatment group.

effect of the latter. The recovery of the inhibitory effect of NADPH by ochratoxin A was dose-dependent (Figure 4A). Moreover, the observed increase in calcium uptake correlated with the decrease in lipid peroxide levels, as measured by the formation of malondialdehyde (Figure 4B). These results are consistent with our previous finding that the levels of malondialdehyde in kidney homogenates of rats dosed with ochratoxin A are reduced or unaltered. Furthermore, the alterations in the rate of calcium uptake brought about by ochratoxin A treatment *in vivo* and *in vitro* correlate well with the levels of the phosphoenzyme intermediate (Table 1).

We also observed a 30–35% reduction in succinate and pyruvate/malate-dependent state-3 respiration rate of renal mitochondria isolated from rats dosed 24 h earlier with ochratoxin A (10 mg/kg body weight orally). Figure 5 shows that ochratoxin A inhibited succinate-dependent mitochondrial calcium uptake in a dose-dependent manner.

Discussion

The free calcium concentration in the cytosol of mammalian cells in usually very low (about 0.1 μM) compared with the concentration of free

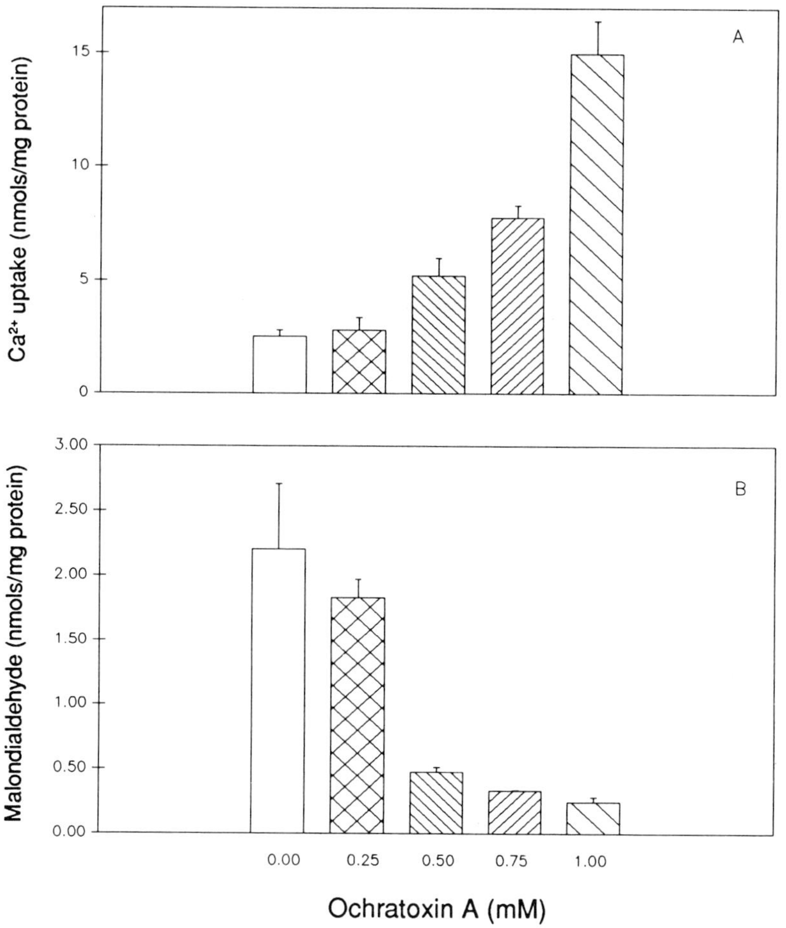

Figure 4. Effect of increasing concentrations of ochratoxin A on (A) calcium uptake activity and (B) malondialdehyde levels in renal cortical microsomes preincubated for 10 min with NADPH

Results are means ± SD of quadruplicate determinations.

calcium in extracellular fluids (about 1.3 mM). In renal tubular cells, the passive influx of calcium at the brush border, driven by its electrochemical gradient, is normally balanced by active calcium efflux at the basolateral membrane by Ca^{2+}-ATPase(s) and possibly by a Na/Ca antiporter (Rouse & Suki, 1990). In addition, the cytosolic calcium concentration is controlled by active sequestration into intracellular stores, which include mitochondria and endoplasmic reticulum, and by calcium binding to intracellular proteins (calmodulin). Mitochondria have a high capacity but a low affinity for calcium uptake, thus preventing them from lowering calcium below 10^{-6} M. In contrast, the endoplasmic

Table 1. Effect of ochratoxin A on phosphorylation of kidney microsomes

Treatment	Bound ^{32}P (nmol/mg protein x 10^6)
In vitro	
Control + NADPH	0.3
Ochratoxin A (1 mM) + NADPH	0.8
Control – NADPH	1.2
Ochratoxin A – NADPH	0.6
In vivo	
Control	0.7
Ochratoxin A (10 mg/kg intraperitoneally, 2 h)	1.7
Control	7.0
Ochratoxin A (4 mg/kg intraperitoneally, daily for 4 days)	2.9

Incubations were carried out for 1 min at 0 °C in 0.2 ml of 37 mM Hepes buffer (pH 6.8) and contained 0.1 M KCl, 2.5 mM $MgCl_2$, 50 μg microsomal protein and 0.6 pmol [γ-^{32}P]ATP. Proteins were separated by sodium dodecyl sulfate-polyacrylamide gel electrophoresis, and radiolabel was determined in the Mr 118 000 band as described by Heilmann *et al.* (1985).

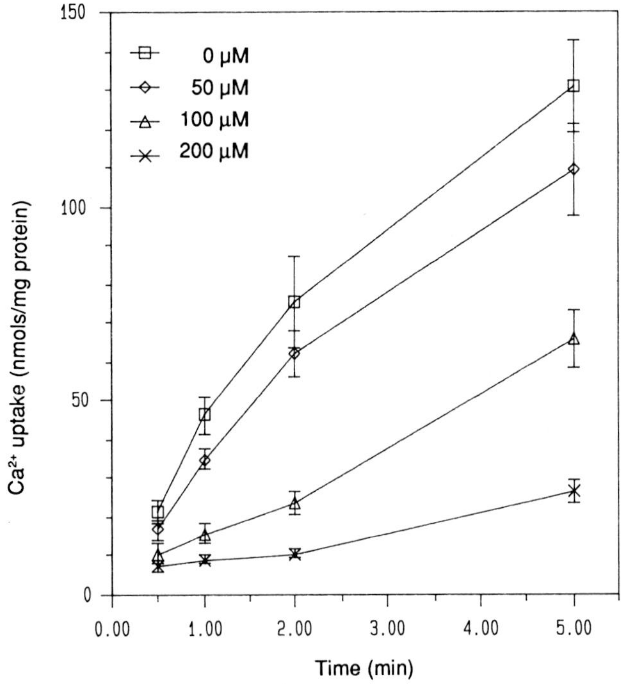

Figure 5. Dose-dependent inhibition of renal mitochondrial calcium uptake by ochratoxin A in the presence of succinate

Results are means ± SD of quadruplicate determinations.

reticulum has a high affinity and a low capacity Ca^{2+}-ATPase (Parys *et al.*, 1985) and is believed to be responsible for maintaining cytosolic calcium in the proper range (Rouse & Suki, 1990).

Berndt *et al.* (1984) first showed that calcium uptake by renal cortical slices was substantially enhanced within 5 min of adding ochratoxin A. We have shown that both mitochondrial respiration and calcium uptake are inhibited by ochratoxin A in a dose-dependent manner. This may lead to an increase in cytosolic calcium levels, and the increase in microsomal calcium uptake observed after dosing *in vivo* may be an attempt to restore calcium homeostasis. Moreover, the increase in calcium pump activity in renal microsomes appears to be the earliest enzymatic change reported in the kidney after administration of ochratoxin A. De Witt *et al.* (1988) observed a similar increase in calcium pump activity in renal endoplasmic reticulum after dosing with cisplatin. Repeated administration of moderate doses of ochratoxin A resulted in an eventual decrease in the rate of calcium uptake, and this could lead to even greater cytosolic concentrations of calcium, with its attendant consequences.

Microsomal calcium pump activity has been shown to be very sensitive to oxidative damage and lipid peroxidation (Waller *et al.*, 1983). The increase in the rate of microsomal calcium uptake observed subsequent to ochratoxin A administration or its addition to microsomes (in the presence of NADPH) is inconsistent with the role of lipid peroxidation. Lipid peroxidation may, however, play a role in the inhibition of calcium uptake observed on repeated dosing of rats with moderate doses of ochratoxin A.

References

Berndt, W.O., Hayes, A.W. & Baggett, J.M. (1984) Effects of fungal toxins on renal slice calcium balance. *Toxicol. Appl. Pharmacol.*, **74**, 78–85

Cain, K. & Skilleter, D.N. (1987) Preparation and use of mitochondria in toxicological research. In: Snell, K. & Mullock, B., eds, *Biochemical Toxicology: A Practical Approach*, Oxford, IRL Press, pp. 217–254

De Witt, L.M., Jones, T.W. & Moore, L. (1988) Stimulation of renal endoplasmic reticulum calcium pump: a possible biomarker for platinate toxicity. *Toxicol. Appl. Pharmacol.*, **92**, 157–169

Gunter, T.E., Gunter, K.K., Puskin, J.S. & Russell, P.R. (1978) Efflux of Ca2+ and Mn2+ from rat liver mitochondria. *Biochemistry*, **17**, 339–345

Heilmann, C., Spamer, C. & Gerok, W. (1985) Reaction mechanism of the calcium-transport ATPase in endoplasmic reticulum of rat liver. *J. Biol. Chem.*, **260**, 788–794

Moore, L. (1982) 1,1-Dichloroethylene inhibition of liver endoplasmic reticulum calcium pump function. *Biochem. Pharmacol.*, **31**, 1463–1465

Orrenius, S., McConkey, D.J., Bellomo, G. & Nicotera, P. (1989) Role of Ca2+ in toxic cell killing. *Trends Pharmacol. Sci.*, **10**, 281–285

Parys, J.B., de Smedt, H., Vendenberghe, P. & Borghgraef, R. (1985) Characterization of ATP-driven calcium uptake in renal basolateral and renal endoplasmic reticulum membrane vesicles. *Cell Calcium*, **6**, 413–429

Rahimtula, A.D., Zachariah, P.K. & O'Brien, P.J. (1979) Differential effects of antioxidants, steroids and other compounds on benzo[a]pyrene 3-hydroxylase activities in various tissues of rat. *Br. J. Cancer*, **40**, 105–120

Rouse, D. & Suki, W.N. (1990) Renal control of extracellular calcium. *Kidney Int.*, **38**, 700–708

Uchiyama, M. & Mihara, M. (1978) Determination of malonaldehyde precursor in tissues by thiobarbituric acid test. *Anal. Biochem.*, **86**, 271–278

Waller, R.L., Glende, E.A., Jr & Recknagel, R.O. (1983) Carbon tetrachloride and bromotrichloromethane toxicity: dual roles of covalent binding of metabolic cleavage products and lipid peroxidation in depression of microsomal calcium sequestration. *Biochem. Pharmacol.*, **32**, 1613–1617

Mycotoxins, Endemic Nephropathy and Urinary Tract Tumours
Ed. M. Castegnaro, R. Pleština, G. Dirheimer, I.N. Chernozemsky & H. Bartsch
Lyon, International Agency for Research on Cancer, pp. 215-227

A MOLECULAR BASIS FOR TARGET-CELL TOXICITY AND UPPER UROTHELIAL CARCINOMA IN ANALGESIC ABUSERS AND PATIENTS WITH BALKAN ENDEMIC NEPHROPATHY

P.H. Bach

School of Science, Polytechnic of East London, London, United Kingdom

Summary

Ochratoxin A is ubiquitous in regions where Balkan endemic nephropathy is common. It damages the kidney cortex in a range of experimental animals and induces renal parenchymal carcinoma in mice, but it is not a potent carcinogen, nor is there experimental evidence to link it to upper urothelial carcinoma (UUC). A model UUC can be induced experimentally in rodents by urothelial initiation, followed by an acutely induced papillary necrosis. This two-stage experimental model may help to clarify the role of ochratoxin A in initiating or promoting upper urothelial cells and increase our understanding of the development of UUC in patients with Balkan endemic nephropathy.

Introduction

Upper urothelial carcinoma (UUC) has a low incidence (1:183 000 to 156 000) (Bengtsson *et al.*, 1978), but analgesic abuse contributed more than 2900 new cases of UUC per 100 000 inhabitants in Switzerland in 1982 (Mihatsch & Knusli, 1982). The incidence of UUC is 1:200 to 1:1000 in areas where Balkan nephropathy is endemic (Chernozemsky *et al.*, 1977; Sattler *et al.*, 1977), but analgesic abuse has never been implicated in its etiology.

As UUC and Balkan endemic nephropathy have marked clinical and pathological differences, it is unlikely that they share a common mechanism. Renal papillary necrosis (RPN) is the primary lesion associated with analgesic abuse (Bach & Bridges, 1985), and cortical degeneration is secondary. Balkan nephropathy is associated with exposure to mycotoxins (WHO, 1979, 1990) and is manifested by degenerative changes in the renal cortex but not, apparently, in the medulla (Hall & Dammin, 1978). A model of RPN and UUC has been developed recently, which provides a mechanistic basis for RPN and UUC in humans. This paper provides a review of the molecular mechanism of RPN and the associated UUC in animal models and extends this concept to increase understanding of the role of ochratoxin A in Balkan endemic nephropathy.

Analgesic Abuse and Renal Papillary Necrosis

The selective targeting of many carcinogens and toxins depends on their physicochemical properties and the biochemical characteristics of the affected cells. RPN is commonly believed to be the key etiological factor in analgesic nephropathy because similar changes can be induced in experimental animals (Bach & Hardy, 1985). Rodents are resistant to the papillotoxic effects of many analgesics and non-steroidal anti-inflammatory drugs, and this has limited attempts to define their role in the genesis of UUC (Bach & Hardy, 1985). Several model papillotoxins, however, cause selective, dose-related necrosis within days (Bach *et al.*, 1983). The 2-bromoethanamine hydrobromide model results in predictable, reproducible functional and morphological changes to the kidney that are almost identical to those caused by chronic exposure to analgesics and non-steroidal anti-inflammatory drugs in animals and humans (Bach & Hardy, 1985).

Early morphological changes in rats are particularly useful for understanding the progression of RPN in humans; these include loss of medullary mucopolysaccharide matrix staining (Bach *et al.*, 1983), similar to that reported in human analgesic abusers (Gloor, 1978). The medullary interstitial cells at the tip of the papilla undergo early degenerative change, followed by damage to the loops of Henle. Microvascular changes occur late in the development of the acute lesion (Gregg *et al.*, 1990a). Active repair at the necrotic papilla, the collecting ducts and papillary and pelvic urothelia causes marked hyperplasia. Following acute papillary necrosis, there is progressive degenerative change that affects the cortex (Gregg *et al.*, 1990a,b), but the factors involved in this cascade are poorly understood.

The probable mechanism of RPN

Several hypotheses have been put forward to explain the mechanistic basis of RPN (Shelley, 1978; Bach & Bridges, 1985). The most attractive hypothesis for selective targeting of the papilla, upper ureter and pelvis involves local metabolic activation of compounds within the medulla. Early investigations concentrated on cytochrome P450-mediated metabolism (Hinson, 1983) as the basis for the formation of biologically reactive intermediates, but the absence of such enzymes in the medulla precludes this mechanism. Prostaglandin synthase has both hydroperoxidase and cyclo-oxygenase activity. Prostaglandin hydroperoxidase oxidizes chemicals similarly to mixed-function oxidases (Marnett & Eling, 1984), including metabolic activation of many analgesics and non-steroidal anti-inflammatory drugs (see Bach & Bridges, 1984, 1985) such as paracetamol (acetaminophen) and *para*-phenetidine (Ross *et al.*, 1985) to reactive intermediates. The medullary interstitial cells, endothelial cells, arteries and arterioles and collecting ducts (Zenser *et al.*, 1983) are very rich in prostaglandin hydroperoxidase. The sensitivity of the medullary interstitial cells has been explained on the basis of high levels of polyunsaturated fatty acids (Bojesen, 1980) which would predispose these cells to lipid peroxidation if reactive intermediates were generated locally (Bach & Bridges, 1984). The fact that neither paracetamol nor *para*-phenetidine causes RPN in rats has limited the development of this concept.

N-Phenylanthranilic acid (an analogue of the fenamate non-steroidal anti-inflammatory drugs) causes a dose-related RPN and changes in renal function and morphology (Hardy & Bach, 1984) that are almost identical to those caused by analgesics, non-steroidal anti-inflammatory drugs and 2-bromoethanamine. *N*-Phenylanthranilic acid is activated by horseradish peroxidase, and prostaglandin synthase (including hydroperoxidase activity) from medullary microsomes produces biologically reactive intermediate(s) that bind to macromolecules (M. Feldman & P.H. Bach, 1987, unpublished data).

Urothelial hyperplasia following RPN

Urothelial hyperplasia and dysplasia are found in analgesic abusers (Blohme & Johansson, 1981), and it is generally agreed that there may be progression from hyperplasia to dysplasia and thence malignancy; however, the interrelationship between proliferative repair, hyperplasia and dysplasia is far less clear.

Urothelial proliferation is a common consequence of chemically induced (2-bromoethanamine, aspirin, paracetamol, *N*-phenylanthranilic acid, ethyleneimine, indeneacetic acid analogue, Aroclor 1242 and diphenylamine) RPN, spontaneous RPN or experimentally induced (ageing mice, those with amyloid deposits, rats fed a fat-free diet over a long period and Gunn rats) RPN, but authors often failed to draw attention to it (see Bach & Bridges, 1985 for full references). Papillary urothelial hyperplasia has also been reported to develop in the absence of RPN following administration of phenacetin (Johansson, 1981) or feeding high levels of sodium chloride, monosodium glutamate (Lalich *et al.*, 1974) or saccharin (Murasaki *et al.*, 1982).

Changes to urothelial cells in animals with 2-bromoethanamine-induced RPN closely parallel those seen in hyperplastic and dysplastic epithelial cells and malignancies (Bannasch, 1986). Alkaline phosphatase changes in proliferating urothelia are reminiscent of premalignant changes (Gregg *et al.*, 1990a,b). There is progressive deposition of neutral lipid in the capillaries and marked accumulation in the epithelial cells (Bach *et al.*, 1991), as seen in human analgesic abusers (Munck *et al.*, 1970). Urothelial cells also accumulate periodic acid-Schiff-positive inclusion bodies (Gregg *et al.*, 1990a,b) similar to those reported in malignant bladder carcinomas (Hukill & Vidone, 1965).

The role of mechanical injury in urothelial hyperplasia

Chronic mechanical injury underlies cell proliferation, hyperplasia and experimental bladder tumours (Mobley *et al.*, 1966). Recently, Cohen and Ellwein (1990) demonstrated that the microabrasion due to silicate crystals produced as a consequence of high-dose, long-term exposure to saccharin caused proliferative changes. 2-Bromoethanamine causes an intense but short-lived magnesium ammonium phosphate crystalluria (Bach, 1990) that could have a similar mechanical effect on the urothelia.

Urothelial proliferation following RPN

The acute nature of the 2-bromoethanamine-induced lesions allows assessment of urothelial kinetics following papillary injury. Each cell type in the kidney has a baseline for renal and urothelial cell proliferation, with a particularly slow turnover in glomerular endothelia and interstitial cells (Mattingley *et al.*, 1991). After administration of 2-bromoethanamine, cell turnover was greatest in the proximal and distal tubules and in the collecting duct epithelium in the mid-medullary region. The turnover of cells decreased progressively from the pelvic fornix, at the origin of the ureter, to the pelvis opposite the tip of the papilla and opposite the zone of papillary injury, and was least in the area of papillary injury. The very marked pelvic response to 2-bromoethanamine in the absence of major ureter or bladder change highlights the focal nature of the injury and the urothelial response; however, the pelvis reverts to near normal within six months (N.J. Gregg & P.H. Bach, unpublished observation). Johansson *et al.* (1989) reported a similar increase in the labelling index of bladder and kidney urothelium after six weeks' administration of paracetamol, antipyrine or phenacetin, although RPN was reported in only some of the kidneys.

The Role of Analgesics in Upper Urothelial Carcinoma

The statistical association between analgesic abuse and UUC is convincing (McCredie *et al.*, 1983), but a causal relationship has never been proven. The IARC (1980, 1982) concluded that the carcinogenic potential of single or mixed analgesics could not be established on the basis of the results of short-term tests or studies in humans, as much of the old epidemiological data (Hultengren *et al.*, 1965) was biased by selective recording of phenacetin intake only. Similarly, the results of carcinogenicity studies in experimental animals were considered to be equivocal. Analgesic mixtures containing phenacetin are carcinogenic to humans (IARC, 1980, 1982), and phenacetin or phenazone alone have been implicated and may be carcinogenic, but these compounds apparently do not affect the urothelial cells of the pelvis or ureter. Macklin and Szot (1980) have argued that the formulation of phenacetin into pellets or cubes would result in the formation of phenacetin *N*-oxide and *N*-nitroso compounds, which could caused tumours of the ear and nose, liver, mammary gland, bladder and renal parenchyma. Thus, the relevance of extrapolating the results of studies in which pelleted diets containing phenacetin were administered to animals to UUC in humans appears to be questionable.

Models of Upper Urothelial Carcinoma

The bladder has been the focus of most of the research on UUC. The identification of a focal RPN demands painstaking preparation of tissue and careful sectioning to assess the papilla tip (Bach & Hardy, 1985). The absence of a papillary lesion may reflect failure to section the medulla tip, but cortical changes often flag the possibility that such a pathological lesion is present. The study of UUC in animals is especially demanding and labour intensive, and the technical difficulties may not have been overcome in some published investigations.

Spontaneous bladder tumours occur in old male brown Norway (Bolhuis *et al.*, 1978) and DA/Han rats (Deerberg *et al.*, 1985), but no data are available on changes to the ureter or pelvis, and variable tumour development, long latent periods and low incidence (50% at best) make such investigations difficult to perform. Chemical models circumvent many of these disadvantages.

Total chemical carcinogens that cause UUC

A number of urinary tract carcinogens are transported to the urothelium *via* the urine (Cohen *et al.*, 1983). *N*-4-(5-Nitro-2-furyl)-2-thiazolyl]formamide (FANFT) is a complete bladder carcinogen in female

Sprague-Dawley and male Fischer 344 rats (Cohen *et al.*, 1982). Early publications reported a few cases of FANFT-induced renal pelvic carcinoma (that had invaded the kidney) and severe renal pelvic epithelial hyperplasia (Erturk *et al.*, 1969). Similarly, feeding of FANFT causes bladder urothelial hyperplasia in Sprague-Dawley (Erturk *et al.*, 1969), Fischer (Tiltmann & Friedell, 1971) and Fischer 344 rats (Jacobs *et al.*, 1979), followed by mitotic activity and squamous metoplasia, which progresses to microscopic papillae throughout the bladder and tumours that protrude into the lumen and invade the stroma. At about 45–50 weeks, there was also late development of tumours in the ureter and renal pelvis with ureteral obstruction and hydronephrosis in Sprague-Dawley but not Fischer rats. Anderström and Johansson (1983) demonstrated a 72% incidence of renal pelvic tumours (52% of which were bilateral) and a high frequency of bladder tumours in male Sprague-Dawley rats fed 0.2% FANFT for 11 weeks. This finding indicates a shift away from the bladder, the most commonly evaluated target site for FANFT-induced carcinogenicity in female Sprague-Dawley rats, and suggests that both renal and extra-renal factors could modulate the selective toxicity of these carcinogens.

Two-stage models of UUC

Significant evidence indicates that a classical two-stage model is at the basis of UUC. Chronic urinary tract infection by *Escherichia coli* (strain 06K 13H1) in FANFT-initiated animals promotes a hyperplastic response in the pelvis and ureter and produces tumours in about 30% of animals (Johansson *et al.*, 1987). A number of factors, such as urinary tract infection, bacterial toxins, metabolic products, mechanical injury, bacterial enzymes and nitrosylation of secondary amines could each contribute to promotion in the urothelium (Claude *et al.*, 1988). Bacterial deconjugation of glucuronide analgesic metabolites could facilitate the presence of the parent compound, which would then undergo local metabolism (Furman *et al.*, 1981) and exacerbate the development of UUC. Analgesic abuse appears to predispose humans to urinary tract infection (Johansson & Wahlqvist, 1977), and animals with experimentally induced RPN are also prone to such infections (Thiele, 1974). Surgical incision of the left renal pelvis stimulates urothelial proliferation, and this intervention has been used after subcarcinogenic urothelial initiation (by six weeks of feeding 0.2% FANFT) to double the incidence of unilateral renal pelvic tumours (Anderström & Johansson, 1983).

Phenacetin on its own always causes marked urothelia hyperplasia, but UUC has rarely been reported, suggesting that phenacetin may be a urothelial promoter. Anderström and Johansson (1983) reported that animals fed 0.2% FANFT for six weeks (a subcarcinogenic exposure period) and then 0.535% phenacetin in the diet for six weeks had enhanced development of well-differentiated non-invasive and poorly differentiated invasive renal pelvic tumours, but not of bladder tumours. Metastases were also present in the lung of FANFT-induced and phenacetin-treated animals. Johansson and Anderström (1988) showed that, after FANFT initiation and antipyrine, one-third of the animals developed tumours, half of which were in the pelvis and the rest in the bladder.

The disadvantages of administering FANFT in the diet include risk to personnel over the six weeks of feeding the rats and uncertainty about the exact dose of carcinogen eaten by each animal. *N*-Nitrosobutyl-*N*-(4-hydroxybutyl)amine (NBHBA) is a specific, complete bladder carcinogen (Fujita *et al.*, 1988) that can be administered by oral gavage in divided doses over a few weeks, thus circumventing the logistic drawbacks of FANFT. NBHBA is a directly acting bladder carcinogen, with a dose-related effect (Ito & Fukushima, 1986). Unilateral ligation of a ureter in male Wistar rats treated with NBHBA for 20 weeks caused hyperplasia of the pelvic and ureteral urothelium and induced papillomas and carcinomas in these regions (Ito *et al.*, 1971). Initiation with NBHBA and long-term feeding with phenacetin plus caffeine or phenacetin on its own (Nakanishi *et al.*, 1978) enhanced the development of urothelial carcinoma. We have developed a model of UUC in which a subcarcinogenic regimen of NBHBA (twice weekly for five weeks to a total dose of 800 mg per 200-g rat) initiates the urothelium; this is followed by a week's respite, before a single dose of 2-bromoethanamine at 75 mg/kg, which causes papillary necrosis. Sustained hyperplasia followed, leading to a specific, localized upper urothelial dysplasia by six weeks in all animals. The earliest malignancies, representing both nodular and flat carcinoma *in situ*, were present in NBHBA-treated animals 13 weeks after dosing with 2-bromoethanamine (Gregg *et al.*, 1991); by 40 weeks, macroscopic tumours were present in the renal pelvis and ureter in about 30% of animals. No change was seen in animals treated with similar doses of NBHBA at 30 weeks, but bladder carcinomas occurred. 2-Bromoethanamine alone caused only transient hyperplasia. These data suggest that a two-stage model could apply to UUC that follows RPN caused by any papillotoxic compound.

In this model, the frequency of macroscopic tumours following treatment with NBHBA and 2-bromoethanamine is still rather low and the time necessary to produce UUC is long. Preliminary evidence (N.J. Gregg & P.H. Bach, 1989, unpublished observation) suggests that dosing with additional NBHBA during the period of high urothelial cell proliferation that follows 2-bromoethanamine-induced RPN increases the frequency and severity of upper urothelial tumours and also accelerates the production of bladder tumours.

The Mechanistic Basis of Upper Urothelial Carcinoma in Humans

The mechanism of UUC in human analgesic abusers may therefore be one of promotion. This suggestion does not, in itself, address the question of how the urothelial tract of analgesic abusers could be initiated. There is very little evidence to suggest that analgesics are in themselves genotoxic (IARC, 1980, 1982). Tobacco is widely accepted as a risk factor in bladder cancer (IARC, 1986), and it is a high risk factor for human analgesic abusers who develop UUC (McCredie *et al.*, 1983). The relevance of the experimental mechanism to human analgesic abusers is not clear, but smoking could be a key factor in initiating the upper urothelium. Furthermore, peroxidative activation of papillotoxins (Bach & Bridges, 1985) has already been implicated in the metabolic activation of *N*-arylamines (Zenser *et al.*, 1983). Polyaromatic hydrocarbons, including 7,8-dihydro-7,8-dihydroxybenzo[*a*]pyrene, are excreted *via* the urine of smokers, and prostaglandin hydroperoxidase activates these compounds to the highly genotoxic 7,8-diol epoxide (Marnett *et al.*, 1978). Analgesic abuse would then give rise to RPN, urinary debris and increased risk of urinary tract infection (see Bach & Bridges, 1985), all of which would contribute to promotion. This hypothesis does not exclude the possibility that metabolites of, e.g., analgesics, non-steroidal anti-inflammatory drugs and caffeine, also contribute to initiation or promotion but suggests that analgesic-associated RPN may lead to the promotion of initiated urothelia in individuals at high risk, especially those who are smokers. This idea is supported by Ross *et al.* (1989).

Nephrotoxic effects of ochratoxin A

Ochratoxin A is regarded as a proximal tubule toxin (WHO, 1979, 1990), but experimental data also indicate that it causes oedematous and sclerotic glomeruli and glomerular basement membrane thickening (Delacruz & Bach, 1991a). Changes to the glomerular basement membrane are common in animals exposed to ochratoxin A and are present in patients with Balkan endemic nephropathy (Hall & Dammin, 1978). Isolated rat glomeruli incorporate 10 times more aromatic amino acid into protein than do proximal tubules. The finding that this incorporation is inhibited by ochratoxin A (Delacruz & Bach, 1991b) could explain the ochratoxin A-related glomerular changes seen *in vivo*.

Role of Ochratoxin A in Balkan Endemic Nephropathy

Whether there is a risk that a similar lesion will develop in other areas of the world where ochratoxin A is ubiquitous remains uncertain. Thus, it is important to understand the mechanism that underlies this lesion and to develop a suitable animal model in order to investigate factors that exacerbate or ameliorate the condition.

The interrelationship between cortical injury in Balkan endemic nephropathy and the development of UUC is difficult to explain unless it is linked indirectly by a two-stage model. If such a model applies, exposure to nitrite, nitrate, polyaromatic hydrocarbons, ochratoxin A and/or unidentified carcinogens could initiate the upper urothelium. Cellular and other debris from ochratoxin A-related proximal tubular or glomerular injury could (together with urinary tract infection and inorganic solids such as silicate) provide the mechanical stimulus that leads to upper urothelial proliferation and carcinoma.

If ochratoxin A can initiate upper urothelial cells, selective promotion with, for example, an acute renal papillary necrosis will produce local dysplasia or neoplasia. Alternatively, a number of factors could initiate the urothelia of individuals in the Balkan regions; the most prominent of these is a high intake of nitrite and nitrate in preserved foods and of polyaromatic hydrocarbons from cigarette smoking and smoke-cured foods. If the upper urothelial cells have already been initiated by genetic or environmental factors, cortical nephrotoxicity or any other nephropathy could act as a promoter. For example, proximal tubular injury often leads to cell exfoliation, proteinuria, casts and urinary changes such as hypotonicity and crystalluria. The resulting deluge of urinary debris would produce an apparently low level of morphological injury, which could, however, be very important in causing upper urothelial proliferation. The resulting hyperplasia could be assessed using histochemistry, morphological and autoradiographic methods.

Prospectives in Research on Balkan Endemic Nephropathy

The demonstration that an animal model of UUC can be induced using a two-stage model indicates that the role of ochratoxin A in inducing or initiating urothelial hyperplasia, dysplasia and malignancy can be assessed. In order to understand the possible role of ochratoxin A in Balkan endemic nephropathy, studies should be carried out to:

- —develop an animal model of Balkan endemic nephropathy as a basis for better therapeutic management;
- —establish a threshold level below which exposure to ochratoxin A and other mycotoxins is safe; and
- —devise early, non-invasive diagnostic markers of renal changes.

The kidneys and urinary tracts of animals with porcine and avian nephropathy should be examined more closely to identify the presence of proliferative urothelial changes.

The human population at risk offers an important resource in which to assess the role of smoking and host factors (genetic susceptibility, dietary nitrate, nitrite, other potential genotoxic carcinogens in individuals with Balkan endemic nephropathy) in initiation of the upper urothelial tract. More kidneys, ureters and bladders should be obtained *post mortem* from people in endemic areas to establish the presence of early hyperplasia and dysplasia and of histochemical changes that indicate whether cells are initiated.

The problem of establishing the role of ochratoxin A in the etiology of Balkan endemic nephropathy is too large and diverse for a single group or institute. A network of collaborative researchers (BEN-NET) should be established to investigate the significance of Balkan endemic nephropathy. There should be an international focus of expertise and resources which could address the primary objectives by exchanging specimens, personnel, research data, etc. Other important strategies might include attempts to improve the animal model of UUC using DA/Han, BN/BiRij or Sprague-Dawley rats administered a more selective or more potent carcinogen (perhaps FANFT rather than NBHBA) to initiate the upper urinary tract, and to exacerbate local proliferation by superimposing bacterial infection and/or urinary microcrystals to induce mechanical injury to the renal pelvis.

Acknowledgements

The author's research was supported by the Cancer Research Campaign, the International Agency for Research on Cancer, the Wellcome Trust, the Kidney Research Fund of Great Britain, the Johns Hopkins Center for Alternatives to Animals in Testing, the Smith-Kline Foundation, the British Council and the Commission of the European Communities. He is indebted to Monique Rivet, Gail Sutherland and Mimps E. van Ek for preparing the manuscript and to Neill J. Gregg and Ligia Delacruz for providing unpublished data.

References

Anderström, C. & Johansson, S.L. (1983) The combined effect of mechanical trauma and phenacetin or sodium saccharin on the rat urinary bladder. *Acta Pathol. Microbiol. Immunol. Scand. Sect. A*, **91**, 381–389

Bach, P.H. (1990) Crystalluria, medullary matrix crystal deposits and bladder calculi associated with an acutely-induced renal papillary necrosis. *Br. J. Urol.*, **66**, 463–470

Bach, P.H. & Bridges, J.W. (1984) The role of prostaglandin synthase mediated metabolic activation of analgesics and non-steroidal anti-inflammatory drugs in the development of renal papillary necrosis and upper urothelial carcinoma. *Prostaglandins Leukotrienes Med.*, **15**, 251–274

Bach, P.H. & Bridges, J.W. (1985) Chemically induced renal papillary necrosis and upper urothelial carcinoma. *CRC Crit. Rev. Toxicol.*, **15**, 217–439

Bach, P.H. & Hardy, T.L. (1985) The relevance of animal models to the study of analgesic associated renal papillary necrosis in man. *Kidney Int.*, **28**, 605–613

Bach, P.H., Grasso, P., Molland, E.A. & Bridges, J.W. (1983) Changes in the medullary glycosaminoglycan histochemistry and microvascular filling during the development of 2-bromoethanamine hydrobromide-induced renal papillary necrosis. *Toxicol. Appl. Pharmacol.*, **69**, 333–344

Bach, P.H., Scholey, D.J., Delacruz, L., Moret, M. & Nichol, S. (1991) Renal and urinary lipid changes associated with an acutely induced renal papillary necrosis. *Food Chem. Toxicol.*, **29**, 211–219

Bannasch, P. (1986) Preneoplastic lesions as end points in carcinogenicity testing. II. Preneoplasia in various non-hepatic tissues. *Carcinogenesis*, **7**, 849–852

Bengtsson, U., Johansson, S. & Angervall, L. (1978) Malignancies of the urinary tract and their relation to analgesic abuse. *Kidney Int.*, **13**, 107–113

Blohme, I. & Johansson, S. (1981) Renal pelvic neoplasms and atypical urothelium in patients with end-stage analgesic nephropathy. *Kidney Int.*, **20**, 671–675

Bojesen, I.N. (1980) Fatty acid composition and depot function of lipid droplet triacylglycerols in renomedullary interstitial cells. In: Mandal, A.K. & Bohman, S.-O., eds, *The Renal Papilla and Hypertension*, New York, Plenum Press, pp. 121–148

Bolhuis, R.L., Klein, J.C. & Kruisbeek, A.M. (1978) Spontaneous urinary bladder and ureter tumours in the Brown Norway rat. *Natl Cancer Inst. Monogr.*, **49**, 301–304

Chernozemsky, I.N., Stoyanov, I.S., Petkova-Bocharova, T.K., Nicolov, I.G., Draganov, I.V., Stoichev, I.I., Tanchev, Y., Naidenov, D. & Kalcheva, N.D. (1977) Geographic correlation between the occurrence of endemic nephropathy and urinary tract tumours in Vratza district, Bulgaria. *Int. J. Cancer*, **19**, 1–11

Claude, J.C., Frentzel-Beyme, R.R. & Kunze, E. (1988) Occupation and risk of cancer of the lower urinary tract among men: a case-control study. *Int. J. Cancer*, **41**, 371–379

Cohen, S.M. & Ellwein, L.B. (1990) Cell proliferation in carcinogenesis. *Science*, **249**, 1007–1011

Cohen, S.M., Murasaki, G., Fukushima, S. & Greenfield, R.E. (1982) Effect of regenerative hyperplasia on the urinary bladder; carcinogenicity of sodium saccharin and N-[4-(5-nitro-2-furyl)-2-thiazolyl]formamide. *Cancer Res.*, **42**, 65–71

Cohen, S.M., Greenfield, R.E. & Ellwein, L.B. (1983) Multi-stage bladder carcinogenesis. *Environ. Health Perspect.*, **49**, 209–215

Deerberg, F., Rehm, S. & Jostmeyer, H.H. (1985) Spontaneous urinary bladder tumours in DA/Han rats: a feasible model of human bladder cancer. *J. Natl Cancer Inst.*, **75**, 1113–1121

Delacruz, L. & Bach, P.H. (1991a) The effects of single and multiple doses of ochratoxin A and aflatoxin B1 on renal function and pathology in the rat. *Toxicol. Lett.*, submitted

Delacruz, L. & Bach, P.H. (1991b) The synergistic toxicity of ochratoxin A and aflatoxin B1 on the proximal tubule and glomeruli in the rat. *Arch. Toxicol.*, submitted

Erturk, E., Cohen, S.M., Price, J.M. & Bryan, G.T. (1969) Pathogenesis, histology and transplantability of urinary bladder carcinomas induced in albino rats by oral administration of N-[4-(5-nitro-2-furyl)-2-thiazolyl]formamide. *Cancer Res.*, **29**, 2219–2228

Fujita, J., Ohuchi, N., Ito, H., Yoshida, O., Nakayama, H., & Kitamura, Y. (1988) Activation of H-ras oncogene in rat bladder tumours induced by N-butyl-N-(4-hydroxybutyl)nitrosamine. *J. Natl Cancer Inst.*, **80**, 37–43

Furman, K.I., Kundig, H. & Lewin, J.R. (1981) Experimental paracetamol nephropathy and pyelonephritis in rats. *Clin. Nephrol.*, **16**, 271–275

Gloor, F.J. (1978) Changing concepts in pathogenesis and morphology of analgesic nephropathy as seen in Europe. *Kidney Int.*, **13**, 27–33

Gregg, N.J., Courtauld, E.A. & Bach, P.H. (1990a) Enzyme histochemical changes in an acutely-induced renal papillary necrosis. *Toxicol. Pathol.*, **18**, 39–46

Gregg, N.J., Courtauld, E.A. & Bach, P.H. (1990b) High resolution light microscopic morphological and microvascular changes in an acutely-induced renal papillary necrosis. *Toxicol. Pathol.*, **18**, 47–55

Gregg, N.J., Ijomah, P., Courtauld, E.A. & Bach, P.H. (1991) Two stage model of upper urothelial carcinoma using N-butyl-N-(4-hydroxybutyl)nitrosamine initiation and promotion by an acutely-induced renal papillary necrosis. *Cancer Res.*, submitted

Hall, P.W. & Dammin, G.J. (1978) Balkan nephropathy. *Nephron*, **22**, 281–300
Hardy, T.L. & Bach, P.H. (1984) The effect of N-phenylanthranilic acid-induced renal papillary necrosis on urinary acidification and renal electrolyte handling. *Toxicol. Appl. Pharmacol.*, **75**, 265–277
Hinson, J.A. (1983) Reactive metabolites of phenacetin and acetaminophen: a review. *Environ. Health Perspect.*, **49**, 71–79
Hukill, P.B. & Vidone, R.A. (1965) Histochemistry of mucus and other polysaccharides in tumours. I. Carcinoma of the bladder. *Lab. Invest.*, **14**, 1624–1635
Hultengren, N., Lagergren, C. & Ljungquivst, A. (1965) Carcinoma of the renal pelvis in renal papillary necrosis. *Acta Chir. Scand.*, **130**, 314–320
IARC (1980) *IARC Monographs on the Evaluation of the Carcinogenic Risk of Chemicals to Humans*, Vol. 24, *Some Pharmaceutical Drugs*, Lyon, pp. 47–49
IARC (1982) *IARC Monographs on the Evaluation of the Carcinogenic Risk of Chemicals to Humans*, Suppl. 4, *Chemicals, Industrial Processes and Industries Associated with Cancer in Humans (IARC Monographs, Volumes 1 to 29)*, Lyon, pp. 47–49
IARC (1986) *IARC Monographs on the Evaluation of the Carcinogenic Risk of Chemicals to Humans*, Vol. 38, *Tobacco Smoking*, Lyon
Ito, N. & Fukushima, S. (1986) Carcinogenesis, urinary tract, rat. In: Jones, T.C., Mohr, U. & Hunt, R.D., eds, *Urinary System, Monograph on the Pathology of Laboratory Animals*, Heidelberg, Springer-Verlag, pp. 317–321
Ito, N., Makiura, S., Yokota, Y., Kamamoto, Y., Hiasa, Y. & Sugihara, S. (1971) Effect of unilateral ureter ligation on development of tumors in the urinary system of rats treated with N-butyl-N-(4-hydroxybutyl)nitrosamine. *Jpn. J. Cancer Res. (Gann)*, **62**, 359–365
Jacobs, J.B., Arai, M., Cohen, S.M. & Friedell, G.H. (1979) Early lesions in experimental bladder cancer: scanning electron microscopy of cell surface markers. *Cancer Res.*, **36**, 2512–2517
Johansson, S.L. (1981) Carcinogenecity of analgesics: long-term treatment of Sprague-Dawley rats with phenacetin, phenazone, caffeine and paracetamol (acetaminophen). *Int. J. Cancer*, **27**, 521–529
Johansson, S.L. & Anderström, C. (1988) The influence of antipyrene on N-[4-(5-nitro-2-furyl)-2-thiazolyl]formamide-induced urinary tract carcinogenesis. *Carcinogenesis*, **9**, 783–787
Johansson, S. & Wahlqvist, L. (1977) Tumours of urinary bladder and ureter associated with abuse of phenacetin-containing analgesics. *Acta Pathol. Microbiol. Immunol. Scand. Section A*, **85**, 768–774
Johansson, S.L., Anderström, C., von Schultz, L. & Larsson, P. (1987) Enhancement of N-[4-(5-Nitro-2-furyl)-2-thiazolyl]formamide-induced carcinogenesis by urinary tract infection in rats. *Cancer Res.*, **47**, 559–562
Johansson, S.L., Radio, S.J., Saidi, J. & Sakata, T. (1989) The effects of acetaminophen, antipyrine and phenacetin on rat urothelial cell proliferation. *Carcinogenesis*, **10**, 105–111
Lalich, J.J., Paik, W.C.W. & Pradhan, B. (1974) Epithelial hyperplasia in the renal papilla of rats. Induction in animals fed excess sodium chloride. *Arch. Pathol.*, **97**, 29–32
Macklin, A.W. & Szot, R.J. (1980) Eighteen month oral study of aspirin, phenacetin and caffeine in C57BL/6 mice. *Drug Chem. Toxicol.*, **3**, 135–163
Marnett L.J. & Eling T.E. (1984) Co-oxidation during prostaglandin bio-synthesis; a pathway for the metabolic activation of xenobiotics. *Rev. Biochem. Toxicol.*, **5**, 135–172
Marnett, L.J., Reed, G.A. & Dennison, D.J. (1978) Prostaglandin synthase dependent activation of 7,8-dihydro-7,8-dihydroxy-benzo[a]pyrene to mutagenic derivatives. *Biochem. Biophys. Res. Commun.*, **82**, 210–216

Mattingley, G., Gregg, N.J. & Bach, P.H. (1991) The response of the pelvic and ureteric epithelial cells to an acutely-induced renal papillary necrosis. *Toxicol. Pathol.*, submitted

McCredie, M., Stewart, J.H. & Ford, J.M. (1983) Analgesics and tobacco as risk factors for cancer of the ureter and renal pelvis in men. *J. Urol.*, **130**, 28–30

Mihatsch, M.J. & Knusli, C. (1982) Phenacetin abuse and malignant tumours: an autopsy study covering 25 years (1953–1977). *Klin. Wochenschr.*, **60**, 1339–1349

Mobley, T.L., Coyle, J.K., Al-Hussaini, M. & McDonald, D.F. (1966) The role of chronic mechanical irritation in experimental urothelial tumorigenesis. *Invest. Urol.*, **3**, 325–333

Munck, A., Lindlar, F. & Masshoff, W. (1970) Pigmentation of the kidney papillae and of the mucous membranes of the discharging urinary tract in the case of chronic sclerosing interstitial nephritis (phenacetin kidneys). *Virchows Archiv. Section A: Pathol. Anat. Histopathol.*, **349**, 323–331 (in German)

Murasaki, G., Greenfield, R.E. & Cohen, S.M. (1982) Alterations in the rat kidney associated with sodium saccharin feeding. *Toxicol. Lett.*, **12**, 251–258

Nakanishi, K., Fukushima, S., Shibata, M., Shirai, T., Ogiso, T. & Ito, N. (1978) Effect of phenacetin and caffeine on the urinary bladder of rats treated with N-butyl-N-(4-hydroxybutyl)nitrosamine. *Jpn. J. Cancer Res. (Gann)*, **69**, 395–400

Ross, D., Larsson, R., Andersson, B., Nilsson, U., Lindqust, T., Lindeke, B., & Moldéus, P. (1985) The oxidation of p-phenetidine by horseradish peroxidase and prostaglandin synthetase and the fate of glutathione during such oxidations. *Biochem. Pharmacol.*, **34**, 343–351

Ross, R.K., Paganini-Hill, A., Landolph, J., Gerkins, V. & Henderson, B.E. (1989) Analgesics, cigarette smoking, and other risk factors for cancer of the renal pelvis and ureter. *Cancer Res.*, **49**, 1045–1048

Sattler, T.A., Dimitrov, T. & Hall, P.W. (1977) Relation between endemic (Balkan) nephropathy and urinary-tract tumours. *Lancet*, **i**, 278–280

Shelley, J.H. (1978) Pharmacological mechanism of analgesic nephropathy. *Kidney Int.*, **13**, 15–26

Thiele, E.H. (1974) An *in vitro* pyelonephritis assay for screening therapeutic antibiotics. *J. Antibiot.*, **27**, 31–41

Tiltmann, A.J. & Friedell, G.H. (1971) The histogenesis of experimental bladder cancer. *Invest. Urol.*, **9**, 218–266

WHO (1979) *Mycotoxins* (Environmental Health Criteria 11), Geneva

WHO (1990) *Selected Mycotoxins* (Environmental Health Criteria 105), Geneva

Zenser, T.V., Cohen, S.M., Mattammal, M.B., Wise, R.W., Rapp, N.S. & Davis, B.B. (1983) Prostaglandin hydroperoxidase-catalyzed activation of certain N-substituted aryl renal and bladder carcinogens. *Environ. Health Perspect.*, **49**, 33–41

Mycotoxins, Endemic Nephropathy and Urinary Tract Tumours
Ed. M. Castegnaro, R. Pleština, G. Dirheimer, I.N. Chernozemsky & H. Bartsch
Lyon, International Agency for Research on Cancer, pp. 229-244

CARCINOGENICITY OF OCHRATOXIN A IN EXPERIMENTAL ANIMALS

J.E. Huff

National Institute of Environmental Health Sciences, Research Triangle Park, North Carolina, USA

Summary

The carcinogenicity of ochratoxin A, a naturally occurring mycotoxin of the fungal genera *Aspergillus* and *Penicillium*, was evaluated in three strains of mice and in one strain of rats. The kidney, and in particular the tubular epithelial cells, was the major target organ for ochratoxin A-induced lesions. In male ddY and DDD mice, atypical hyperplasia, cystadenomas and carcinomas of the renal tubular cells were induced, as were neoplastic nodules and hepatocyte tumours of the liver. In $B6C3F_1$ mice, tubular-cell adenomas and carcinomas of the kidneys were induced in male mice, and the incidences of hepatocellular adenomas and carcinomas were increased in male and female mice. In male and female F344 rats, ochratoxin A induced nonneoplastic (degeneration, karyomegaly, proliferation, cytoplasmic alteration, hyperplasia) and neoplastic effects (adenomas, and carcinomas with metastases) in the kidneys; the incidence of fibroadenomas of the mammary glands was also increased in female rats. Other studies on ochratoxin A were considered inadequate for evaluating the presence or absence of a carcinogenic effect; however, these are mentioned and referenced below. The collective experimental findings, together with accumulating evidence in humans, forecast further toxic and carcinogenic effects in humans exposed to ochratoxin A, mainly *via* foodstuffs.

Introduction

Naturally occurring chemicals and other agents that have been part of the human environmental milieu since the beginning of time continue to be prime candidates for evaluating potential long-term toxic and carcinogenic hazards (Huff & Hoel, 1991; Huff *et al.*, 1991a). In many cases however, 'natural chemicals' cease to be truly natural once human ingenuity has brought about increased exposure because of population needs or profit incentives, using technology to excavate (asbestos), extract (benzene), distil (ethanol), grow (tobacco), imitate (pesticides) or synthesize (allyl isothiocyanate, reserpine) these 'natural' chemicals. Ochratoxins are, however, indeed natural, as they are made by several species of the fungal genera *Aspergillus* and *Penicillium*. These fungi are found virtually everywhere, and contamination of animal feed and foodstuffs for human consumption is widespread. Yet human activities play a major role in the spread even of this purely natural chemical (as with aflatoxin). Ochratoxin A, the hallmark mycotoxin, has been found mainly in cereals, and to a lesser degree in beans (coffee, soya, cocoa) (WHO, 1990). For these reasons, and later because of the reported association with Balkan endemic nephropathy and cancer of the urinary tract, long-term carcinogenesis studies were conducted by exposing laboratory animals to ochratoxin A. An overview of these studies forms the basis of this paper.

Experimental Protocols and Results

To allow some historical perspective, the carcinogenesis studies are summarized chronologically. Only key details of the experiments, together with the essence of the results, are presented. The cited papers should be consulted for more information.

Kanisawa & Suzuki (1978)

Groups of 10 male ddy mice were offered diets containing ochratoxin A at 0 or 40 mg/kg for 44 weeks. Other groups of 20–21 animals were given aflatoxin B_1 at 20 mg/kg by gavage, followed a week later by basal diet or one containing ochratoxin A at 40 mg/kg. Experiments were ended at week 50, after five weeks of no exposure. Histopathological findings were reported only for liver and kidney (Table 1). Despite the small numbers of animals and the single exposure level, ochratoxin A obviously caused tumours of the kidney and of the liver. Equally interesting were the results of the combination experiments with aflatoxin, a combined exposure that humans might encounter. Because dimethyl sulfoxide had no influence on the effects of ochratoxin A, data from groups 1 and 2 can be combined so that the carcinogenicity of ochratoxin A can be appreciated more fully. A comparison of data for

Table 1. Effects of ochratoxin A and aflatoxin B_1 on liver and kidney of ddy male mice (0 or 40 mg/kg ochratoxin A in diet for 44 weeks)[a]

Group[b]	No.	Liver[c]		Kidney tubule[d]		
		Nodule	Tumour	Atypical hyperplasia	Cyst-adenoma	Renal-cell tumour
1. Control	10	2	0	0	0	0
2. DMSO	10	1	0	0	0	0
1 + 2	20	3	0	0	0	0
3. Ochratoxin A	9	1	5	4	9	2
4. Ochratoxin A + DMSO	10	2	3	6	9	3
3 + 4	19	3	8	10	18	5
5. Aflatoxin B_1	18	1	2	0	1	0
6. Aflatoxin B_1 + ochratoxin A	20	8	15	11	19	3

[a]From Kanisawa & Suzuki (1978), except for combined data
[b]DMSO, dimethyl sulfoxide to administer 20 mg/kg aflatoxin B_1 by single gavage one week before ochratoxin A
[c]Nodule, hyperplastic nodule with some atypia; tumour, well-differentiated hepatocellular trabecular adenoma
[d]Atypical hyperplasia, of tubular epithelial cells; renal-cell tumour, malignant renal-cell tumour

groups 1 and 2 with those for group 5 shows a possible, albeit small, effect of aflatoxin on the liver and kidney and a substantial synergistic impact of the combination (group 6) on the liver: a comparison of the results for group 6 with those for groups 3 and 4 shows an increase in the prevalence of nodules from 16% (3 of 19 animals) to 40% (8 of 20) and of liver tumours from 40% (8 of 19) to 75% (15 of 20). To determine whether the combination has any influence on the kidney, experiments must be conducted with both higher and lower concentrations, because the effects of ochratoxin A alone were sizable for tubular-cell atypia and for cystic adenomas; the marginal decrease in the incidence of tubular-cell tumours, from 25% (5/19, groups 3 and 4) to 15% (3/20, group 6), is insufficient to conclude that this compound had an inhibitory effect.

Imaida *et al.* (1982)

Ochratoxin A and citrinin (and three other mycotoxins) were evaluated for initiating and promoting activity using groups of 20 male F344 rats in a 10-week model involving partial hepatectomy and administration of carbon tetrachloride and 2-acetylaminofluorene; the number and area of hyperplastic nodules were then measured. Citrinin at 1000 mg/kg or ochratoxin A at 50 mg/kg was given in the diet for six weeks (partial hepatectomy at week 1), followed by one week on basal diet and two weeks of 2-acetylaminofluorene at 200 mg/kg of diet, with carbon tetrachloride given by gavage once at week 8. The opposite design was used to test for promoting capability: 2-acetylaminofluorene during weeks 0–2, carbon tetrachloride at week 1, followed by one week on basal diet then six weeks of citrinin or ochratoxin A, with partial hepatectomy at week 4. Ochratoxin A exhibited both initiating and promoting activity, whereas citrinin had about one-half as much initiating activity and no promoting effect (Table 2).

Table 2. Summary results of short-term initiation-promotion assays in male Fischer 344 rats *in vivo*[a]

Mycotoxin	Initiating activity	Promoting activity
Ochratoxin A	++	++
Sterigmatocystin	++	++
Patulin	++	–
(+) Rugulosin	+	+
Citrinin	+	–

[a]From Imaida *et al.* (1982)
++, significantly different from respective control in number and area of hyperplastic nodules
+, significantly different from respective control in number of hyperplastic nodules
–, no difference from respective control

Kanisawa (1984)

Groups of 16 (or 17) male ddY mice were given diets containing ochratoxin A at 50 mg/kg of diet for 0, 5, 10, 15, 20, 25 and 30 weeks, followed by basal diet through 70 weeks. 'Hepatomas' were seen in 2 of 14 mice in the 20-week group, in 5/15 in the 25-week group and in 6/17 in the 30-week group; none was reported in the groups exposed for shorter periods. The cumulative amounts of ochratoxin A received by these groups were 33, 40 and 50 mg/mouse, respectively. Renal-cell tumours occurred in groups exposed for 15 weeks or longer (3/15 for 15 weeks; 1/14 for 20 weeks; 2/15 for 25 weeks; 4/17 for 30 weeks) to cumulative amounts of 29 mg/mouse and more. No kidney tumour was observed in controls. Lung tumours were found in all groups but were discounted because no dose-response relationship was seen: 4/15 controls and a mean per exposed group of 6.5/15. These findings provide a better notion of the durations and total exposure combinations necessary for the induction of cancer by ochratoxin A.

In another set of experiments, groups of 20 male DDD mice received diets containing ochratoxin A at 0 or 25 mg/kg; citrinin at 100 or 200 mg/kg; ochratoxin A at 25 mg/kg plus citrinin at 100 or 200 mg/kg; ochratoxin A at 25 mg/kg for 25 weeks then citrinin at 200 mg/kg for 45 weeks; or citrinin at 200 mg/kg for 25 weeks followed by ochratoxin A at 25 mg/kg for 45 weeks. The experiments were ended at 70 weeks. The results are given in Table 3. The carcinogenicity of ochratoxin A was confirmed for both liver and kidney; citrinin did not cause tumours in either organ. Unfortunately, the unique design of this study appears to have been compromised by the undeniable effects of ochratoxin A

Table 3. Effects of ochratoxin A and citrinin on liver and kidney of DDD male mice (70-week experiments)[a]

Group[b]	No.	Liver		Kidney tubule		
		Nodule	Tumour	Atypical hyperplasia	Cyst-adenoma	Renal-cell tumour
1. Control	17	4	1	0	0	0
2. Ochratoxin A	20	16	8	18	20	6
3. Citrinin 100	17	0	0	0	0	0
4. Citrinin 200	19	2	0	0	0	0
5. Ochratoxin A + citrinin 100	19	11	10	18	19	0
6. Ochratoxin A + citrinin 200	18	7	7	18	17	10
7. Ochratoxin A— citrinin 200	19	9	3	17	16	0
8. Citrinin 200— ochratoxin A	18	4	2	16	2	0

[a] From Kanisawa (1984)
[b]+, both chemicals in diet; —, sequential chemical exposure: groups 7 and 8, 1st chemical for 25 weeks, 2nd for 45 weeks. Numbers are mg/kg of diet. For other definitions, see footnotes to Table 1.

alone. The dose of 25 mg/kg ochratoxin A reflects a reduction from the 40 mg/kg used earlier (Kanisawa & Suzuki, 1978), but even lower exposures appear to be needed to obtain dose-response curves and to determine any synergism. The experiments with combinations showed little other than a possible inhibitory effect on the induction of renal-cell tumours (no renal-cell tumour was observed with ochratoxin A at 25 mg/kg and citrinin at 100 mg/kg, yet 6 of 20 animals given ochratoxin A alone had renal-cell tumours) or a possible enhancement of the induction of renal-cell tumours (10 of 18 mice had renal-cell tumours with ochratoxin A at 25 mg/kg and citrinin at 200 mg/kg, but 0 of 19 did with ochratoxin A at 25 mg/kg and citrinin at 100 mg/kg, and 6 of 20 did with ochratoxin A alone). This 'bimodel mechanism' of action of citrinin on the effects of ochratoxin A on the kidney is surprising, especially since both have initiating avtivity (Table 2). Kanisawa (1984) stated that the 'simultaneous administration... enhanced extent of these renal lesions with the dose-response to concentration of citrinin'. Data are not given, but this conclusion is supported by the comment that 'difference between both groups was much more evident' if compared for 'the tumour size and nodule number induced in the bilateral kidneys'. The results of the experiments in which the two compounds were given sequentially might imply that administration of ochratoxin A first and for a shorter period has a greater impact than when it is given later and longer, when more cyst adenomas of the kidney (16/19 *versus* 2/18) were found; in either case, no renal-cell tumour was induced. This result may reflect a 'stronger' initiating activity of ochratoxin A than citrinin (Table 2), yet the latter experiments were done using male rat liver. Whether citrinin can be considered to enhance ochratoxin A-induced lesions of the kidney and thus presents a larger risk to humans remains to be confirmed.

Bendele *et al.* (1985)

Diets containing ochratoxin A at 0, 0.8 or 33.6 mg/kg were supplied to groups of 50 male and female B6C3F1 mice for two years. Tubular-cell tumours of the kidney were induced in male mice (adenomas, 0/50 controls *versus* 0/47 and 26/49 at the two dose levels, respectively; carcinomas, 0/50 *versus* 0/47 and 14/49). Cysts were present in the kidneys of most of the males given 33.6 mg/kg of diet; nephropathy was also seen, characterized by dilatation of renal tubules, often with hyperplasia of the lining epithelium (data on incidence were not given). The incidences of hepatocellular adenomas and carcinomas were increased in male (1/50 in controls and 8/47 and 10/50 at the two dose levels, respectively) and female (0/47 *versus* 2/45 and 7/49) mice. These experiments involved use of the lowest dietary exposure concentration of ochratoxin A that has been shown to cause cancer in laboratory animals: liver tumours were induced at less than 1 mg/kg of diet.

Boorman (1989)

Male and female F344 rats were given ochratoxin A in corn oil by gavage on five days per week for nine months (15 per gender per group), 15 months (15 per gender per group) and two years (50 per gender per group) at concentrations of 0, 21, 70 or 210 μg/kg body weight per day (Table 4). The kidney was the major organ in which chemically induced responses were seen. With one exception (karyomegaly in 8/51 females exposed for two years to 21 μg/kg), the lesions induced in the kidney were induced by the two highest doses. In the 9- and 15-month exposure studies, karyomegaly was seen in all rats fed 70 and 210 μg/kg, whereas epithelial hyperplasia was found in 15/60 males and in 4/60 females in these groups; an adenoma was found in one male given 210 μg/kg. Degeneration of renal tubular epithelium was diagnosed at 15 months in all rats given the highest dose, and two adenomas and three carcinomas were observed in 30 males given the two highest doses.

Table 4. Design and summary results of two-year experiments with ochratoxin A in male and female Fischer 344/N rats[a]

Exposure concentrations: 0, 21, 70 or 210 μg/kg per day five times/week
Route of exposure: oral intubation in corn oil vehicle
Mean body weights: highest dose groups exhibited 4–7% lower body weights between weeks 18–77 for males and weeks 6–89 for females.
Survival at 104 weeks: males, 39/50 *vs* 26/51; 26/51; 23/50 females, 32/50 *vs* 23/51; 35/50; 34/50
Nonneoplastic effects in renal tubular epithelium: degeneration, karyomegaly, proliferation, cytoplasmic alteration (males only) and hyperplasia
Neoplastic effects: renal tubular cell: cyst, adenoma, carcinoma, metastases (males only); mammary gland: fibroadenoma (females only)

[a]From Boorman (1989)

In rats of each sex exposed to ochratoxin A for two years, considerable nonneoplastic (degeneration, karyomegaly, proliferation, cytoplasmic alteration, hyperplasia) and neoplastic effects (cysts, adenomas, carcinomas) were observed in the kidneys. Metastases occurred in 17 males (with a total of 55 sites affected) and in one female rat. Degeneration and karyomegaly occurred in all animals given the two highest doses. Other kidney lesions (except for cysts) affected males more commonly than females (Table 5). Apart from karyomegaly in female rats, no lesion seen in the group given 21 μg/kg could be attributed to ochratoxin A. The incidences of both single and multiple fibroadenomas of the mammary gland were increased in female rats (single tumours were found in 17/50 controls and in 23/51, 22/50 and 28/50 rats in the exposed groups and multiple tumours in 4/50 and 4/51, 5/50 and 14/50). Carcinomas of the liver were found in one male control and in one male given 210 μg/kg; none was seen in females.

Summary of experimental results

Considering these long-term experiments collectively, the exposure levels associated with carcinogenicity in the liver (0.8 mg/kg for two years in male and female B6C3F1 mice) were considerably lower than those for the kidney (25 mg/kg for 70 weeks in male DDD mice). Even if total amount consumed is considered, the differences remain. This difference was not seen in male DDD mice, in which liver tumours were seen in groups with a total exposure of 33 mg/mouse (50 mg/kg for 20 weeks), and kidney tumours occurred with 29 mg/mouse (50 mg/kg for 15 weeks). In Fischer rats, liver tumours did not occur, and 70 μg/kg body weight per day was the lowest concentration that caused tumours of the kidney in males; for females, this level was 210 μg/kg. One adenoma of the kidney was seen, however, at nine months in a male

Table 5. Kidney lesions observed in two-year experiments with ochratoxin A in male and female Fischer 344/N rats[a]

	Males			Females		
	Controls	70 µg/kg	210 µg/kg	Controls	70 µg/kg	210 µg/kg
No. examined	50	51	50	50	50	50
Cytoplasmic alterations	0	3	8	0	1	2
Karyomegaly	0	51	50	0	50	50
Degeneration	0	50	49	0	49	49
Proliferation	0	10	26	0	3	16
Hyperplasia	1	16	24	0	12	13
Cyst	0	0	10	0	1	31
Adenoma	1	5	10	0	1	3
multiple	0	1	0	0	0	2
Carcinoma	0	12	20	0	1	3
bilateral/multipl	0	4	10	0	0	0
Metastases	0	4	13	0	1	0
Adenoma + carcinoma	1	20	36	0	2	8
Age-adjusted	1.3	26.7	43.6	0	2.2	11.5
Percentage	2.6	53.4	87.2	0	4.3	22.9

[a]From Boorman (1989)

Numerical incidence data indicate the number of animals with the stated lesion. The groups given 21 µg/kg are not included because most had no diagnosed lesion, with the exception of one male each with karyomegaly, hyperplasia and adenoma and eight females with karyomegaly. The age-adjusted incidence figures indicate the number of animals that would have had these tumours if all animals had survived to the end of week 104 of the experiment.

given 210 µg/kg; after 15 months, one benign and one malignant tumour were found in males given 70 µg/kg, and one adenoma and two carcinomas were seen in males receiving 210 µg/kg. Except in a single male fed 21 µg/kg, no hyperplasia was seen at doses below 70 µg/kg.

Thus, after exposure to ochratoxin A, mice appear to develop tumours of the liver at considerably lower doses and more readily than do rats, while rats are more responsive to lower exposure levels for tumours of the kidney; for example, mice receiving diets containing ochratoxin A at 40 mg/kg would have received about 4.5 mg/kg body weight per day or nearly 20–60 times the dose administered to rats by gavage. Furthermore, males are more prone than females to develop either tumour type. Ochratoxin A seems to fit the typical male–female ratio pattern, since chemicals that cause tumours of the kidney affect primarily male rats (Barrett & Huff, 1991; Huff *et al.*, 1991b).

Experiments considered to be inadequate

Several experiments conducted decades ago have limitations that make them inappropriate for evaluating the carcinogenicity of ochratoxin A (IARC, 1976, 1983, 1987); in every case cited, the findings indicated no evidence of carcinogenicity. All studies in which positive responses were obtained, even with similar limitations (e.g., too few animals or too short exposure), were considered to be relevant and were summarized above.

Dickens and Waynforth (1968) gave 10 male and 10 female mice 10-mg subcutaneous injections of ochratoxin A twice a week for 36 weeks; no tumour was reported at the end of 81 weeks.

Purchase and van der Watt (1971) gave groups of five male and five female Wistar rats ochratoxin A by gavage at doses of 0, 100 or 300 μg dissolved in a solution of sodium bicarbonate, on five days/week for 50 weeks, with an observation period up to 110 weeks. One rat [sex not stated] given 300 μg was reported to have a hamartoma of the kidney. The rarity of this tumour type might indicate a positive effect. Further, under these experimental conditions (the exposures were approximately equal to 6000 μg/kg body weight per day at the start of the experiment, since reportedly 50-g rats were used), tumours might have been expected to be induced by ochratoxin A. Another group of 10 female Wistar rats was given ochratoxin A at 2.5 mg/kg body weight by subcutaneous injection twice a week for 35 injections. Only a few injection-site fibrosarcomas were reported.

Discussion and Conclusions

Chronic nephropathy in humans has been associated with long-term exposure to ochratoxin A, on the basis of evidence of consumption of ochratoxin A-contaminated foods grown in areas endemic for Balkan (Bulgaria, Romania and Yugoslavia) nephropathy and of blood levels in people with this disease. Balkan endemic nephropathy is a chronic and uniformly fatal disease characterized morphologically by extensive tubulointerstitial damage (Cotran, 1981). Its incidence appears to exceed 350/100 000. Further, about one-third of the people who have died from Balkan endemic nephropathy have been found to have papillomas and/or carcinomas of the renal pelvis, ureter or urinary bladder (Castegnaro *et al.*, 1987). The disease occurs typically in people between the ages of 30 and 60 years, and women are more often affected at younger ages, although this difference in sex-associated incidence is thought to disappear later on (Cotran, 1981). Nikolov and Chernozemsky (1990) reported the most current figures for this disease and provided separate incidence rates per 100 000 population: these range from 96 in moderately affected regions to 315 in hyperendemic areas for men and from 208 to 506 for women. For transitional-cell carcinoma, the rates in the different areas are 10.2–43.5 for men and 21.3–74.2 for women.

In laboratory rodents, ochratoxin A causes cancers of the kidney, liver and mammary gland. Toxic, preneoplastic and neoplastic lesions of the kidney predominate following exposure to this commonly occurring mycotoxin (Kuiper-Goodman & Scott, 1989; Kuiper-Goodman, this volume). The primary location of these induced effects is the renal proximal convoluted tubular epithelium in the inner cortex and outer medulla. The tissue distribution of ochratoxin A, measured after a single intravenous injection, does not explain why the kidney is more responsive than other organs which receive as much or more ochratoxin A (e.g., liver, myocardium, bone marrow). Moreover, larger amounts of ochratoxin A appear in the renal medulla than in the kidney cortex (Emanuelsson *et al.*, this volume). Ochratoxin A has been shown to be mutagenic to *Salmonella typhimurium* strains TA1535, 1538, and 100 (but not to TA1537 or 98) following incubation with primary cultures of hepatocytes from uninduced female Wistar rats, and caused sister chromatid exchange in human peripheral lymphocytes exposed to hepatocyte-converted ochratoxin A (Hennig *et al.*, this volume). On the basis of SOS-DNA repair-inducing activity in *Escherichia coli* PQ37 strain, Malaveille *et al.* (this volume) propose that an ochratoxin A-derived free radical is the genotoxic intermediate, rather than a reduced oxygen species. Using ^{32}P-postlabelling, Pfohl-Leszkowicz *et al.* (this volume) observed DNA adducts in liver, kidney and spleen of mice. Much remains to be accomplished before we can decipher the mechanism of action of ochratoxin A (Dirheimer & Creppy, this volume) and of the greater majority of other carcinogenic chemicals (Vainio *et al.*, 1992).

In most species and strains, chemically associated responses in the kidney are more pronounced in males than in females (Barrett & Huff, 1991; Huff *et al.*, 1991b); the same is true in humans, whereby men have higher incidences of and greater mortality rates from kidney cancer than women (Ries *et al.*, 1990). Interestingly, in some contrast to the largely male-specific kidney responses in experimental animals, the humans most affected by the 'Balkan' disease are middle-aged women, with fewer men and almost no adolescents (Chernozemsky, this volume). In laboratory experiments, however, the longer durations of exposure may allow fuller expression of neoplastic lesions in females, given that tumours of the kidney appear late and the longest study reported lasted only two years. This length of time corresponds roughly to two-thirds to three-fourths of the average lifespan of Fischer rats, and this is roughly equivalent to 55–65 years of age in humans. Another possibility (discussed later) concerns the value of taking and examining multiple histopathological sections for better and more complete detection of these neoplastic lesions (frequently microscopic in size).

Nonetheless, the kidney lesions observed in these experiments appear to have substantiated and forecast the effects increasingly being seen in humans in regions where high levels of ochratoxin A have been consumed and are causally implicated (Castegnaro *et al.*, 1990). A potential confounding factor that requires further investigation is the apparent difference in histogenesis between ochratoxin A-induced epithelial kidney cancers in rodents (tubular cell) and the ochratoxin A-associated tumours in humans (transitional cell). However, the consistency of observations among species and strains and of experimental designs likewise indicates that ochratoxin A represents an inherent potential human cancer hazard. The responses in the kidney are significantly greater than those to most other chemicals that have been identified as causing cancer of the kidney in laboratory rodents (Barrett & Huff, 1991; Huff *et al.*, 1991b), and in particularly at such very low exposure levels.

Although the kidney receives 25% or more of the cardiac output and is a major site for the metabolism and excretion of xenobiotics, a surprisingly small number of chemicals actually cause cancer of the kidney; of the 394 chemicals evaluated in 1394 experiments conducted by the National Cancer Institute/National Toxicology Program, only 29 (7%) chemicals were considered to be carcinogenic for the kidney in at least one sex of one species. Only one chemical caused tumours of the kidney in female mice (nitrilotriacetic acid), four chemicals did so in male mice, nine did in female rats, and 25 in male rats (Huff *et al.*, 1991b). Thus, in less than 3% of the experiments (39 single sex–species effects out of 1394 units) was there a carcinogenic response in the kidney. Further, of these 29 chemicals, only one caused tumours of the kidney in animals of each sex in both species (nitrilotriacetic acid), three caused tumours of the kidney in three of the four sex–species experiments (bromodichloromethane, ochratoxin A [Bendele *et al.*, 1985, where female mice had no tumour of the kidney] and tris(2,3-dibromopropyl)-phosphate). Other chemicals that induce kidney tumours belong to an equally diverse group of structurally unrelated chemicals: metals, aromatic and heterocyclic amines and amides, furyls, nitroso and related compounds, mycotoxins, pharmaceuticals, organohalides (especially alkyl and alkenyl) and miscellaneous compounds (hormones, inorganic chemicals) (Kluwe *et al.*, 1984; Hard, 1987).

Regarding the importance of tissue sampling, the possibility clearly exist that the comparative significance of the numbers of animals with tumours of the kidney or the incidences of multiple tumours would have been enhanced (more in exposed groups) or diminished (more in control groups) if more tissue sections had been taken. This possibility is particularly important for risk assessment, in which the most 'sensitive' site of chemically induced carcinogenesis forms the basis of quantitative estimates of human risks. Typically, in the US National Toxicology Program studies, one section is taken from each kidney as well as from grossly visible lesions. Kurokawa *et al.* (1990), for example, observed a sizable enhancement in the already dramatic effects of potassium bromate on kidney cancer after examining 10–15 additional sections of kidney. Another interesting finding was confirmation (despite multiple sectioning) of the very low background rates for tubular-cell tumours in F344 rats (from 1.1% up to 6% with multiple sections), and in particular in females (0.1%) (Haseman *et al.*, 1985; J. Haseman, personal communication). Importantly, however, taking more kidney sections does not mean automatically that more microscopic lesions will be discovered. In a series of studies on nine chemicals, 4–6 additional sections per group were evaluated; the results showed no consistent pattern or trend. Thus, in some experiments the chemically associated effects were confirmed while in others the incidences were enhanced or decreased (e.g., more tumours were found in exposed groups or enough tumours were found in the control group to negate the effect in exposed groups) (see Huff, 1990; Kari, 1990; G. Boorman, personal communication).

The major lesson to be learned from taking additional sections from any organ, and in particular the kidney, is that one cannot predict with equal confidence whether the findings based on single sections will be confirmed, enhanced or eliminated. Of course, the results revealed by single sections are sometimes obvious (e.g., in the case of ochratoxin A), but at other times they may not be clear (e.g., toluene [Huff, 1990] and any chemical that induces a marginal increase in the incidence of tumours of the kidney). When neoplastic effects seen in long-term chemical carcinogenesis studies are confined to a single system, organ or tissue (e.g., with ochratoxin A, especially for female rats), one should consider obtaining additional sections to confirm the carcinogenic response. More knowledge can be gained about the true effect of the chemical, and this will permit better characterization of dose–response curves (at lower exposures) and of inherent potency, as well as allowing more precise quantitative risk estimations. For example, in the studies in rats reported by Boorman (1989), tumours might have been observed in the group given 21 μmg/kg or more tumours might have been found in female rats if more sections had been evaluated. This possibility is particularly important in the case of chemicals that cause cancer at a single exposure level in animals of one sex of one strain of one species. Moreover, such experimental enhancement would allow better ascertainment of a 'no-observable-carcinogenic-effect level'.

Another oft-made statement about lesions of the kidney (and of other organs) concerns the association between toxic effects and tumours (Huff, 1992). This hypothesis is based on the notion that chemically induced toxicity leads to an eventual carcinogenic effect rather than being an inherent activity of the chemical. In the kidney, for example, a spectrum of 'spontaneous' degenerative changes, termed 'nephropathy' or 'protein-overload nephropathy', is seen in virtually all rats at variable severity (Montgomery & Seely, 1990). Results from the study of Boorman (1989) are typical: chronic diffuse nephropathy occurred in 96–100% of male rats and in 70–92% of female rats, both control and exposed; graded severity was not reported. Thus, the spectrum of lesions was extant, yet not all animals exhibited tumours (or hyperplasia). Of the 201 rats with toxic lesions after exposures to the two upper doses, 'only' 56/101 males and 10/100 females had neoplasia. Further, there was no evidence of hyaline droplet nephropathy. Granted, the magnitude of the kidney cancer response in males in this study is unprecedented in National Toxicology Program studies, but not 'all' males and only 10% of females had tumours, even though all rats in these two exposure groups had cellular degeneration, karyomegaly and nephropathy. Other studies show this disparity more dramatically. Thus, even though chronic nephropathy occurs in all male and most female rats, the background rates of kidney tumours are extremely low. Nonetheless, overlying this issue is the potential biological impact of the typically observed chronic 'protein overload nephropathy'.

To overcome or minimize this key 'protein' confounding factor in long-term studies, Rao *et al.* (1991) proposed to reduce the protein content of the basal laboratory diet formulations by at least one-half, from the current level of 23–25% to about 12–15%. After feeding these diets to Fischer rats for two years, they observed a lowering of severity in control animals, from 3.3 (moderate to marked nephropathy) to 1.4 (minimal to mild) in male rats and from 1.6 (minimal to mild) to 1.1 (minimal) in female rats. Perhaps other confirmatory studies can now be done with this lower protein content, exposing animals to chemicals known to cause tumours of the kidney. In any event, control male rats typically show moderate-to-marked nephropathy (grades 3 to 4) at two years and rarely exhibit kidney tumours: the background rates for kidney tumours are approximately 1 in every 100 control male and 1 in 1000 control female Fischer rats.

The cardinal findings from the experiments with combinations of chemicals evoke a troublesome dilemma for public health, which is not often taken into account in risk management decisions. That is, most such efforts are directed at single chemicals and their potential for harm, whereas humans are exposed to an ocean of intentional and unintentional chemical risks that act with varying degrees of synergism. Studies are needed to better decipher the role and carcinogenic impact of habitual exposure to commonly encountered chemicals and life-style factors. Kanisawa and Suzuki (1978), Imaida *et al.* (1982) and Kanisawa (1984) designed the types of experiments that should be attempted to help answer these important public health issues. Few have been done (Arcos *et al.*, 1988). More are needed.

Renal cancers in humans have been associated with tobacco smoking, certain environmental and occupational factors (e.g., coke oven emissions and possibly rubber industry by-products) and therapeutic agents, particularly analgesic agents containing phenacetin (IARC,1987).

Arsenic has been suspected of causing kidney cancer (Tomatis *et al.*, 1989). Serious consideration might now be given to adding ochratoxin A to the list (although a causal relationship has not been established; Stefanovic & Polenakovic, 1991), and this chemical can likewise be included in the expanding list of chemicals first shown to cause cancer in laboratory animals and subsequently in humans (Tomatis *et al.*, 1989; Huff & Rall, 1991; IARC, 1991). Most of the causes of kidney cancer are not known, and the 20 000 new cases diagnosed each year in the USA alone represent an important human health problem (Barrett & Huff, 1991). The global importance of renal cancer to public health multiplies this concern, especially in areas of endemic renal diseases that may predispose to or act as cofactors with other agents to cause cancer. Thus, when natural and synthetic chemicals have been shown unequivocally to cause cancer in laboratory animals, human exposure to these chemicals should be reduced to the lowest levels that are technologically and agriculturally feasible, or, when expedient and necessary, exposure to these carcinogenic agents should be eliminated (Huff *et al.*, 1991a). Given the undeniable fact of Balkan endemic nephropathy and the associated tumorigenesis observed in humans, together with the nearly identical toxicity and carcinogenicity induced in laboratory animals by ochratoxin A at extremely low doses, the necessity for public health of keeping ochratoxin A (and other carcinogenic mycotoxins) out of foods consumed by humans is especially urgent.

Acknowledgements

I appreciate the helpful comments and valuable suggestions made by John Bucher, J. Carl Barrett, Ronald Melnick and Michael Elwell, and I am grateful to Elisabeth Heseltine for editing this paper.

References

Arcos, J.C., Woo, Y-T. & Lai, D.Y. (1988) Database on binary combination effects of chemical carcinogens. *Environ. Carcinogen. Rev.*, **C6**, 1–150

Barrett, J. C. & Huff J.E. (1991) Cellular and molecular mechanisms of chemically induced renal carcinogenesis. In: Bach, P.H., Gregg, N.J., Wilks, M.F. & Delacruz, L., eds, *Nephrotoxicity. Mechanisms, Early Diagnosis, and Therapeutic Management*, Chapter 45, New York, Marcel Dekker, pp. 287–306

Bendele, A.M., Carlton, W.W., Krogh, P. & Lillehoj, E.B. (1985) Ochratoxin A carcinogenesis in the (C57Bl/6J X C3H)F1 mouse. *J. Natl Cancer Inst.*, **75**, 733–742

Boorman, G.A., ed. (1989) *Toxicology and Carcinogenesis Studies of Ochratoxin A (CAS No. 303-47-9) in F344/N Rats (Gavage Studies)* (NTP Technical Report Series No. 358), Research Triangle Park, NC, National Toxicology Program, National Institute of Environmental Health Sciences

Castegnaro, M., Bartsch, H. & Chernozemsky, I. (1987) Endemic nephropathy and urinary tract tumours in the Balkans. *Cancer Res.*, **47**, 3608–3609

Castegnaro, M., Chernozemsky, I., Hietanen, E. & Bartsch, H. (1990) Are mycotoxins risk factors for endemic nephropathy and associated urothelial cancers? *Arch. Geschwulstforsch.*, **60**, 295–303

Cotran, R.S. (1981). Tubulointerstitial diseases. In: Brenner, B.M. & Rector, F.C., Jr, eds, *The Kidney*, 2nd ed., Chapter 29, Philadelphia, Saunders, pp. 1633–1667

Dickens, F. & Waynforth, H.B. (1968) Studies on carcinogenesis by lactones and related substances. *Rep. Br. Emp. Cancer Campaign*, **46**, 108 (Abstract)

Hard, G.C. (1987) Chemically induced epithelial tumours and carcinogenesis of the renal parenchyma. In: Bach, P.H. & Lock, E.A., eds, *Nephrotoxicity in the Experimental and Clinical Situation*, Chapter 7, Dordrecht, Martinus Nijhoff, pp. 211–250

Haseman, J.K., Huff, J.E., Rao, G.N., Arnold, J.E., Boorman, G.A. & McConnell, E.E. (1985) Neoplasms observed in untreated and corn oil gavage control groups of F344/N rats and (C57BL/6N x C3H/HeN)F1 (B6C3F1) mice. *J. Natl Cancer Inst.*, **75**, 975–984

Huff, J.E., ed. (1990) *Toxicology and Carcinogenesis Studies of Toluene (CAS No. 108-88-3) in F344/N Rats and B6C3F1 Mice (Inhalation Studies)*, Research Triangle Park, NC, National Toxicology Program, National Institute of Environmental Health Sciences

Huff, J.E. (1992) Chemical toxicity and chemical carcinogenesis. Is there a causal connection? In: Vainio, H., Magee, P., McGregor, D. & McMichael, A.J., eds, *Mechanisms of Carcinogenesis in Risk Evaluation* (IARC Scientific Publications No. 116), Lyon, IARC (in press)

Huff, J.E. & Hoel, D.G. (1991) Hazard identification. Perspective and overview on the concepts and value of the initial phase in the risk assessment process of cancer and human health. *Scand. J. Work Environ. Health* (in press)

Huff, J.E. & Rall, D.P. (1991) Relevance to humans of carcinogenesis results from laboratory animal toxicology studies. In: Last, J.M., ed., *Maxcy-Rosenau's Public Health and Preventive Medicine*, 13th Ed., Norwalk, CT, Appleton-Century-Crofts, pp. 433–440, 453–457

Huff, J.E., Haseman, J.K. & Rall, D.P. (1991a) Scientific concepts, value, and significance of chemical carcinogenesis studies. *Ann. Rev. Pharmacol. Toxicol.*, **31**, 621–652

Huff, J.E., Cirvello, J., Haseman, J.K. & Bucher, J.R. (1991b) Chemicals associated with site-specific neoplasia in 1394 long-term carcinogenesis experiments in laboratory rodents. *Environ. Health Perspect.*, **93**, 247–271

IARC (1976) *IARC Monographs on the Evaluation of the Carcinogenic Risk of Chemicals to Man*, Vol. 10, *Some Naturally Occurring Substances*, Lyon, pp. 191–197

IARC (1983) *IARC Monographs on the Evaluation of the Carcinogenic Risk of Chemicals to Humans*, Vol. 31, *Some Food Additives and Naturally Occurring Substances*, Lyon, pp. 191–206

IARC (1987) *IARC Monographs on the Evaluation of Carcinogenic Risks to Humans*, Suppl. 7, *Overall Evaluations of Carcinogenicity: An Updating of* IARC Monographs *Volumes 1 to 42*, Lyon, pp. 271–272

IARC (1991) Preamble. In: *IARC Monographs on the Evaluation of Carcinogenic Risks to Humans*, Vol. 53, *Occupational Exposures in Insecticide Spraying; and Some Pesticides*, Lyon, pp. 13–33

Imaida, K., Hirose, M., Ogiso, T., Kurata, Y. & Ito, N. (1982) Quantitative analysis of initiating and promoting activities of five mycotoxins in liver carcinogenesis in rats. *Cancer Lett.*, **16**, 137–143

Kanisawa, M. (1984) Synergistic effect of citrinin on hepatorenal carcinogenesis of ochratoxin A in mice. *Dev. Food Sci.*, **7**, 245–254

Kanisawa, M. & Suzuki, S. (1978) Induction of renal and hepatic tumours in mice by ochratoxin A, a mycotoxin. *Gann*, **69**, 599–600

Kari, F.W., ed. (1990) *Toxicology and Carcinogenesis Studies of Phenylbutazone (CAS No 50-33-9) in F344/N Rats and B6C3F1 Mice (Gavage Studies)*, Research Triangle Park, NC, National Toxicology Program, National Institute of Environmental Health Sciences

Kluwe, W.M., Abdo, K.M. & Huff, J.E. (1984) Chronic kidney disease and organic chemical exposures: evaluations of causal relationships in humans and experimental animals. *Fundam. Appl. Toxicol.*, **4**, 889–901

Kuiper-Goodman, T. & Scott, P.M. (1989) Risk assessment of the mycotoxin ochratoxin A. *Biomed. Environ. Sci.*, **2**, 179–248

Kurokawa, Y., Maekawa, A., Takahashi, M. & Hayashi, Y. (1990) Toxicity and carcinogenicity of potassium bromate—a new renal carcinogen. *Environ. Health Perspect.*, **87**, 309–335

Montgomery, C.A., Jr & Seely, J.C. (1990) Kidney. In: Boorman, G.A., Eustis, S.L., Elwell, M.R., Montgomery, C.A., Jr & MacKenzie, W.F., eds, *Pathology of the Fischer Rat*, Chapter 10, San Diego, Academic Press, pp. 127–153

Nikolov, I.G. & Chernozemsky, I.N. (1990) Balkan endemic nephropathy and transitional cell carcinoma: two fatal chronic diseases and the environment. *J. Environ. Pathol. Toxicol. Oncol.*, **10**, 317–320

Purchase, I.F.H. & van der Watt, J.J. (1971) The long-term toxicity of ochratoxin A to rats. *Food Cosmet.Toxicol.*, **9**, 681–682

Rao, G.N., Edmonson, J. & Elwell, M.R. (1991) Influence of dietary protein concentration on severity of nephropathy in Fischer 344 rats. *Toxicologist*, **11**, 139 (Abstract No. 481)

Ries, L.A.G., Hankey, B.F. & Edwards, B.K., eds (1990) *Cancer Statistics Review 1973–1987*, Bethesda, MD

Stefanović, V. & Polenaković, M.H. (1991) Balkan nephropathy. Kidney disease beyond the Balkans? *Am. J. Nephrol.*, **11**, 1–11

Tomatis, L., Aitio, A., Wilbourn, J. & Shuker, L. (1989) Human carcinogens so far identified. *Jpn. J. Cancer Res.*, **80**, 795–807

Vainio, H., Magee, P., McGregor, D. & McMichael, A.J., eds (1992) *Mechanisms of Carcinogenesis in Risk Evaluation* (IARC Scientific Publications No. 116), Lyon, IARC (in press)

WHO (1990) *Selected Mycotoxins: Ochratoxins, Trichothecenes, Ergot* (Environmental Health Criteria 105), Geneva

Mycotoxins, Endemic Nephropathy and Urinary Tract Tumours
Ed. M. Castegnaro, R. Pleština, G. Dirheimer, I.N. Chernozemsky & H. Bartsch
Lyon, International Agency for Research on Cancer, pp. 245-253

DNA ADDUCT FORMATION IN MICE TREATED WITH OCHRATOXIN A

A. Pfohl-Leszkowicz[1], K. Chakor[1], E.E. Creppy[2] & G. Dirheimer[1,3]

[1]Institute of Molecular and Cellular Biology, National Centre for Scientific Research, Strasbourg; and [2]Laboratory of Toxicology, University of Bordeaux II, Bordeaux, France

Summary

Several authors have reported the occurrence of renal and hepatic tumours in mice and rats exposed to ochratoxin A in long-term studies. The compound was not mutagenic, however, in various microbial and mammalian gene mutation assays, either with or without metabolic activation. Contradictory results were obtained for induction of unscheduled DNA synthesis and sister chromatid exchange. We showed previously that ochratoxin A causes DNA damage, manifested as single-strand breaks in mouse spleen cells and *in vivo*. These findings, which suggest that ochratoxin A is weakly genotoxic to mammalian cells, prompted us to search for DNA adducts using a modified ^{32}P-postlabelling method, the sensitivity of which was improved by treatment with nuclease P1. DNA was isolated from liver, kidney and spleen excised from mice 24, 48 and 72 h after oral treatment with ochratoxin A at 0.6, 1.2 and 2.5 mg/kg body weight. Several adducts were found in the DNA of the three organs, the levels varying greatly. After administration of 2.5 mg/kg body weight, 40 adducts per 10^9 nucleotides were found in kidney DNA and 7 adducts per 10^9 nucleotides in liver after 72 h. The levels of most of the adducts increased from 24 to 72 h, but those of others diminished after 24 or 48 h. Adducts were found in spleen only at 24 and 48 h. These results confirm the genotoxicity of ochratoxin A.

[3]To whom correspondence should be addressed

Introduction

Ochratoxin A is a mycotoxin produced by *Aspergillus ochraceus* and other moulds, notably *Penicillium viridicatum.* Ochratoxin A consists of a dihydroisocoumarin moiety linked through its 7-carboxyl group by an amide bond to one molecule of L-ß-phenylalanine. It has been shown to have a number of toxic effects, the most marked of which is nephrotoxicity (Krogh *et al.*, 1979). It has been implicated in Balkan endemic nephropathy, a disease characterized by a high incidence of end-stage renal failure, as it contaminates feedstuffs in areas of Romania, Bulgaria and Yugoslavia where the disease occurs (Chernozemsky *et al.*, 1977; Petkova-Bocharova & Castegnaro, 1985).

The first evidence for the carcinogenicity of ochratoxin A came from long-term carcinogenicity studies in mice in which liver and renal tumours were observed (Kanizawa & Suzuki, 1978; Kanizawa, 1984; Bendele *et al.*, 1985a). In a recent two-year study in rats (Boorman, 1989), renal adenomas and carcinomas were found at combined incidences of 20/51 in males and 2/50 in females at a dose of 70 µg/kg body weight per day. The significance of these ochratoxin A-induced renal carcinomas is increased by the high frequency of metastases.

The carcinogenicity of ochratoxin A was also suspected in humans because of the high incidence of kidney, pelvis, ureter and urinary bladder carcinomas among patients suffering from Balkan endemic nephropathy (Chernozemsky *et al.*, 1977; Castegnaro *et al.*, 1987; Petkova-Bocharova *et al.*, 1988).

The genotoxicity of ochratoxin A has been reviewed (Bendele *et al.*, 1985b; Kuiper-Goodman & Scott, 1989). It was shown to be nonmutagenic in various microbial and mammalian gene mutation assays, either with or without exogenous metabolic activation (Ueno & Kubota, 1976; Umeda *et al.*, 1977; Kuczuk *et al.*, 1978; Wehner *et al.*, 1978). Contradictory results were found for induction of unscheduled DNA synthesis and sister chromatid exchange (Cooray, 1984; Mori *et al.*, 1984; Bendele *et al.*, 1985b; Boorman, 1989). Some of our previous results drew attention to the possible genotoxicity of ochratoxin A, as expressed by DNA single-strand breaks (Creppy *et al.*, 1985; Kane *et al.*, 1986).

To understand better the carcinogenicity of ochratoxin A, we looked for ochratoxin A-DNA adducts, even though other authors had failed to find any. For this purpose, a modified ^{32}P-postlabelling method was used, the sensitivity of which was improved by nuclease P1 treatment (Reddy & Randerath, 1986).

Material and Methods

Ochratoxin A was purified from a culture of *Aspergillus ochraceus* strain NRRL/3174 on wheat, according to the method of Bunge *et al.*

(1979). Concentrations of the toxin were determined using an extinction coefficient of e = 5500 at 333 nm in methanol. After the methanol had been evaporated, ochratoxin A was dissolved in sterile 0.1 M $NaHCO_3$ pH 7.4.

Swiss male mice weighing 25 ± 2 g, aged seven weeks, were given ochratoxin A at 0.6, 1.2 and 2.5 mg/kg body weight by gastric intubation in 0.1 M $NaHCO_3$. They were sacrificed by decapitation after 24, 48 and 72 h, and spleen, kidney and liver were excised.

To isolate DNA, tissues were homogenized with polytron in 10 ml of 10 mM Tris-HCl, 50 mM ethylenediaminetetraacetic acid (EDTA), 0.5 M NaCl and 10 mM ß-mercaptoethanol pH 7. Then, 0.75 ml of 20% SDS was added, and the homogenate was incubated for 30–60 min at 65 °C; subsequently, 3.75 ml of 6 M potassium acetate were added. The reaction mixture was stored at 0 °C, which resulted in precipitation of most proteins within 30 min. After centrifugation for 30 min at 4 °C (2200 *g*), the DNA was found in the supernatant. It was precipitated overnight at –20 °C by adding two volumes of ethoxyethanol, collected on a glass rod and dissolved immediately in 0.1 M NaCl, 20 mM EDTA and 50 mM Tris-HCl pH 8 (SET). In order to remove all RNA from the DNA samples, 10 μg pancreatic RNase A and five units of T1 RNase were added to 150–300 μg of absorbing material (A 260 nm) after previous boiling of the RNases for 15 min to destroy DNases. After digestion for 2–3 h at 37 °C, 300 μl phenol saturated with 0.1 M Tris-HCl pH 7 were added, shaken for 30 min at room temperature and centrifuged. The aqueous supernatant was taken, and two volumes of ethoxyethanol were added. After 2 h at –20 °C, the precipitate was collected by centrifugation and redissolved in 300 μl SET. A second precipitation was done with 600 μl ethoxyethanol for 2 h at –20 °C. The DNA pellet was then washed twice with SET:ethoxyethanol (1:9, v/v).

The method used for ^{32}P-postlabelling was that previously described by Randerath *et al.* (1981), modified by Reddy and Randerath (1986). Briefly, digested DNA is treated with nuclease P1 before ^{32}P-postlabelling. Nuclease P1 dephosphorylates 3′-monophosphate of normal nucleotides, whereas adducts protect this binding. Only adducted nucleotides are labelled by T4 polynucleotide kinase. Normal nucleosides and pyrophosphate were removed by chromatography on polyethyleneimine-cellulose in 2.3 M NaH_2PO_4 pH 5.7 (D1) overnight. Origin areas containing labelled adducted nucleotides were cut and transferred to another polyethyleneimine-cellulose plate in 4.25 M lithium formate and 7.5 M urea pH 3.35 (D2). Two further migrations (D3 and D4) were performed perpendicular to D2. The solvent for D3 was 0.6 M NaH_2PO_4 and 5.9 M urea pH 6; and the solvent for D4 was 1.6 M NaH_2PO_4 pH 6. Autoradiography was carried out at –80 °C for three days' exposure in the presence of an intensifying screen (Cronex). Spots were scraped off, and their radioactivity was counted by the Cerenkov technique.

Results

DNA adducts were found in kidney, liver and spleen of mice treated with ochratoxin A. The pattern of these adducts in kidney, the main target organ, is shown in Figure 1. The predominant spots, designated A to J, were grossly dose-dependent and time-related, especially in liver and kidney but with great variation for individual spots. In spleen, DNA adducts were detected only at 24 h at all doses and were still present at 48 h only with the highest dose.

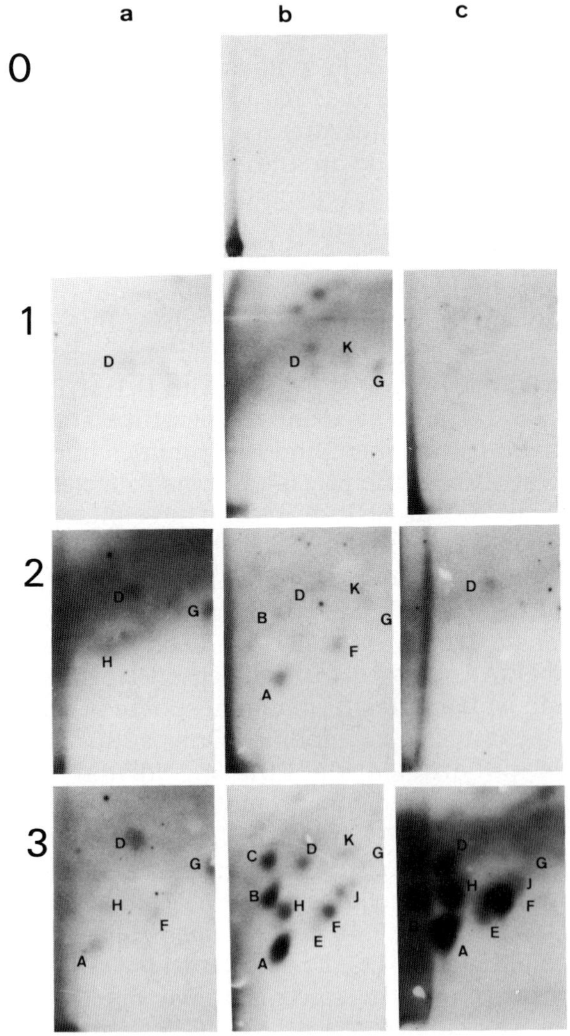

Figure 1. Autoradiograph of mouse kidney DNA after treatment with ochratoxin A at (1) 0.6 mg/kg; (2) 1.2 mg/kg or (3) 2.5 mg/kg at (a) 24 h, (b) 48 h and (c) 72 h

The total amounts of DNA adducts obtained with the different doses of ochratoxin A in the three organs are shown in Figure 2. After administration of 0.6 or 1.2 mg/kg, most of the adducts had disappeared from kidney by 72 h (Figure 2A), indicating that DNA repair had taken place. This repair mechanisms did not seem to operate at higher doses of ochratoxin A (2.5 mg/kg), implying that ochratoxin A impairs the repair. In liver (Figure 2B), adducts had already appeared after 24 h with all doses, but the level increased up to 72 h. Adduct levels were similar at all doses. In this organ, no repair appeared to occur; nevertheless, it should be noted that the total amount of adducts was 5.3 times lower in liver than in kidney after 72 h with the highest dose of ochratoxin A. In spleen (Figure 2C), adducts also appeared at 24 h. The repair mechanism appears to be very effective, as no more adducts were detected after 48 h.

Quantitative estimates of the different adducts obtained after administration of 2.5 mg/kg ochratoxin A are shown in Table 1. The major adducts are not the same in the three organs: In kidney, the predominant adduct is A, in liver it is C and in spleen X1. Some adducts predominate in two organs, e.g., B in kidney and liver (about 20%). The adducts that represent individually more than 10% of the total adducts after 72 h are A > B > C > F > H in kidney and C > B > D > G > K in liver. Thus, only B and C are major adducts in both liver and kidney. It is striking that the amount of individual adducts is 2–20 fold higher in the kidney than in the liver.

In kidney, some adducts (C, F, J and K) were not detected after 24 h; the levels of some increased regularly from 24 to 72 h, and some increased exponentially (A, B, F and H). One was apparent only after 48 h (K), and a small one reached a maximal level at 24 h and decreased thereafter (G). One was detected only in kidney (J), appearing at 48 h and increasing at 72 h.

A similar phenomenon was seen in liver, but, in contrast to kidney, adduct G appeared only after 72 h and adduct K was found already after 24 h. This was the major adduct present at 24 and 48 h. This adduct occurred at a higher level in liver than in kidney, in contrast to all other adducts. It was, however, maximal at 48 h, as in the kidney, and the level decreased from 48 to 72 h whereas the levels of most of the other adducts (except F) increased from 48 to 72 h.

Only a few adducts were seen in spleen, each with a maximum at 24 h. The two major adducts in spleen, X1 and X2, were not found in liver or kidney.

Discussion

DNA adducts were found for the first time in this study after ochratoxin A treatment. This may be due to improvement of the ^{32}P-postlabelling technique by Reddy and Randerath (1986), which

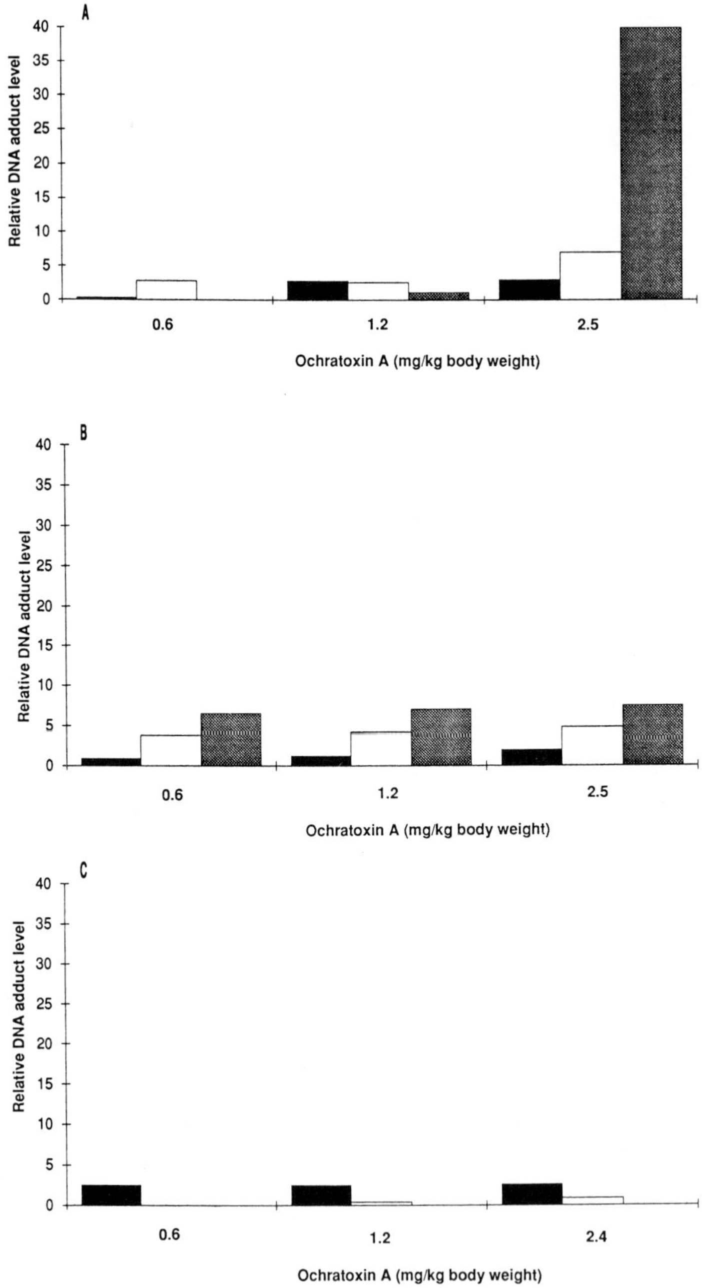

Figure 2. Total DNA adduct levels in kidney (A), liver (B) and spleen (C) after treatment with three doses of ochratoxin A

Levels were determined at three times: black bars, 24 h; white bars, 48 h; hatched bars, 72 h

Table 1. Relative DNA adduct levels in mouse kidney, liver and spleen after administration of a single dose (2.5 mg/kg body weight) of ochratoxin A

Organ	Time (h)	Adduct (no./10^9 nucleotides)											
		A	B	C	D	E	F	G	H	J	K	X1	X2
Kidney	24	0.31	0.3	0	1.47	0.2	0	0.5	0.24	0	0	0	0
	48	1.02	1.06	1.71	1.78	0.3	0.59	0.2	0.62	0.45	0.38	0	0
	72	13.7	7.55	5.21	2.45	1.48	4.56	0.1	3.88	1.4	0	0	0
Liver	24	0.2	0.2	0.2	0.48	0.1	0.2	0	0.1	0	0.59	0	0
	48	0.6	0.63	0.2	0.6	0.14	0.14	0	0.4	0	1.2	0	0
	72	0.7	1.57	1.83	0.87	0.34	0.1	0.79	0.5	0	0.74	0	0
Spleen	24	0.1	0	0.2	0.5	0	0	0	0	0	0	1	0.8
	48	0	0	0.1	0.1	0	0	0	0	0	0	0.5	0.2
	72	0	0	0	0	0	0	0	0	0	0	0	0

permits detection of 1–2 adducts/10^9 nucleotides, whereas the classical method of Randerath *et al.* (1981) allows detection of only one adduct/10^7 nucleotides. As the maximal level we found for an individual adduct was 14/10^9 nucleotides (i.e., 0.14/10^7 nucleotides)(spot A in kidney DNA), it is not surprising that the classical method cannot show these adducts. The DNA adducts induced by ochratoxin A did not occur in control animals, indicating that there was no interference with the feed intake of other genotoxic mycotoxins. The structures of these DNA adducts remain to be determined, and this will not be a easy task because of the high diversity and low amounts of adducts. These results confirm the genotoxicity of ochratoxin A, suggested previously by the finding of single-strand breaks both *in vitro* and *in vivo* (Creppy *et al.*, 1985; Kane *et al.*, 1986). The finding of many more adducts in kidney than in liver or spleen correlates well with the results of carcinogenicity studies, as stated above. The finding that the adducts are not the same in the three organs tested might be due to different metabolism in the three organs, leading to different ultimate carcinogens.

Some of the adducts disappeared early, whereas others appeared only after 24 h (C, F, J, K, for example, in kidney). This sequential appearance might be due to metabolism of the adduct on the DNA or to early repair of some adducts, so that only some of the ultimate carcinogens would be formed and would act at a late stage. With the highest dose used (2.5 mg/kg), the adduct levels in kidney and liver continued to rise from 24 to 72 h. It will be interesting to follow up the adduct level after 72 h. More investigations *in vivo* and *in vitro* are now needed to identify how metabolism in different tissues leads to different adducts, the nature of the modified bases and the nature of the compounds bound to the DNA in order to better understand the mechanism of ochratoxin A-induced genotoxicity and carcinogenicity.

Acknowledgments

This research was supported by grants from the Région Aquitaine, University of Bordeaux II, Direction de la Recherche, and the Ligue Nationale contre le Cancer, Comités Départementaux de la Gironde et du Haut-Rhin, Ministère de la Recherche et de la Technologie and INSERM 89/2007.

References

Bendele, A.M., Carlton, W.W., Krogh, P. & Lillehoj, E.B. (1985a) Ochratoxin A carcinogenesis in the (C57BL/6J x C3H)F1 mouse. *J. Natl Cancer Inst.*, **75**, 733–742

Bendele, A.M., Neal, S.B., Oberly, T.J., Thompson, C.Z., Bewsey, B.J., Hill, L.E., Rexroat, M.A., Carlton, W.W. & Probst, G.S. (1985b) Evaluation of ochratoxin A for mutagenicity in a battery of bacterial and mammalian cell assays. *Food Chem. Toxicol.*, **23**, 911–918

Boorman, G., ed. (1989) *NTP Technical Report on the Toxicology and Carcinogenesis Studies of Ochratoxin A (CAS NO. 303-47-9) in F344/N rats (Gavage Studies)* (NIH Publication No. 89-2813), Research Triangle Park, NC, US Department of Health and Human Services, National Institutes of Health

Bunge, I., Dirheimer, G. & Röschenthaler, R. (1978) *In vivo* and *in vitro* inhibition of protein synthesis in *Bacillus stearothermophilus* by ochratoxin A. *Biochem. Biophys. Res. Commun.*, **83**, 398–405

Castegnaro, M., Bartsch, H. & Chernozemsky, I. (1987) Endemic nephropathy and urinary tract tumors in the Balkans. *Cancer Res.*, **47**, 3608–3609

Chernozemsky, I.N., Stoyanov, I.S., Petkova-Bocharova, T.K., Nocolov, I.G., Draganov, I.G., Stoichev, I., Tanchev, Y., Naidenov, D. & Kalcheva, N.D. (1977) Geographic correlation between the occurrence of endemic nephropathy and urinary tract tumours in Vratza district, Bulgaria. *J. Cancer*, **19**, 1–11

Cooray, R. (1984) Effects of some mycotoxins on mitogen-induced blastogenesis and SCE frequency in human lymphocytes. *Food Chem. Toxicol.*, **22**, 529–534

Creppy, E.E., Kane, A., Dirheimer, G., Lafarge-Frayssinet, C., Mousset, S. & Frayssinet, C. (1985) Genotoxicity of ochratoxin A in mice: DNA single-strand break evaluation in spleen, liver and kidney. *Toxicol. Lett.*, **28**, 29–35

Kane, A., Creppy, E.E., Roth, A., Röschenthaler, R. & Dirheimer, G. (1986) Distribution of the [3H]-label from low doses of radioactive ochratoxin A ingested by rats, and evidence for DNA single-strand breaks caused in liver and kidneys. *Arch. Toxicol.*, **58**, 219–224

Kanizawa, M. (1984) Synergistic effect of citrinin on hepatorenal carcinogenesis of ochratoxin A in mice. In: Kurata, H. & Ueno, Y., eds, *Toxigenic Fungi, Their Toxins and Health Hazard* (Developments in Food Science No. 7), Amsterdam, Elsevier, pp. 245–254

Kanizawa, M. & Suzuki, S. (1978) Induction of renal and hepatic tumors in mice by ochratoxin A, a mycotoxin. *Gann*, **69**, 599–600

Krogh, P., Elling, F., Friis, C.H.R., Mald, B., Larsen, A.E., Lillehoj, E.B., Madsen, A., Mortensen, H.P., Rasmussen, F. & Ravuskou, U. (1979) Porcine nephropathy induced by long-term ingestion of ochratoxin A. *Vet. Pathol.*, **16**, 466–475

Kuczuk, M.H., Benson, P.M., Heath, H. & Hayes, W. (1978) Evaluation of the mutagenic potential of mycotoxins using *Salmonella typhimurium* and *Saccharomyces cerevisiae*. *Mutat. Res.*, **53**, 11–20

Kuiper-Goodman, T. & Scott, P.M. (1989) Risk assessment of the mycotoxin ochratoxin A. *Biomed. Environ. Sci.*, **2**, 179–248

Mori, H., Kawai, K., Ohbayashi, F., Kuniyasu, T., Yamazaki, M., Hamasaki, T. & Williams, G.M. (1984) Genotoxicity of a variety of mycotoxins in the hepatocyte primary culture/DNA repair test using rat and mouse hepatocytes. *Cancer Res.*, **44**, 2918–2923

Petkova-Bocharova, T. & Castegnaro, M. (1985). Ochratoxin A contamination of cereals in an area of high incidence of Balkan endemic nephropathy in Bulgaria. *Food Addit. Contam.*, **2**, 267–270

Petkova-Bocharova, T., Chernozemsky, I.N. & Castegnaro, M. (1988) Ochratoxin A in human blood in relation to Balkan endemic nephropathy and urinary system tumours in Bulgaria. *Food Addit. Contam.*, **5**, 299–301

Randerath, K., Reddy, M.V. & Gupta, R.C. (1981) ^{32}P-Labeling test for DNA damage. *Proc. Natl Acad. Sci. USA*, **78**, 6126–6129

Reddy, M.V. & Randerath, K. (1986) Nuclease P1-mediated enhancement of sensitivity of ^{32}P-postlabeling test for structurally diverse DNA adducts. *Carcinogenesis*, **7**, 1543–1551

Ueno, Y. & Kubota, K. (1976) DNA-attacking ability of carcinogenic mycotoxins in recombination-deficient mutant cells of *Bacillus subtilis*. *Cancer Res.*, **36**, 445–451

Umeda, M., Tsutsui, T. & Saito, M. (1977) Mutagenicity and inducibility of DNA single-strand breaks and chromosome aberrations by various mycotoxins. *Gann*, **68**, 619–625

Wehner, F.C., Thiel, P.G., van Rensburg, S.J. & Demasius, I.P.C. (1978) Mutagenicity to *Salmonella typhimurium* of some *Aspergillus* and *Penicilium* mycotoxins. *Mutat. Res.*, **58**, 193–203

Mycotoxins, Endemic Nephropathy and Urinary Tract Tumours
Ed. M. Castegnaro, R. Pleština, G. Dirheimer, I.N. Chernozemsky & H. Bartsch
Lyon, International Agency for Research on Cancer, pp. 255-260

MUTAGENICITY AND EFFECTS OF OCHRATOXIN A ON THE FREQUENCY OF SISTER CHROMATID EXCHANGE AFTER METABOLIC ACTIVATION

A. Hennig[1], J. Fink-Gremmels[2] & L. Leistner[1]

[1]Federal Institute for Meat Research, Kulmbach, Germany; and [2]Faculty of Veterinary Medicine, Department of Veterinary Pharmacology, Pharmacy and Toxicology, Utrecht, The Netherlands

Summary

Primary cultures of hepatocytes derived from untreated rats were incubated in the presence of ochratoxin A for 24 h. Five different strains of histidine auxotroph *Salmonella typhimurium* were exposed to conditioned cell culture medium before being tested for mutagenicity. A clear hepatocyte-mediated mutagenic response was observed in TA1535, TA1538 and TA100. In addition, sister chromatid exchange frequency was increased in human peripheral lymphocytes that had been incubated in the presence of conditioned medium derived from ochratoxin A-exposed hepatocytes.

Introduction

Previous investigations using various short-term tests gave inconsistent results on the genotoxicity of ochratoxin A, and its biotransformation has not been elucidated in detail; therefore, the possible contribution of ochratoxin A metabolites to its genotoxicity and carcinogenicity remains unclear. As primary cultures of rat hepatocytes have proven to be a convenient model for studying metabolism-mediated toxicity, we conducted two selected short-term genotoxicity tests *in vitro*—the *Salmonella* mutagenicity assay and the sister chromatid exchange test in human lymphocytes—using the cell culture medium of ochratoxin A-exposed hepatocytes.

Experimental

Chemicals

Crystalline ochratoxin A with a purity of 99.8% was obtained from Sigma (Munich). Chemicals used as reference compounds for the *Salmonella*/microsome assay—*N*-methyl-3-nitro-1-nitrosoguanidine, 9-aminoacridine and fluorene-2-amine—were obtained from Aldrich (Steinheim) and Schuchard (Munich). RPMI 1640 cell culture medium, histopaque 1077, 5-bromo-2´-deoxyuridine and colchicine were purchased from Sigma (Munich); HEPES and fetal calf serum from Biochrom (Frankfurt), phytohaemagglutinin M from Gibco (New York), Aroclor 1254 from Supelco (Bellefonte) and Hoechst 33258 bisbenzimide-fluorescence dye from Hoechst (Frankfurt).

Salmonella/microsome assay

The histidine auxotrophs of *S. typhimurium* strains TA1535, TA1537, TA1538, TA98 and TA100 (BAE, Karlsruhe) were checked prior to testing for *rfa* and *uvrB* mutation and for the presence of the pKM 101 plasmid in TA98 and TA100. The mutability of the selected strains was tested with positive reference compounds with and without metabolic activation, as described by Ames *et al.* (1973). Overnight cultures containing 1–2 x 10^8 microorganisms prepared from stock cultures were used for all assays. Ochratoxin A was dissolved in 0.1 ml dimethyl sulfoxide to final concentrations of 0.5, 5.0, 50 and 1000 nmol. The assay was carried out as described by Ames *et al.* (1975).

For the microsome assay, ochratoxin A (0.5, 5.0, 50 and 1000 nmol/plate) was preincubated for 30 min with 0.5 ml of S9 homogenate derived from Aroclor 1254-pretreated male Wistar rats and diluted with a NADPH-generating system, as described elsewhere (Ames *et al.*, 1973). The amount of microsomal protein added per plate was 1.5 mg.

Primary hepatocyte cultures

Hepatocytes were isolated from livers of uninduced female Wistar rats by the method of Butterworth *et al.* (1987). Cells were maintained in RPMI 1640 medium supplemented with 20 mmol/l HEPES, 2 mmol/l L-glutamine and 10 vol% of heat-inactivated fetal calf serum. Cell viability was checked by means of trypan blue exclusion. Only preparations with a cell viability exceeding 95% were used. Cells were seeded at a density of 3 x 10^6/10 ml in cell culture flasks and maintained at 37 °C in a 5% CO_2 air atmosphere. Ochratoxin A predissolved in dimethyl sulfoxide was added at various concentrations as described in the results. After 24 h of exposure to the toxin, cell-free medium was collected from the hepatocyte monolayer.

Salmonella/hepatocyte mutation test

For the modified *Salmonella*/hepatocyte mutation test, 0.2 ml of overnight cultures of the different *Salmonella* strains were diluted with 2 ml of medium derived from hepatocytes exposed to ochratoxin A (200 nmol/2 ml) and were incubated at 37 °C for 2 h. The bacteria were separated by centrifugation and reconstituted in 0.1 ml sterile nutrient broth before being added to the top agar, as in the standard plate incorporation assay. Results were read after 48 h.

Sister chromatid exchange assay

Peripheral lymphocytes were separated from venous blood from healthy nonsmokers of each sex, aged 25–40 years, using histopaque 1077 according to the method described by Boyum (1968). Cells were suspended at a density of 10^6 cells/ml in 10 ml RPMI 1640 culture medium containing 20 mmol/l HEPES, 2 mmol/l L-glutamine, 2 mmol/l glutathione, 10 vol% of heat-inactivated fetal calf serum and 2 vol% of reconstituted phytohaemagglutinin M and incubated in the presence of 10^{-5} mol/l 5-bromo-2′-deoxyuridine for 24 h in the dark (37 °C, 5% CO_2 atmosphere). Ochratoxin A was added to the culture at concentrations of 0.001, 0.01, 0.1, 1 and 10 μmol/l, followed by an incubation period of 48 h to allow two cycles of cell replication. For sister chromatid exchange analysis, 2 μg/ml colchicine were added to the culture for metaphase arrest 4 h prior to cell harvesting. Chromosomes were prepared as described by Moorhead *et al.* (1960), and chromatids were stained differentially using the fluorescence plus Giemsa technique of Wolff and Perry (1974). A total of 50 mitotic figures in the secondary metaphase from each culture stage were scored for sister chromatid exchange. In addition, the mitotic index was estimated from 10^3 cells, by comparing the rates of the first and second cell cycles.

In a second series of experiments, we measured the effect of hepatocyte-converted ochratoxin A on the induction of sister chromatid exchange. Thus, 5 ml of conditioned medium derived from ochratoxin A-exposed hepatocytes were added to 5 ml lymphocyte medium after the 24 h of culture time necessary to stimulate lymphocyte blastogenesis. The mycotoxin concentrations tested represent 0.1, 1 and 10 μmol/l of ochratoxin A or its metabolites.

Results

Bacterial mutagenicity

Ochratoxin A was not mutagenic, with or without metabolic activation, as reported previously (Bendele *et al.*, 1985). When ochratoxin A was metabolized by primary rat hepatocytes, however, the conditioned culture medium induced a clear mutagenic response in strains TA1535, TA1538 and TA100, indicating that hepatocytes convert ochratoxin A into bacterial mutagens (Table 1). This effect was not due to any acute toxicity of ochratoxin for the hepatocytes, as seen by measurements of lactic dehydrogenase release.

Table 1. Mutation induced in *Salmonella typhimurium* strains after activation of ochratoxin A by primary cultures of rat hepatocytes[a]

Dose of ochratoxin A-converted hepatocytes (nmol/plate)	*Salmonella* strain				
	TA1535	TA1537	TA1538	TA98	TA100
0	70.5	12.5	87.5	86.5	110.0
200	427.5	19.0	303.0	87.5	420.0

[a]*Salmonella* strains were preincubated for 2 h in the presence of cell-free supernatant of ochratoxin A-exposed hepatocytes, as described. Values are means of *his*+ revertants per plate of two separate experiments ($r > 0.98$).

Induction of sister chromatid exchange in human lymphocytes

The frequency of sister chromatid exchange in phytohaemagglutinin M-stimulated peripheral blood lymphocytes was elevated by a concentration of 0.1 µmol/l ochratoxin A. Increasing the concentration to 1 µmol/l reduced the mitotic index, and complete cytotoxicity was seen at a concentration of 10 µmol/l (Table 2).

When the assay was conducted in the presence of culture medium from hepatocytes exposed to 0.1 µmol/l ochratoxin A, the frequency of sister chromatid exchange was considerably increased. Conditioned medium from hepatocytes exposed to 1 µmol/l ochratoxin A had a mitostatic effect, since only cells in the first mitotic cycle were observed after 72 h culture time. Medium from hepatocytes exposed to 10 µmol/l ochratoxin A was completely cytotoxic (Table 2).

Discussion

Several protocols have been designed to study the influence of metabolic activation of xenobiotics on their genotoxicity. In the present study, we compared the mutagenic effect of microsomally activated ochratoxin A and that of hepatocyte-converted ochratoxin A in a modified

Table 2. Induction of sister chromatid exchange in human peripheral lymphocytes by ochratoxin A and by conditioned medium derived from hepatocyte monolayers exposed to ochratoxin A[a]

Ochratoxin A (µmol/L)	Without conversion by hepatocytes	Ratio (%) 1:2 mitosis	With conversion by hepatocytes	Ratio (%) 1:2 mitosis
0	0	32:68	0	44:56
0.001	0.5 ± 0.4	40:60	ND	ND
0.01	1.0 ± 0.7**	42:58	ND	ND
0.1	1.4 ± 0.8***	28:72	2.2 ± 0.5***	76:24
1.0	1.0 ± 0.5	35:65	Mitostatic	100:0
10.0	Cytotoxic	–	Cytotoxic	–

[a]Spontaneous sister chromatid exchange rates of individual controls have been subtracted. Values are means ± SD from a total of 100 metaphases in two independent experiments. $^{**}p < 0.01$; $^{***}p < 0.001$; ND, not determined

Salmonella assay. As found previously, ochratoxin A was not mutagenic after microsomal activation; however, the culture medium of rat hepatocytes exposed to 100 µmol/l ochratoxin A induced a clear mutagenic response in *Salmonella* type strains sensitive to frameshift mutagens and to base-pair substitutions. In addition, an increase in sister chromatid exchange rate and a mitostatic effect were observed in human peripheral lymphocytes that had been cultured in the presence of medium conditioned by a hepatocyte monolayer exposed to ochratoxin A.

The chemical nature of the ochratoxin A metabolites generated by hepatocytes has not yet been elucidated. However, these metabolites apparently escape cellular detoxification and are excreted by cultured cells into the culture medium.

Previous experiments demonstrated that ochratoxin A is metabolized by liver microsomes from various species to 4(R)- and 4(S)-hydroxyochratoxin A as well as to 10-hydroxyochratoxin A and ochratoxin α (through cleavage of the amide bond) (Stormer *et al.*, 1981, 1983). 4(R)-4-hydroxyochratoxin A was also the first metabolite to be generated by isolated hepatocytes (Hansen *et al.*, 1982). As ochratoxin A has given negative results in the *Salmonella*/mammalian microsome assay, however, as reported both by us and others, it seems unlikely that these metabolites are responsible for the observed mutagenic effects and for the induction of sister chromatid exchange in human lymphocytes in the presence of hepatocyte-converted ochratoxin A.

Inclusion in genotoxicity screening tests of an 'activating' system consisting of intact hepatocytes makes it possible for biotransformation of xenobiotics to proceed under more physiological conditions than with S9 mix. This system might therefore provide additional information which might contribute to a closer understanding of the carcinogenicity induced by ochratoxin A in rats *in vivo* (Boorman, 1989).

References

Ames, B.N., Lee, F.D. & Durston, W.E. (1973) An improved bacterial test system for the detection and classification of mutagens and carcinogens. *Proc. Natl. Acad. Sci. USA*, **70**, 782–786

Ames, B.N., McCann, J. & Yamasaki, E. (1975) Methods for detecting carcinogens and mutagens with the Salmonella/mammalian-microsome mutagenicity test. *Mutat. Res.*, **31**, 347–364

Bendele, A.M., Neal, S.B., Oberly, T.J., Thompson, C.Z., Bewsey, B.J., Hill, L.E., Rexroat, M.A., Carlton, W.W. & Probst, G. (1985) Evaluation of ochratoxin A for mutagenicity in a battery of bacterial and mammalian cell assays. *Food Chem. Toxicol.*, **23**, 911–918

Boorman, G., ed. (1989) *Toxicology and Carcinogenesis Studies of Ochratoxin A in F344/N Rats (Gavage Studies)* (NTP Technical Report 358), Research Triangle Park, NC, National Toxicology Program

Boyum, A. (1968) Isolation of leucocytes from human blood. Further observations. *Scand. J. Clin. Invest.*, **21** (Suppl. 97), 31–50

Butterworth, B.E., Ashby, J., Bermudez, E., Casciano, D., Mirsalis, J., Probst, G. & Williams, G. (1987) A protocol and guide for the in vitro rat hepatocyte DNA-repair assay. *Mutat. Res.*, **189**, 113–121

Hansen, C.E., Dueland, S., Drevon, C.A. & Stormer, F.C. (1982) Metabolism of ochratoxin A by primary cultures of rat hepatocytes. *Appl. Environ. Microbiol.*, **43**, 1267–1271

Moorhead, P.S., Nowell, P.C., Mellman, W.J., Battips, D.M. & Hungerford, D.A. (1960) Chromosome preparations of leucocytes cultured from human peripheral blood. *Exp. Cell Res.*, **20**, 613–616

Stormer, F.C., Hansen, C.E., Pedersen, J.I., Hvistendahl, G. & Aasen, A.J. (1981) Formation of (4R)- and (4S)-4-hydroxyochratoxin A and 10-hydroxyochratoxin A from ochratoxin A by liver microsomes from various species. *Appl. Environ. Microbiol.*, **42**, 1051–1056

Stormer, F.C., Storen, O., Hansen, C.E., Pedersen, J.I. & Aasen, A.J. (1983) Formation of (4R)- and (4S)-4-hydroxyochratoxin A and 10-hydroxyochratoxin A by rabbit liver microsomes. *Appl. Environ. Microbiol.*, **45**, 1183–1187

Wolff, S. & Perry, P. (1974) Differential Giemsa staining of sister chromatids and the study of sister chromatid exchanges without radiography. *Chromosoma*, **48**, 341–353

Mycotoxins, Endemic Nephropathy and Urinary Tract Tumours
Ed. M. Castegnaro, R. Pleština, G. Dirheimer, I.N. Chernozemsky & H. Bartsch
Lyon, International Agency for Research on Cancer, pp. 261-266

GENOTOXICITY OF OCHRATOXIN A AND STRUCTURALLY RELATED COMPOUNDS IN *ESCHERICHIA COLI* STRAINS: STUDIES ON THEIR MODE OF ACTION

C. Malaveille, G. Brun & H. Bartsch

International Agency for Research on Cancer, Lyon, France

Summary

Ochratoxin A, ochratoxin α (its major metabolite in rodents) and seven structurally related substances were assayed for SOS DNA repair inducing activity in *Escherichia coli* PQ37 strain. At a concentration range of 0.1–4 mM, ochratoxin A, chloroxine, 5-chloro-8-quinolinol, 4-chloro-*meta*-cresol and chloroxylenol were found to induce SOS-DNA repair in the absence of an exogenous metabolic activation system. Ochratoxin B, ochratoxin α, 5-chlorosalicylic acid and citrinin were inactive, but all except ochratoxin α were cytotoxic. Thus, the presence of a chlorine at C-5 in ochratoxin A and in other analogues appears to be one determinant of their genotoxicity. In order to ascertain whether this reactivity involves a bacterial glutathione conjugation reaction, we investigated the modifying effect on the genotoxicity of ochratoxin A of amino oxyacetic acid, an inhibitor of cysteine conjugate ß-lyase. Amino oxyacetic acid decreased the cytotoxicity of ochratoxin A but did not alter its genotoxic activity, suggesting the formation of a cytotoxic thiol-containing derivative. The way in which ochratoxin A and some of its active analogues induce SOS DNA repair activity was further investigated in *E. coli* PQ37 and in three derived strains (PQ300, OG100 and OG400, containing deletions within the *oxy R* regulon). The response in PQ37 strain was measured in the absence and presence of Trolox C, a hydrosoluble form of vitamin E. Trolox C completely quenched the genotoxicity of ochratoxin A, which was no greater in mutated than in wild type strains. These results implicate an ochratoxin A-derived free radical rather than reduced oxygen species as genotoxic intermediate(s) in bacteria.

Introduction

Ochratoxin A, a dehydroisocoumarin-containing mycotoxin, is produced by members of the *Aspergillus* and *Penicillium* species. Its occurrence in food and feed is widespread (IARC, 1983), and it has been found to induce nephropathy in several species (reviewed by Krogh, 1978); furthermore, it is hepatotoxic, an immunosuppressor and a teratogen (Haubeck *et al.*, 1981; Hayes, 1981). Dietary feeding of ochratoxin A has been shown to induce renal adenomas and hepatocellular carcinomas in female mice (Bendele *et al.*, 1985), and it induced single-strand breaks in the DNA of liver, kidney and spleen of mice (Creppy *et al.*, 1985). At variance with these findings, ochratoxin A was reported to be non-genotoxic in various short-term tests *in vitro* (IARC, 1987). We report here on the SOS-DNA repair-inducing activity in *E. coli* PQ37 of ochratoxin A, of its major metabolite in rodents ochratoxin α and of seven structurally related substances and propose a mode of action of ochratoxin A as a genotoxin in bacteria.

Genotoxicity of Ochratoxin A and Related Substances

A modified SOS chromotest was used to measure genotoxicity (Quillardet *et al.*, 1982; Malaveille *et al.*, 1989), to avoid interference of the test compound with the enzyme activities that reflect the level of SOS-DNA repair synthesis and surviving bacteria. Genotoxicity is expressed by the induction factor, which is an index of the induction of the SOS-DNA repair function adjusted for cytotoxicity.

Ochratoxin A, chloroxine, 5-chloro-8-quinolinol, 4-chloro-*meta*-cresol and chloroxylenol induced SOS-DNA repair synthesis in *E. coli* PQ37 at concentrations ranging from 0.1 to 4 mM (Table 1). Ochratoxin B, 5-chlorosalicylic acid and citrinin exerted only cytotoxicity; ochratoxin α, at concentrations up to 1 mM, was neither toxic nor genotoxic. Addition of a rat liver or kidney metabolic activation system (up to 20% by volume of a post-mitochondrial supernatant, fortified with an NADPH-generating system) decreased the genotoxicity of the active test substances.

The SOS inducers were active only at cytotoxic concentrations; ochratoxin A was the most active when compared to other compounds leaving less than 30% survivors. No association was observed between the induction factor and survival; thus, at similar cytotoxicity, ochratoxin A was an SOS inducer while ochratoxin B and 5-chloro-

Table 1. SOS-DNA repair-inducing activity and cytotoxicity of ochratoxin A, ochratoxin α and seven structurally related substances in *Escherichia coli* PQ37 strain

Substance	Concentration (mM)	Genotoxicity[a] (induction factor)	Toxicity[a] (% survivors)
Dimethyl sulfoxide (solvent)		1	100
Ochratoxin A 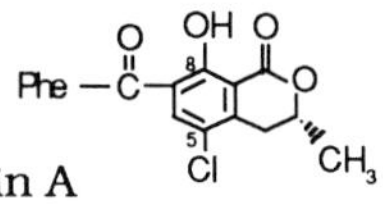	1	1.2	63
	2	2.3 ± 0.14 (n=8)	30 ± 1
	4	2.7 ± 0.16 (n=8)	25 ± 1
Ochratoxin B 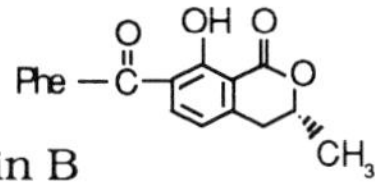	2	1.1	77
	4	1.4 ± 0.0 (n=2)	37 ± 6
Ochratoxin α 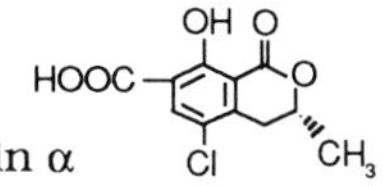	0.1	0.9	100
	0.5	0.8	100
	1	0.8	95
5-Chlorosalicylic acid 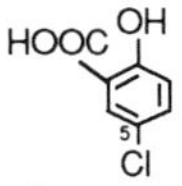	0.1	0.8	100
	0.5	0.8	65
	1	0.9	34
	2	0.01	25
Chloroxine 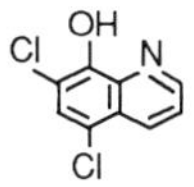	0.1	1.6 ± 0.6 (n=2)	25 ± 14
	0.3	2.3 ± 0.4 (n=3)	15 ± 4
	0.5	2.9 ± 0.2 (n=2)	11 ± 2
	1	0.8	18
5-Chloro-8-quinolinol 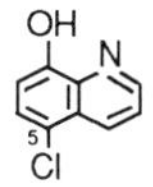	0.1	1.1	63
	0.3	1.2	27
	1	1.8	11
	2	1.2	10
4-Chloro-*meta*-cresol 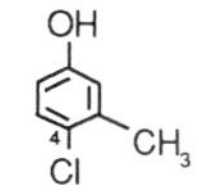	0.1	1.5 ± 0.2 (n=2)	70 ± 10
	0.3	3.6 ± 0.0 (n=2)	21 ± 1
	1	2.0 ± 0.2 (n=2)	10 ± 0
	2	1.6	9
Chloroxylenol 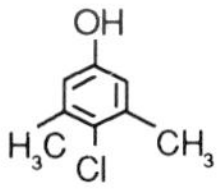	0.1	1.6 ± 0.4 (n=2)	54 ± 1.4
	0.3	2.8 ± 0.1 (n=2)	11 ± 0
	1	1.2 ± 0.0 (n=2)	10 ± 0
Citrinin 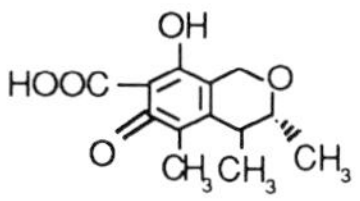	0.26	0.9	81
	0.5	1.1	42
	1	1.0	18

[a]Mean value of one duplicate experiment or mean value ± SE of 2–8 duplicate experiments. A concurrent positive control using methylmethane sulfonate accompanied each experiment.

salicylic acid were not (Table 1). To substantiate our findings, we manipulated the experimental conditions in an attempt to distinguish the cytotoxic and genotoxic actions of the nine compounds under study.

Possible Mode of Action of Ochratoxin A and Related Substances as Genotoxic and Cytotoxic Agents

Our results indicate that there are certain structural requirements for the genotoxicity of these compounds. A carboxyl group in the *ortho* position was associated negatively with activity (for ochratoxin α, 5-chlorosalicylic acid and citrinin), and a chlorine atom in the position *para* to the phenolic group (for ochratoxin A, chloroxine, 4-chloro-*meta*-cresol, chloroxylenol and 5-chloro-8-quinolinol) was associated positively with activity. Thus, the presence of a chlorine at C-5 in ochratoxin A appears to be one determinant of its genotoxic action. Genotoxic, nephrotoxic species of halogenated alkenes and bromobenzene are produced by a bacterial glutathione/cysteine conjugation reaction, followed by enzymatic processing of the conjugate(s) by ß-lyase (Elfarra & Anders, 1984). To ascertain whether a similar reaction is involved, we investigated the modifying effect of amino oxyacetic acid, an inhibitor of cysteine conjugate ß-lyase (Elfarra *et al.*, 1986). This compound had no effect on the genotoxicity of ochratoxin A, expressed as the SOS induction factor (Figure 1A), and ochratoxin A exerted such activity only at cytotoxic concentrations, as shown by reduction of SOS-DNA repair activity in bacterial survivors as a function of ochratoxin A concentration (Figure 1B). Amino oxyacetic acid counteracted this effect. Thus, inhibition of ß-lyase activity decreased ochratoxin A cytotoxicity but not genotoxic activity, suggesting intra-bacterial formation of a cytotoxic, but not genotoxic, thiol containing an ochratoxin A derivative. These findings confirm that ochratoxin A is genotoxic to bacteria, since assay conditions exist in which the SOS induction factor increases in parallel with SOS-DNA repair activity in surviving bacteria.

Ochratoxin A forms a strong complex with ferric ion, leading to stimulation of lipid peroxidation *in vitro* (Omar *et al.*, 1990). We therefore addressed the possibility that the bacterial genotoxicity was due to an iron-based oxidative stress within bacteria that ultimately yields radical species. The response of *E. coli* PQ37 strain was measured in the presence and absence of Trolox C, a hydrosoluble form of vitamin E. For concentrations of 1–2.5 mM, this radical scavenger completely quenched the genotoxicity of ochratoxin A at 2-4 mM, of 4-chloro-*meta*-cresol at 0.3 mM and of chloroxine at 0.25 mM. The way in which ochratoxin A induces SOS-DNA repair was further investigated by comparing the response of PQ37 with that of PQ300, OG 100 and OG 400. These PQ37-derived strains have a partially deleted *oxy R* gene and are thus more sensitive to oxidative DNA damage (Goerlich *et al.*, 1989; P. Quillardet,

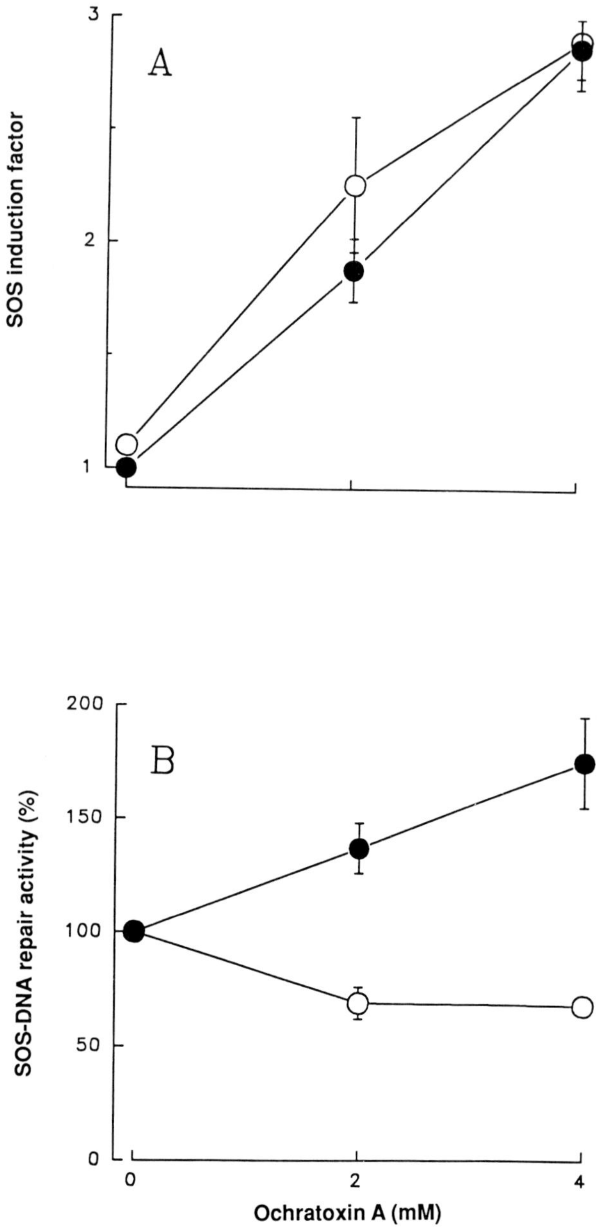

Figure 1. Concentration-dependent genotoxic activity of ochratoxin A in *Escherichia coli* PQ37 in the absence (open circles) and presence (filled circles) of amino oxyacetic acid

A, SOS induction factor; B, SOS-DNA repair activity. Each point represents mean ± SE of two duplicate experiments.

unpublished data). The four strains displayed very similar sensitivity to the genotoxic effect of ochratoxin A, whereas they showed differential sensitivity to cumene hydroperoxide, used as a positive control (data not shown).

Our results do not indicate the participation of reduced oxygen species and suggest that an ochratoxin A-derived free radical (possibly an alkoxy radical at C-8) is a genotoxic intermediate in bacteria.

Acknowledgments

We are grateful to Dr P. Quillardet and Dr M. Hofnung (Institut Pasteur, Paris, France) for providing the *E. coli* strains and for giving helpful advice. We thank Ms Y. Granjard for secretarial help.

References

Bendele, A.M., Carlton, W.W., Krogh, P. & Lillehoj, E.B. (1985) Ochratoxin A carcinogenesis in the (C57BL/6J x C3H) F1 mouse. *J. Natl Cancer Inst.*, **75**, 733–739

Creppy, E.E., Kane, A., Dirheimer, G., Lafarge-Frayssinet, C., Mousset, S. & Frayssinet, C. (1985) Genotoxicity of ochratoxin A in mice: DNA single-strand break evaluation in spleen, liver and kidney. *Toxicol. Lett.*, **28**, 29–35

Elfarra, A.A. & Anders, M.W. (1984) Commentary: renal processing of glutathione conjugates role in nephrotoxicity. *Biochem. Pharmacol.*, **33**(23), 3729–3732

Elfarra, A.A., Lash, L.H. & Anders, M.W. (1986) Metabolic activation and detoxification of nephrotoxic cysteine and homocysteine S-conjugates. *Proc. Natl Acad. Sci. USA*, **83**, 2667–2671

Goerlich, O., Quillardet, P. & Hofnung, M. (1989) Induction of the SOS response by hydrogen peroxide in various *Escherichia coli* mutants with altered protection against oxidative DNA damage. *J. Bacteriol.*, **171**, 6141–6147

Haubeck, H.D., Lorkoweski, G., Kolsh, E. & Roschenthaler, R. (1981) Immunosuppression by ochratoxin A and its prevention by phenylalanine. *Appl. Environ. Microbiol.*, **41**, 1040–1042

Hayes, A.W. (1981) *Mycotoxin Teratogenicity and Mutagenicity*, Boca Raton, FL, CRC Press

IARC (1983) *IARC Monographs on the Carcinogenicity of Chemicals to Humans*, Vol. 31, *Some Food Additives, Feed Additives and Naturally Occurring Substances*, Lyon, pp. 191–206

IARC (1987) *IARC Monographs on the Evaluation of Carcinogenic Risks to Humans*, Suppl. 6, *Genetic and Related Effects: An Updating of Selected* IARC Monographs *from Volumes 1–42*, Lyon, pp. 434–436

Krogh, P. (1978) Ochratoxins. In: Rodricks, J.V., Hesseltine, C.W. & Mehlman, M.A., eds, *Mycotoxins in Human and Animal Health*, New York, Pathotox Publishers, pp. 489–498

Malaveille, C., Vineis, P., Estève, J., Ohshima, H., Brun, G., Hautefeuille, A., Gallet, P., Ronco, G., Terracini, B. & Bartsch, H. (1989) Levels of mutagens in the urine of smokers of black and blond tobacco correlate with their risk of bladder cancer. *Carcinogenesis*, **10**(3), 577–586

Omar, R.F., Hasinoff, B.B., Mejilla, F. & Rahimtula, A.D. (1990) Mechanism of ochratoxin A stimulated lipid peroxidation. *Biochem. Pharmacol.*, **40**(6),1183–1191

Quillardet, P., Huisman, O., D'Ari, R. & Hofnung, M. (1982) SOS chromotest, a direct assay of induction of an SOS function in *Escherichia coli* K-12 to measure genotoxicity. *Proc. Natl Acad. Sci. USA*, **79**, 5971–5975

Mycotoxins, Endemic Nephropathy and Urinary Tract Tumours
Ed. M. Castegnaro, R. Pleština, G. Dirheimer, I.N. Chernozemsky & H. Bartsch
Lyon, International Agency for Research on Cancer, pp. 267-272

CHROMOSOMAL ALTERATIONS IN LYMPHOCYTES OF PATIENTS WITH BALKAN ENDEMIC NEPHROPATHY AND OF HEALTHY INDIVIDUALS AFTER INCUBATION *IN VITRO* WITH OCHRATOXIN A

G. Manolov[1], Y. Manolova[1], M. Castegnaro[2,3] & I.N. Chernozemsky[1]

[1]*National Centre of Oncology, Sofia, Bulgaria; and*
[2]*International Agency for Research on Cancer, Lyon, France*

Summary

The possible involvement of mycotoxins in chromosomal alterations in patients with Balkan endemic nephropathy (BEN) was investigated cytogenetically. Lymphocyte cultures from patients with BEN and from individuals from a nonendemic region were examined and compared with cultures from healthy people which had been incubated *in vitro* with noncytotoxic doses of ochratoxin A. Significantly increased numbers of various numerical and structural anomalies were found in patients with BEN. Chromosome X in female patients occurred in both monosomic and polysomic forms. A 'prosomization' effect was seen along the entire length of supernumerary X chromosomes, manifested by retarded contraction resembling early mitotic stages, with a comparably detailed band pattern. No other specific numerical or structural change was found consistently in BEN patients. Incubation of the lymphocytes of healthy people with ochratoxin A induced similar aberrations and prosomization. These findings may support the hypothesis that ochratoxin A is involved in the pathogenesis of BEN.

---

[3]To whom correspondence should be addressed

Introduction

Bulić (1967) first described genetic links among patients with Balkan endemic nephropathy (BEN). Recently, Toncheva *et al.* (1988) reported that a marker on chromosome 3 was present in 100% of lymphocytes from such patients. Evidence against genetic predisposition is provided by studies that have associated the disease and subsequent renal cancer with environmental exposure to ochratoxin A (Krogh *et al.*, 1977; Pavlovich *et al.*, 1979; Hult *et al.*, 1982; Krogh & Nesheim, 1982; Bendele *et al.*, 1985; Petkova-Bocharova & Castegnaro, 1985; Boorman, 1988; Petkova-Bocharova *et al.*, 1988).

In order to test the two hypotheses at a cytogenetic level, we analysed the chromosomes of stimulated peripheral lymphocytes from patients with BEN and from healthy controls and compared them with chromosomes in lymphocyte cultures from healthy donors that had been incubated *in vitro* with ochratoxin A.

Cytogenetic Findings in Lymphocytes of Patients with Balkan Endemic Nephropathy

In 1986 and 1991 we investigated lymphocytes from 25 patients (17 female and 8 male) with BEN and kidney failure (grade 0 to 3–4), all of whom were inhabitants of Vratza district, Bulgaria. For comparison, lymphocytes were also obtained from six individuals in a nonendemic area for this disease. Chromosomes were prepared from stimulated peripheral lymphocytes by a standard method and by a high-resolution method, to make air-dried preparations. These were then banded by the trypsin–Giemsa technique.

The main results are shown in Table 1. Eighteen times more aberrant cells were found in BEN patients (14.7%) than in control cultures (0.8%), and significant involvement of aneuploid cells (10.8%) was noted. A large number of inconsistent numerical and structural aberrations were seen in the autosomes (chromosomes 1–22) of the patients. The chromosomes predominantly involved were X, 10 and 21; the X chromosome was affected in 12 of the 17 female patients. In one-third of the cells with X trisomy from BEN patients, one or two of the X chromosomes appeared to be longer, to show less contraction and to have more detailed banding, typical of an earlier mitotic stage (a 'prosomization' effect). The more detailed banding of these less condensed X chromosomes corresponded to an additional 200 bands in the haploid set.

Most of the aberrations observed were not repeated. In one terminal BEN patient who had a relatively large number of lymphocytes with aberrant chromosomes, however, two minimal cell clones were found, each consisting of two cells carrying the same marker chromosomes.

Table 1. Chromosomal changes in cells from patients with Balkan endemic nephropathy (BEN) and from controls, and in healthy cells treated in culture with ochratoxin A, in the absence and presence of an exogenous metabolic system, with the metabolic system alone or with the buffer of the metabolic system (containing KCl and K_2EDTA) alone

Cell source and treatment	No. of cells analysed	No. of cells with aberrant autosomes				No. of cells with aberrant X chromosomes					
		Numerical		Structural		Numerical		Structural		Prosomy[a]	
		No.	%	No.	%	No.	%	No.	%	No.	%
16 BEN patients (1986)	731	79	10.8	69	9.4	27	3.7	4	0.5	3/9	33.0
9 BEN patients (1991)	189	20	10.6	8	4.2	7	3.7	–		–	
Cultured cells from six healthy individuals, treated with:											
6 controls (1988)	477	2	0.4	3	0.6	1	0.2	–		–	
Ochratoxin A	465	17	3.7	4	0.9	12	2.6	–		6/8	75.0
Ochratoxin A preincubated with S9	534	17	3.2	4	0.7	10	1.9	1	0.2	5/7	71.5
Metabolic activation (S9)	490	3	0.6	1	0.2	2	0.4	–		–	
Buffer	482	7	1.5	9	1.9	4	0.8	1	0.2	–	

[a] In X trisomies; for definition of 'prosomy', see text

Cytogenetic Findings in Lymphocytes from Healthy Donors Treated in vitro *with Ochratoxin A*

Lymphocytes from six healthy individuals were cultured as such and treated in one of four ways: with ochratoxin A for 48 h, at 6 ng/ml of medium; with ochratoxin A preincubated with an exogenous metabolic system for 2 h before addition; with the metabolic system alone; or with the buffer (containing KCl and K_2EDTA) used for obtaining the metabolic system (see Manolova *et al.*, 1990). A total of about 2500 cells (1971 treated cells and 477 untreated cells) were examined directly under the microscope.

Numerical and structural aberrations similar to those seen in the lymphocytes from patients with BEN were observed in autosomes and X chromosomes in treated cultures. Three-quarters of the cells with X trisomy showed the prosomization effect (Table 1). A cell with a totally prosomized chromosome 4 homologue and a cell with a similarly prosomized partner of a trisomic chromosome 17 were also seen in cultures treated with ochratoxin A. Aberrant chromosomes occurred five times more frequently in cells treated with ochratoxin A than in untreated cells, but they occurred 3.7 times less frequently than in the cells of patients with BEN.

Cultures treated with ochratoxin A preincubated with an exogenous metabolic activation system had slightly fewer chromosomal aberrations, but they still had 4.5 times more anomalies than control cultures, and 71.5% of X trisomic cells showed prosomization of one or two gonosomes. Cells treated with the metabolic system alone had few aberrant chromosomes, but those treated with buffer had 3.5 times more anomalies than the control.

Discussion

We have shown that the lymphocytes of patients with BEN contain significantly more numerical and structural autosomal aberrations than lymphocytes from controls. Our finding that treatment of normal lymphocytes *in vitro* with ochratoxin A, the mycotoxin that has been associated etiologically with this disease, induced similar anomalies adds support to this hypothesis.

The number of aberrations in treated cultures was somewhat smaller than that in lymphocytes from diseased patients. This difference is probably due, however, to the fact that we used a dose of ochratoxin A that would not cause significant damage to the plates. The reduction in the number of aberrations seen in cultures that were treated with ochratoxin A preincubated with a metabolic system suggests that ochratoxin A has a direct action and does not require metabolic activation. When the lymphocytes were treated with the 'buffer' only, the number of cells with chromosomal aberrations increased significantly; in the presence of the metabolic activation system, the number of aberrations decreased to the control level. This finding suggests either an 'anti-aberration' potential of the components of the system other than the buffer (enzymes, proteins and electrolytes), which would counteract the aberration-inducing activity of the buffer components (probably K_2EDTA), or that the metabolic activation system metabolizes the active components of the buffer.

Similar effects were found in the different cell cultures with regard to the X chromosome. A significant involvement of the X chromosome emerges when the results for this single chromosome are compared with those for the 22 autosomal chromosomes: the X chromosome was involved 20 times more frequently than any single autosome in lymphocytes from patients with BEN and more than 50 times more frequently in lymphocytes treated *in vitro* with ochratoxin A.

The finding of prosomization in X chromosomes of lymphocytes from female BEN patients and in normal female lymphocytes after treatment with ochratoxin A is of particular interest. Up to now, this phenomenon has been seen only in translocation regions and in whole tumour markers (Manolov *et al.*, 1979; Manolova *et al.*, 1979). It is obviously associated with the activity of ochratoxin A, as no other substance is known that has similar activity. Furthermore, this phenomenon was seen in none of the cells from control subjects or in cells treated only with the metabolic system or with buffer. Since prosomization was also found in autosomal cells in cultures treated with ochratoxin A, the mycotoxin appears to affect other parts of the genome as well.

The observation that only the X chromosomes, and never the Y chromosomes, show prosomization may be associated with the often remarked prevalence of females among BEN patients and the sex-related pathology of this disease in endemic regions.

Many different types of chromosomal aberration were observed in the BEN patients; in only one case of terminal BEN were two clones found with the same chromosomal aberration. A similar situation is seen in preleukaemic conditions, in the survivors of the nuclear explosions in Japan and in homosexuals with pre-acute immune deficiency syndrome (Manolov *et al.*, 1985). As the presence of diverse chromosomal aberrations thus appears to be a prerequisite of malignant lymphoblastic transformation, neoplasia might be expected to be a consequence. The neoplasms that follow BEN originate in the epithelium of the urinary tract. It would therefore be interesting to investigate this tissue cytogenetically.

Our results strongly support the hypothesis that ochratoxin A is involved in the induction of various chromosomal aberrations and that mycotoxins play a significant role in the etiopathogenesis of BEN and urinary tract tumours. No consistent chromosomal aberration was found in the peripheral lymphocytes of 25 BEN patients. This investigation therefore excludes the existence of a specific chromosomal marker in patients with this disease, as predicted by Toncheva *et al.* (1988) for chromosome 3.

References

Bendele, A.N., Carlton, W.W., Krogh, P. & Lillehoj, E.B. (1985) Ochratoxin A carcinogenesis in B6C3F1 mouse. *J. Natl Cancer Inst.*, **75**, 733–739

Boorman, G. (1988) *Toxicology and Carcinogenesis Studies of Ochratoxin A* (NTP Tech. Rep. 358; NIH Publ. No. 88-2813), Research Triangle Park, NC, National Toxicology Program

Bulić, F. (1967) The possible role of genetic factors in the aetiology of the Balkan nephropathy. In: Ciba Foundation Study Group 30, eds, *The Balkan Nephropathy*, London, Churchill, pp. 17–27

Hult, K., Pleština, R., Habazin-Novak, V., Radić, B. & Čeović, S. (1982) Ochratoxin A in human blood and Balkan endemic nephropathy. *Arch. Toxicol.*, **51**, 313–321

Krogh, P. & Nesheim, S. (1982) Ochratoxin A. In: Stoloff, L., Castegnaro, M., Scott, P., O'Neill, I.K. & Bartsch, H., eds, *Environmental Carcinogens: Selected Methods of Analysis*, Vol. 5, *Some Mycotoxins* (IARC Scientific Publications No. 44), Lyon, IARC, pp. 247–253

Krogh, P., Hald, B., Pleština, R. & Čeović, S. (1977) Balkan (endemic) nephropathy and foodborn ochratoxin A: preliminary results of a survey of foodstuffs. *Acta Pathol. Microbiol. Immunol. Scand. Sect. B*, **85**, 238–240

Manolov, G., Urumov, I., Argirova, R. & Petkova, P. (1979) Cytogenetic study of foetal colon mouse tumour—AKATOL-1-71—cultivated in vitro. *Hereditas*, **90**, 227–236

Manolov, G., Manalova, Y., Sonnabend, J., Lipscomb, H. & Purtilo, D.T. (1985) Chromosome aberrations in peripheral lymphocytes of male homosexuals. *Cancer Genet. Cytogenet.*, **18**, 337–350

Manolova, Y., Manolov, G., Kieler, J., Levan, A. & Klein, G. (1979) Genesis of the 14q+ marker in Burkitt's lymphoma. *Hereditas*, **90**, 5–10

Manolova, Y., Manolov, G., Parvanova, L., Petkova-Bocharova, T., Castegnaro, M. & Chernozemsky, I.N. (1990) Induction of characteristic chromosomal aberrations, particularly X-trisomy, in cultured human lymphocytes treated by ochratoxin A, a mycotoxin implicated in Balkan endemic nephropathy. *Mutat. Res.*, **231**, 143–149

Pavlović, M., Pleština, R. & Krogh, P. (1979) Ochratoxin A contamination of foodstuffs in an area with Balkan (endemic) nephropathy. *Acta Pathol. Microbiol. Immunol. Scand. Sect. B*, **87**, 243–246

Petkova-Bocharova, T. & Castegnaro, M. (1985) Ochratoxin A contamination of cereals in an area of high incidence of Balkan (endemic) nephropathy in Bulgaria. *Food Addit. Contam.*, **2**, 267–270

Petkova-Bocharova, T., Chernozemsky, I. & Castegnaro, M. (1988) Ochratoxin A in relation to endemic nephropathy and urinary system tumours in Bulgaria. *Food Addit. Contam.*, **5**, 299–301

Toncheva, D., Dimitrov, T. & Tzoneva, M. (1988) Cytogenetic studies in Balkan endemic nephropathy. *Nephron*, **48**, 18–21

Mycotoxins, Endemic Nephropathy and Urinary Tract Tumours
Ed. M. Castegnaro, R. Pleština, G. Dirheimer, I.N. Chernozemsky & H. Bartsch
Lyon, International Agency for Research on Cancer, pp. 273-277

EFFECT OF OCHRATOXIN A ON BRUSH BORDER ENZYMES OF RAT KIDNEY

S. Pepeljnjak[1], I. Čepelak[2,3] & D. Juretić[2]

[1]Institute of Microbiology and [2]Institute of Medical Biochemistry, Faculty of Pharmacy and Biochemistry, Zagreb, Yugoslavia

Summary

Ochratoxin A was given orally at 60 µg/kg body weight in neutral olive oil to Fischer rats for 30 days, at which time they were killed. Clinical state, weights of animals and of their organs and urea and creatinine concentrations were not affected during the exposure period. Significant increases in the activity of enzymes in urine were found: 60% increase in alanine aminopeptidase, 45% increase in γ-glutamyltransferase and 90% increase in alkaline phosphatase. These changes indicate early pathological changes in the kidney. Relatively small amounts of the toxin thus affect kidney membrane cells.

Introduction

The frequent occurrence of ochratoxin A in relatively higher concentrations in endemic than in other regions (Pepeljnjak & Cvetnić, 1985) has made this secondary metabolite of *Aspergillus* and *Penicillium* species the focus of etiological studies as a possible causative agent of endemic nephropathy. In order to study the speed and site of action of ochratoxin A in kidney tissue, we determined the activities of alanine aminopeptidase (EC 3.4.11.2), alkaline phosphatase (EC 3.1.3.1) and γ-glutamyltransferase (EC 2.3.2.2) in urine and kidney homogenates from rats given ochratoxin A orally at low concentrations.

[3]To whom correspondence should be addressed

Materials and Methods

Twenty male Fischer rats, weighing approximately 300 g, were divided into four groups and given ochratoxin A in neutral olive oil at 60 μg/kg body weight for 5, 10, 20 or 30 days. A control group of 12 animals received the same amount of neutral olive oil. Animals were sacrificed after each exposure period, and their kidneys were removed and homogenated at a dilution of 10 g/litre. A supernatant was obtained by centrifugation in a MSE Mistral-2L apparatus at 12 000 *g* for 30 min, and enzyme activities and protein concentration were determined.

Urine was collected in metabolic cages for 24 h prior to sacrifice, centrifuged and examined using test tapes, then dialysed for 3 h in running water, using a Visking hose. Ochratoxin A was obtained by biosynthesis on wheat using *Aspergillus sulphureus* and extracted by the method of Balzer *et al.* (1978) at 80% purity.

γ-Glutamyltransferase activity was determined by the method of Szasz *et al.* (1974), alkaline phosphatase by the method of McComb and Bowers (1972) and alanine aminopeptidase by the method of Jung and Scholz (1980); protein concentration was determined by the method of Bradford (1976). The results were evaluated statistically by the parametric Mann-Whitney test (Zar, 1974).

Results and Discussion

Alanine aminopeptidase, γ-glutamyltransferase and alkaline phosphatase are enzymes located mainly on membranes of the brush border epithelium of the proximal kidney tubule (Guder & Ross, 1984). Statistically significant increases ($p < 0.05$) occurred in γ-glutamyltransferase activity in urine (by about 45%) and in alkaline phosphatase activity (by about 52%) after 20 days (Figures 1 and 2) and in alanine aminopeptidase activity (by about 60%) after 30 days (Figure 3), with no concomitant alteration in kidney tissue, except for a slight increase in alkaline phosphatase activity on the 30th day. These results suggest that continuous intake of ochratoxin A at extremely low concentrations over a relatively brief period induces pathological changes in the proximal tubules of the kidney. The differences in the severity of enzymuria might be due to different intramembranous distribution and mode of attachment of these enzymes on the cell membrane (Scherberich, 1989). The volume of urine, concentrations of creatinine and urea, body weight of animals, weight of kidneys and protein concentration in tissue homogenates were similar to those of the control group.

Numerous recent studies suggest a possible effect of ochratoxin A on human and animal health. This is confirmed by our findings of ochratoxin A in commodities consumed in households in areas endemic for nephropathy: in smoked meat at 40–920 μg/kg (Pepeljnjak & Blažević, 1982), beans at 17–53 μg/kg (Pepeljnjak, 1984), cereals at

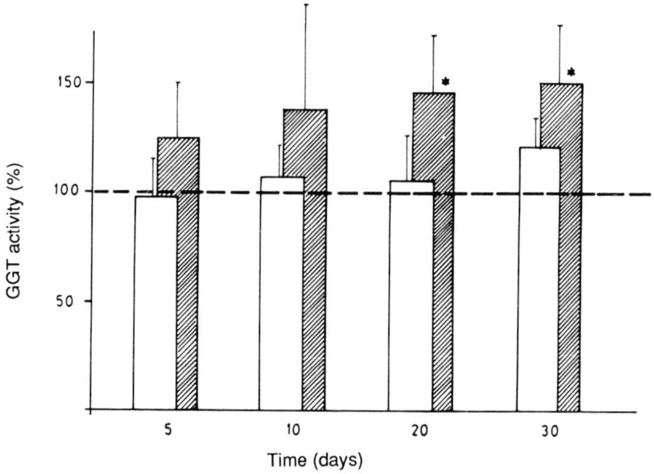

Fig. 1 Effect of ochratoxin A on the activity of γ-glutamyltransferase (GGT)

The relative activity is shown as percent of the activity of the control group. The activities in the control group were 0.58 ± 0.19 IU/24-h volume in urine (shaded bars) and 9.05 ± 4.0 IU/g protein in kidney (open bars); $p < 0.05$

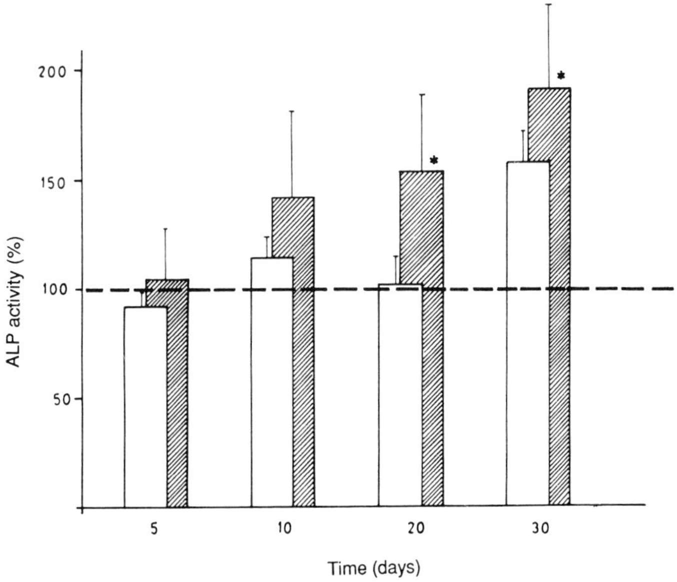

Fig. 2 Effect of ochratoxin A on the activity of alkaline phosphatase (ALP)

The relative activity is shown as percent of the activity of the control group. The activities in the control group were 0.32 ± 0.15 IU/24-h volume in urine (shaded bars) and 1.58 ± 0.45 IU/g protein in kidney (open bars); $p < 0.05$

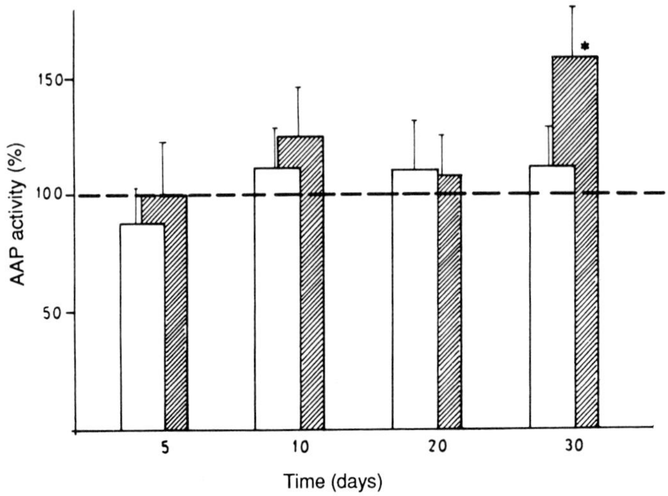

Fig. 3 Effect of ochratoxin A on the activity of alanine aminopeptidase (AAP)
The relative activity is shown as percent of the activity of the control group. The activities in the control group were 0.20 ± 0.05 IU/24-h volume in urine (shaded bars) and 10.4 ± 4.46 IU/g protein in kidney (open bars); $p < 0.05$

0.02–68.9 mg/kg; and in blood from slaughtered pigs at 36–37 µg/kg, in liver at 0–21 µg/kg and in kidneys at 16–27 µg/kg (Pepeljnjak & Cvetnić, 1985). In such areas, 7.5% of samples of cereals contained more than 20 mg/kg ochratoxin A, 12.5% contain more than 5 mg/kg and 25% contained more than 2 mg/kg. In regions with no endemic nephropathy, only 15% of samples had concentrations up to 2 mg/kg (Pepeljnjak & Cvetnić, 1985). In addition, the finding of ochratoxin A in human blood at 3–4 µg/ml (Hult *et al.*, 1982) in areas of endemic nephropathy suggests that it could act on kidney function.

In order to explain fully the role of ochratoxin A in human and animal health, the action of low concentrations on the function of kidney cells must be investigated further.

References

Balzer, I., Bogdanić, Č. & Pepeljnjak, S. (1978) Rapid thin layer chromatographic method for determining aflatoxin B_1, ochratoxin A and zearalenon in corn. *J. Assoc. Off. Anal. Chem.*, **61**, 584–585

Bradford, M.M. (1976) A rapid and sensitive method for the quantitation of microgram quantities of protein utilizing the principle of protein-dye binding. *Anal. Biochem.*, **72**, 248–254

Guder, W.G. & Ross, B.D. (1984) Enzyme distribution along the nephron. *Kidney Int.*, **26**, 101–111

Hult, K., Pleština, R., Čeović, S., Habazin-Novak, V. & Radić, B. (1982) Ochratoxin A in human blood: analytical results and confirmational tests from a study in connection with Balkan endemic nephropathy. In: *Proceedings of the V International IUPAC Symposium on Mycotoxins and Phytotoxins*, Vienna, International Union of Pure and Applied Chemists, pp. 338–341

Jung, K. & Scholz, D. (1980) An optimised assay of alanine-aminopeptidase activity in urine. *Clin. Chem.*, **26**, 1251–1254

McComb, R.B. & Bowers, G.N. (1972) Study of optimum buffer conditions for measuring alkaline phosphatase activity in human serum. *Clin. Chem.*, **18**, 97–104

Pepeljnjak, S. (1984) Mycotoxic contamination of haricot beans in nephropathic areas in Yugoslavia. *Microbiol. Aliments Nutr.*, **2**, 331–336

Pepeljnjak, S. & Blažević, N. (1982) Contamination with moulds and occurrence of ochratoxin A in smoked meat products from endemic nephropathy regions of Yugoslavia. In: *Proceedings of the V International IUPAC Symposium on Mycotoxins and Phytotoxins*, Vienna, International Union of Pure and Applied Chemists, pp. 102–105

Pepeljnjak, S. & Cvetnić, Z. (1985) The mycotoxicological chain and contamination of food by ochratoxin A in Yugoslavia. *Mycopathologia*, **90**, 147–153

Scherberich, E.J. (1989) Immunological and ultrastructural analysis of loss of tubular membrane-bound enzymes in patients with renal damage. *Clin. Chim. Acta*, **185**, 271–282

Szasz, G., Weimann, G., Stähler, F., Wahlfeld, A.W. & Persijn, J.P. (1974) New substrates for measuring gamma-glutamyl transpeptidase activity. *Z. Klin. Chem. Klin. Biochem.*, **12**, 228

Zar, J.H. (1974) *Biostatistical Analysis*, London, Prentice Hall, pp. 109–114

ROLE OF GENETIC FACTORS AND DRUG METABOLISM IN THE NEPHROTOXICITY AND CARCINOGENICITY OF OCHRATOXIN A

Mycotoxins, Endemic Nephropathy and Urinary Tract Tumours
Ed. M. Castegnaro, R. Pleština, G. Dirheimer, I.N. Chernozemsky & H. Bartsch
Lyon, International Agency for Research on Cancer, pp. 281-287

INDIVIDUALITY IN CYTOCHROME P450 EXPRESSION AND ITS ASSOCIATION WITH THE NEPHROTOXIC AND CARCINOGENIC EFFECTS OF CHEMICALS

C.R. Wolf

Imperial Cancer Research Fund, Molecular Pharmacology Group, Edinburgh, Scotland, United Kingdom

Summary

The susceptibility of a tissue to the toxic and/or carcinogenic effects of chemicals is determined by a variety of factors, which include their rate of metabolic activation by the cytochrome P450-dependent monooxygenases. Individual differences in the levels of cytochrome P450 expression would be expected, and are known, to give rise to profound differences in toxicological response. Such effects are almost best exemplified by the sex differences observed in the toxic effects of a variety of nephrotoxins and carcinogens. In recent work, we have shown that in species such as the mouse and rat almost all cytochrome P450 enzymes in the kidney are sexually differentiated. This difference in cytochrome P450 regulation is mediated by testosterone and explains the large differences observed in the metabolic activation, toxicity and carcinogenicity of chloroform and possibly of other compounds such as ochratoxin A. In addition to hormonal or environmental influences on cytochrome P450 expression, genetic factors have also been shown to be important. In man, this is best exemplified by the genetic polymorphism observed in the metabolism of debrisoquine and approximately 25 other drugs. This genetic defect affects approximately 5–10% of the Caucasian population and has been associated with altered susceptibility to cancer.

In this presentation, the development of a simple DNA-based assay to identify affected individuals is described. Use of this assay will allow clarification of the reported association of this genetic polymorphism to susceptibility to Balkan nephropathy and cancer.

Introduction

The cytochrome P450s play a central role in determining our responses to the toxic and carcinogenic effects of environmental chemicals. Probably a direct consequence of this function is that this enzyme system has evolved into a multigene family of proteins, the genes for which are scattered through the human genome. Estimates of the number of P450 genes in man range significantly, from 50 to 200. Although certain cytochrome P450s exhibit overlapping substrate specificities, some specific isozymes are almost exclusively responsible for the metabolism of a particular drug or foreign compound.

The central role of the P450 system in drug and carcinogen metabolism implies that individuality in the levels of specific P450 isozymes may be an important determinant of individual susceptibility to chemical toxicity and cancer (Guengerich, 1990; Wolf *et al.*, 1990). It is therefore important to characterize the factors that determine cytochrome P450 isozyme expression in man.

The cytochrome P450s appear to have evolved as an adaptive response to chemical challenge, as exemplified by the finding that exposure to a chemical often results in the induction of a cytochrome P450 isozyme which is active in its metabolism, thereby facilitating its excretion. Analysis of the cytochrome P450 content in a panel of human liver samples indicates that the relative level of isozyme expression is subject to large individual differences, which could be related to environmental, hormonal or genetic factors.

Molecular Basis of the P450 CYP2D6 Polymorphism

A variety of reports have implicated the cytochrome P450 system in the pathogenesis of Balkan endemic nephropathy at both the biochemical and the genetic level (see other chapters in this volume). Of particular interest in this regard is the reported association between a genetic polymorphism in the cytochrome P450 CYP2D6 gene locus and susceptibility to this disease.

This cytochrome P450 polymorphism was first identified by the inability of certain individuals to metabolize the marker drugs debrisoquine or sparteine. Such individuals have been termed 'poor metabolizers' (see Eichelbaum & Gross, 1990). This polymorphism is inherited as an autosomal recessive trait and affects approximately 5% of the Caucasian population. Thus, in a population of 50 million, approximately 2.5 million individuals will be affected by this metabolic defect.

This observation gains further importance on the realization that the metabolism of over 25 drugs (Table 1) is compromised in affected individuals and in certain instances can give rise to life-threatening drug side-effects.

Table 1. Drugs subject to polymorphic oxidation due to mutations in the cytochrome P450 CYP2D6 gene

Cardiovascular drugs	Psychiatric drugs	Other drugs
Metoprolol	Amitriptyline	Dextromethorphan
Bufuralol	Imipramine	Codeine
Timolol	Desipramine	Methoxyphenamine
Propranolol	Nortriptyline	Phenformin
Perhexilene	Clomipramine	Melatonin (?)
Chlorpropamide (?)	Thioridazine	1-Methyl-4-phenyl-1,2,3,6-tetrahydro-pyridine (MPTP) (?)
Sparteine	Perphenazine	
N-Propylamaline	Amiflamine	
Propafenone	Tomoxitene	
Encainide		
Flecainide		
Mexiletine		

In addition to its association with adverse drug reactions, this genetic polymorphism has also been correlated with altered susceptibility to cancers of the lung, bladder and liver (see Idle & Ritchie, 1983; Gough *et al.*, 1990) as well as to Balkan endemic nephropathy (see Nikolov *et al.*, this volume). Some of these studies are, however, controversial, owing partly to limitations of the phenotyping assay used, in which factors such as drug–drug interactions, disease state and impaired kidney function can all confound the identification of affected individuals.

We therefore decided to try to find mutations in the CYP2D6 gene that could be used to develop a DNA-based assay to identify individuals with this polymorphism. With this in mind, we isolated cDNA clones that had been previously associated with the metabolic defect (Gonzalez *et al.*, 1988) and identified the base-pair differences between these and the known functional CYP2D6 gene (Gough *et al.*, 1990). These are shown in Figure 1. On the basis of these differences we then developed a simple assay to detect these affected individuals.

The assay is based on the base-pair deletion observed at position 506 in the cDNA. This gene-inactivating mutation is due to a base-pair substitution (G to A transition) at the junction of intron 3 and exon 4 of CYP2D6 gene (Gough *et al.*, 1990). This transition deletes a *Bst* NI restriction site in the normal CYP2D6 gene; therefore, amplification of

CTGCACGCTACCCACCAGGCCCCCT .... Normal allele
CTGCACGCTAC**T**CACCAGGCCCCCT .... Mutant allele
▲
100

CGCTGGTGACCCACGGCGAGGACACCGCCG............. Normal
CG**A**TGGTGACCC**G**CGGCGAGGACAC**G**GCCG............ Mutant
▲ ▲ ▲
271 281 294

CCACTCCGGACGC.......... Normal
CCACTCCG-ACGC........... Mutant
▲
506

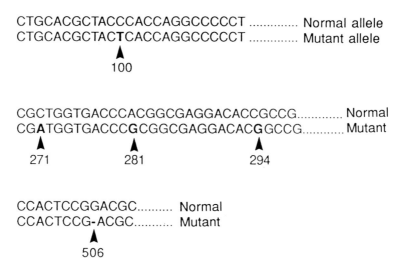

Figure 1. Position of mutations identified in the CYP2D6 cDNA associated with the poor metabolizer phenotype

Of the mutations shown in the CYP2D6 (bold letters), only the base-pair deletion at position 506 will prevent the production of CYP2D6 protein.

the DNA covering this site followed by *Bst* NI digestion will identify individuals homozygous or heterozygous for this mutation (Figure 2). The G to A transition is found in poor metabolizers at an allele frequency of approximately 0.8. In addition to this mutation, a gene deletion has been identified in certain poor metabolizers with an allele frequency of approximately 0.1 (Gough *et al.*, 1990) and a further mutation in exon 5 at a frequency of approximately 0.02 (Kagimoto *et al.*, 1990). Analysis of G to A transitions at position 506 will identify 80% of poor metabolizers, i.e., people homozygous for the G to A transition or heterozygous for the G to A transition and the gene deletion.

The assay has recently been further refined to facilitate the analysis of all known CYP2D6 mutant alleles (approximately 90% of poor metabolizers) (Spurr *et al.*, 1991). Studies are currently in progress using this assay to establish clearly whether there is an association between susceptibility to Balkan endemic nephropathy and the CYP2D6 polymorphism.

Sex Differences in Renal Cytochrome P450 Gene Expression

An intriguing aspect of Balkan endemic nephropathy is the difference in susceptibility between males and females. Interestingly, a significant sex difference is also observed in susceptibility to kidney cancer in general, the incidence being higher in males than in females. A

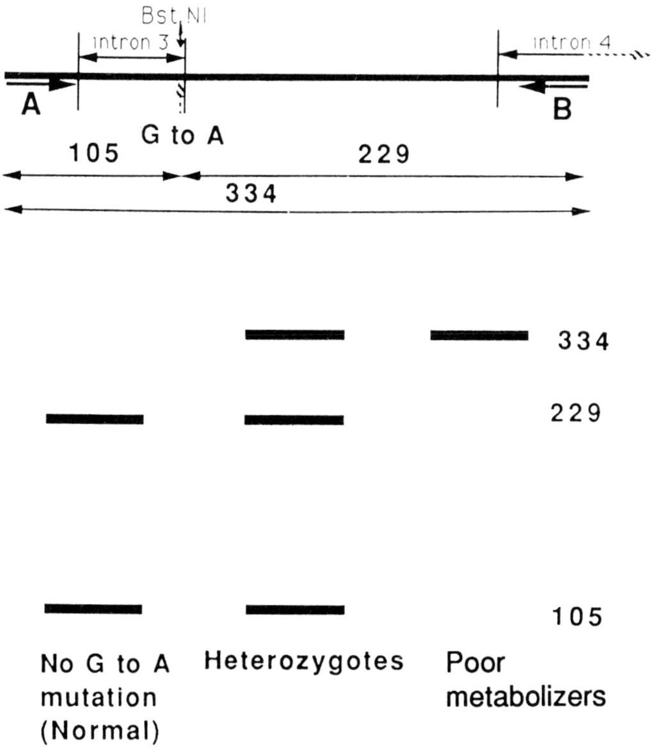

Figure 2. Principle behind the DNA-based assay for identifying poor metabolizers; top, amplification method; bottom, predicted binding pattern

A and B are the primers used for the amplification reaction. For further details, see Gough *et al.* (1990).

profound sex difference has also been observed in the incidence of kidney tumours in animals receiving ochratoxin A (see Huff, this volume).

In view of these observations and the central role of the cytochrome P450s in the metabolism and activation of chemical carcinogens, we were interested in establishing the extent to which these enzymes are sexually differentiated in the kidney. The results of initial studies, together with those of others (see Henderson *et al.*, 1990), demonstrated that male mice have a higher total renal cytochrome P450 content and greater renal microsomal activity towards a wide range of diverse cytochrome P450 substrates. In order to establish which cytochrome P450 isozymes were involved, Western and Northern blots were done on microsomal fractions from male and female C57Bl/6 and DBA/2N mice, using a panel of cytochrome P450 antibodies that react with proteins in the CYP2A, CYP2B, CYP2C, CYP2E, CYP3A and CYP4A gene families.

Intriguingly, all 11 of the cytochrome P450 isozymes detected were sexually differentiated. In most cases, males had a higher isozyme content than females (Henderson *et al.*, 1990), although certain isozymes were detected only in females (Figure 3).

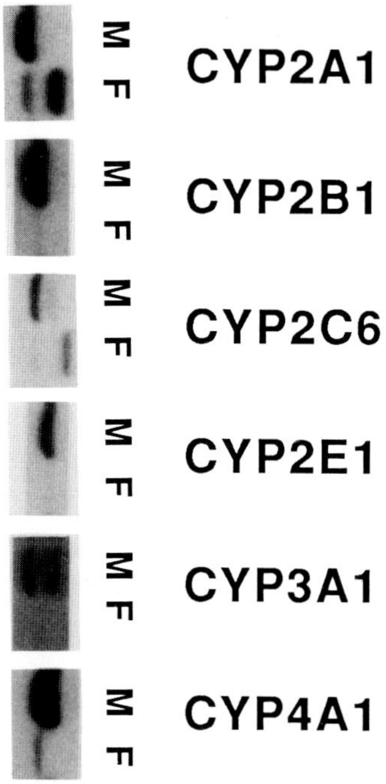

Figure 3. Expression of P450 isozymes in mouse kidney

Mouse kidney microsomal samples were probed for the expression of cytochrome P450 isozymes from a variety of gene families, as described by Henderson *et al.* (1990). The majority of the isozymes are expressed predominantly in males.

Sex differences in cytochrome P450 isozyme expression have been related to the pattern of growth hormone secretion. In order to establish whether it is involved here, experiments were carried out on growth hormone-deficient 'little mice'; it was shown not to be a factor.

We therefore investigated whether steroid hormones are involved. As a consequence of these studies, we could demonstrate unequivocally that the sexual differentiation observed is determined by circulating testosterone levels and mediated by the androgen receptor (Henderson & Wolf, 1991). Further studies were then carried out in order to establish whether this sexual differentiation is seen in other species. This indeed proved to be the case in rats and rabbits, but no sex difference in P450 levels was seen in dog kidney.

Sex differences in cytochrome P450 expression may explain the differences observed in the incidence of renal tumours in human populations. The results of the studies described above would imply that circulating testosterone levels are important in the pathogenesis of renal cancer. It will be interesting to establish whether testosterone can reverse the profound sex difference in the incidence of renal tumours observed in rats and mice following administration of ochratoxin A.

References

Eichelbaum, M. & Gross, A. (1990) The genetic polymorphism of debrisoquine/spartein metabolism—clinical aspects. *Pharmacol. Ther.*, **46**, 377–394

Gonzalez, F.J., Skoda, R.C., Kimura, S., Umeno, M., Zanger, U.M., Nebert, D.W., Gelboin, H.V., Hardwick, J.P. & Meyer, U.A. (1988) Characterization of the common genetic defect in humans deficient in debrisoquine metabolism. *Nature*, **331**, 442–446

Gough, A.C., Miles, J.S., Spurr, N.K., Moss, J.E., Gaedigk, A., Eichelbaum, M. & Wolf, C.R. (1990) Identification of the primary gene defect at the cytochrome P450 CYP2D locus. *Nature*, **347**, 773–776

Guengerich, F.P. (1990) Characterization of the roles of human cytochrome P450 enzymes in carcinogen metabolism. *Asia Pacific J. Pharmacol.*, **5**, 327–345

Henderson, C.J. & Wolf, C.R. (1991) Evidence that the androgen receptor mediates sexual differentiation of mouse renal cytochrome P450 expression. *Biochem. J.*, **278**, 499–503

Henderson, C.J., Scott, A.A., Yang, C.S. & Wolf, C.R. (1990) Testosterone-mediated regulation of cytochrome P450 gene expression will explain sex differences in response to nephrotoxins and carcinogens. *Biochem. J.*, **266**, 675–681

Idle, J.R. & Ritchie, J.C. (1983) Probing genetically variable carcinogen metabolism using drugs. In: Harris, C.C. & Autrup, H.N., eds, *Human Carcinogenesis*, New York, Academic Press, pp. 857–881

Kagimoto, M., Heim, M., Kagimoto, K., Zeugin, T. & Meyer, U.A. (1990) Multiple mutations of the human cytochrome P450IID6 gene (CYP2D6) in poor metabolizers of debrisoquine. *J. Biol. Chem.*, **265**, 17209–17214

Spurr, N.K., Gough, A.C., Smith, C.A.D. & Wolf, C.R. (1991) Genetic analysis of cytochrome P450 gene loci. *Methods Enzymol.*, **206**, 149–166

Wolf, C.R., Miles, J.S., Gough, A. & Spurr, N.K. (1990) Molecular genetics of the human cytochrome P450 system. *Biochem. Soc. Trans.*, **18**, 21–24

Mycotoxins, Endemic Nephropathy and Urinary Tract Tumours
Ed. M. Castegnaro, R. Pleština, G. Dirheimer, I.N. Chernozemsky & H. Bartsch
Lyon, International Agency for Research on Cancer, pp. 289-296

GENETIC PREDISPOSITION TO BALKAN ENDEMIC NEPHROPATHY: ABILITY TO HYDROXYLATE DEBRISOQUINE AS A HOST RISK FACTOR

I. G. Nikolov[1], I.N. Chernozemsky[1] & J.R. Idle[2]

[1]*National Oncological Centre, Medical Academy, Sofia, Bulgaria; and* [2]*Pharmacology Department, St Mary's Hospital Medical School, London, United Kingdom*

Summary

The objective of this study was to examine the association between efficiency of oxidative metabolism and risk for developing Balkan endemic nephropathy (BEN) and/or transitional-cell carcinoma of the urinary tract, using a case-control design controlling for age, gender and socioeconomic factors. Over 900 urine samples were taken from 646 subjects, divided into the following groups: healthy subjects from areas with no BEN; healthy subjects from villages with BEN; subjects suspected of having BEN; and subjects with BEN and/or upper urinary tract tumours (UUT). BEN patients and controls from the same villages were of similar age. The highest urinary recovery of debrisoquine was found among controls from areas with no BEN; recovery in BEN patients was only 50% of that in controls. The most interesting result is that BEN patients did not have impaired debrisoquine metabolism: subjects who metabolized < 25% of the drug represented only 2.9% of BEN patients, 12.4% of controls from BEN villages and 12.7% of controls from outside the BEN area. The very poor metabolizers represented 1.0% of BEN patients and 4.8–5.8% of controls. the percentages of extensive metabolizers in the same groups were 86.3, 64.5 and 67.4%, respectively. The mean metabolic ratio rose progressively from BEN patients < suspected BEN patients < controls from BEN villages < controls from non-BEN villages; the maximum metabolic ratios were 40, 51, 72 and 87, respectively. The cumulative distribution of the 8-h urinary debrisoquine metabolic ratios, presented as a normal probability plot, formed a discrete population with values over 10. The distribution among patients with BEN/UUT indicates a predominance of extensive debrisoquine hydroxylation and a lack of poor metabolizers. These results are consistent with the hypothesis that the efficiency of oxidative metabolism is greater in BEN patients and that it may be one of the key host factors determining predisposition to these diseases.

Introduction

The etiology and pathogenesis of Balkan endemic nephropathy (BEN) and of the associated tumours of the upper urinary tract (UUT) have been the subject of much speculation. The evidence seems to favour the action of an environmental toxicant (on the basis of the rural origins of the patients and the finding that immigrants are affected after a prolonged stay in these areas). A set of specific epidemiological characteristics, however, suggests the modifying effects of certain host factors. These characteristics include:

—the very peculiar mosaic distribution between and within villages, which is difficult to explain as the mosaic distribution of an environmental factor (which would have to be stable for many years);

—the familial character of both diseases (it is well recognized that there are 'ill' and 'healthy' families in the same village and that one of the most important bases for diagnosing BEN is belonging to a 'BEN family');

—the very specific location of tumours in the upper urinary tract, which might well be the result of genetically determined, specific metabolic properties of cells in that part of the urinary tract; and

—the simultaneous appearance of BEN and UUT in 30% of patients, which may reflect a specific and/or a very great metabolic efficiency in these subjects.

Many xenobiotics, including most carcinogens, are metabolized by cytochrome P450-dependent mono-oxygenases to form reactive electrophilic intermediates, through which the carcinogenic, mutagenic and toxic effects of these chemicals are mediated.

When a single isozyme is responsible for a specific metabolic route for a drug, the efficiency of conversion of the parent drug to that metabolite can be used to measure the activity of the isozyme. A well-characterized example is the isozyme responsible for metabolic hydroxylation of the drug debrisoquine, which is responsible for more than 90% of the conversion to 4-hydroxydebrisoquine. The activity of this enzyme is under the control of a single pair of genes with a homozygous recessive trait resulting in defective metabolism. The ability to metabolize debrisoquine to its 4-hydroxy metabolite is known to vary widely among individuals. Genetic polymorphism for debrisoquine was first described by Mahgoub *et al.* (1977), who showed that there are two human phenotypes for this reaction, based on the urinary ratio of drug to metabolite, measured by a simple gas chromatographic procedure (Idle *et al.*, 1979).

Working Hypothesis

The aim of this study was to test the hypothesis that the efficiency of oxidative metabolism of debrisoquine is higher in patients with BEN/UUT than in healthy control subjects from the same endemic villages and that an individual's capacity for metabolic oxidation may thus be a key determinant in the effect of any hypothetical nephrotoxic and/or oncogenic environmental agent. We used a case-control design, in which we controlled for certain socioeconomic factors and for age, gender and certain habits.

This approach also provides an opportunity to speculate about the nature of the environmental factor: If the hypothesis is true and there is a statistically significant difference in oxidative metabolism between BEN patients and healthy controls from the same villages (and thus sharing the same environment and habits), it can be assumed that the hypothetical environmental factor requires metabolic oxidative activation to exert its nephrotoxic and/or carcinogenic effect.

Patients and Methods

Patients were drawn from the population living in the BEN area of Vratza District in north-western Bulgaria. The choice of an adequate control population is critical to the success of noninterventional epidemiological studies. We used the following groups of control subjects:

—a group drawn from the same population at risk as the patients with BEN/UUT, with similar socioeconomic and occupational status; and

—a group drawn from a population in an area with similar geographic, socioeconomic and occupational conditions but free of BEN and with an incidence of UUT no different from that in the rest of the country.

The patients and controls were thus classified as follows:

—subjects with a positive diagnosis of BEN;
—subjects suspected of having BEN;
—healthy controls from villages affected by BEN;
—healthy subjects from areas with no BEN; and
—patients with UUT.

Some of the patients with UUT also had, or were suspected of having, BEN; others had no evidence of BEN. A separate analysis of UUT patients is not given here, and BEN and UUT patients were combined because of the very close epidemiological characteristics of the two diseases. There were 1.7 times more controls than patients. The mean age of all subjects was about 61 years, but slightly lower in controls from the non-BEN area.

The subjects under study gave informed consent to the protocol. All were given a 10-mg tablet of debrisoquine (Declinax, Roche) by mouth after an overnight fast; all other therapy was discontinued 48 h before the study. Urine was collected over the following 8 h; the volume was recorded, and 5-ml aliquots were taken and stored frozen in plastic containers before being shipped to London by air for analysis at the Pharmacology Department, St Mary's Hospital Medical School. Debrisoquine and 4-hydroxydebrisoquine were measured by electron capture gas chromatography after derivatization with hexafluoroacetone.

In studies to define the genetic polymorphism of debrisoquine hydroxylation, it has been customary to give the 8-h urinary metabolic ratio, as this mathematical measure permits ready identification of poor metabolizers. The metabolic ratio for each subject was calculated from the quotient: % dose excreted as unchanged debrisoquine/% dose excreted as 4-hydroxydebrisoquine during the collection period. The mean metabolic ratio was taken for people whose urine was analysed twice.

Differences in the distribution of phenotypic indices of metabolism in BEN/UUT and control subjects are presented graphically as normal probability plots. These provide a visual indication of multiple populations, characterized by gaps in the curve produced by the data points. All data were tested by the chi-square test, Fisher's exact test and the Wilcoxon rank sum test.

Results and Discussion

We analysed over 900 urine samples from patients and control subjects. Preliminary results have already been reported (Ritchie *et al.*, 1982; Nikolov, 1986), and the most important data are given in Table 1. Urinary recoveries of debrisoquine and 4-hydroxydebrisoquine were similar in patients suspected of having BEN and in control subjects from the BEN area but were 50% lower in patients with BEN. The highest recoveries were obtained in controls from the non-BEN area.

The most interesting feature of the data is the lack of patients with BEN who had relatively impaired metabolism of debrisoquine. Subjects who metabolized < 25% of the drug (metabolic ratio, > 3.0) represented only 2.9% of BEN patients, 12.4% of healthy controls from BEN villages and 12.7% of controls from outside the BEN area. The exceptionally poor metabolizers (metabolic ratio, > 10.0) represent 1.0% of BEN patients and 4.8–5.8% of the control groups. The percentages of exceptionally extensive metabolizers (metabolic ratio, > 0.8) were 86.3, 64.5 and 67.4% in these groups, respectively.

BEN patients were significantly faster oxidizers than either the controls from BEN villages ($2p < 0.01$) or the controls from non-BEN areas ($2p < 0.01$) when tested in the Wilcoxon rank test. The maximal metabolic ratios in these groups were 40, 72 and 87, respectively, and the mean metabolic ratio was lower in BEN patients than in either control group (Table 1).

Table 1. Summary of debrisoquine phenotyping of patients with Balkan endemic nephropathy (BEN) and control subjects, Vratza District, Bulgaria

Group	No.	Age (years)	Recovery (%)	Metabolic ratio		No. of subjects with metabolic ratio		
				Mean±SD	Max	< 0.8	> 3.0	> 10.0
BEN patients	102	61.2±8.8	14.1±6.6	1.08±4.0	40.0	88 (86.3%)	3 (2.9%)	1 (1.0%)
Suspected BEN patients	135	61.8±9.6	27.6±9.9	1.67±5.6	51.0	100 (74.1%)	12 (8.9%)	3 (2.2%)
Controls from BEN villages	243	59.6±10.1	25.6±8.8	3.36±10.7	72.0	163 (67.4%)	30(12.4%)	14(5.8%)
Controls from areas with no BEN	166	53.1±12.2	37.6±13.0	3.16±9.2	87.0	107 (64.5%)	21 (12.7%)	8 (4.8%)

The cumulative distribution of metabolic ratios for debrisoquine in 8-h urine samples is presented as normal probability plots in Figure 1. The antimode appears to lie at a metabolic ratio of 10. Among the control subjects, the poor metabolizers formed a discrete population with values over 10. The distribution in patients with BEN/UUT indicates a predominance of people with extensive debrisoquine hydroxylation activity and an absence of poor metabolizers.

A number of subjects in the populations studied were members of the same family, some of whom developed BEN and others who did not or were only suspected of having BEN. We therefore compared the metabolic ratios for debrisoquine in these family members. Since the subjects had a similar mean age at the time of the study and at the time

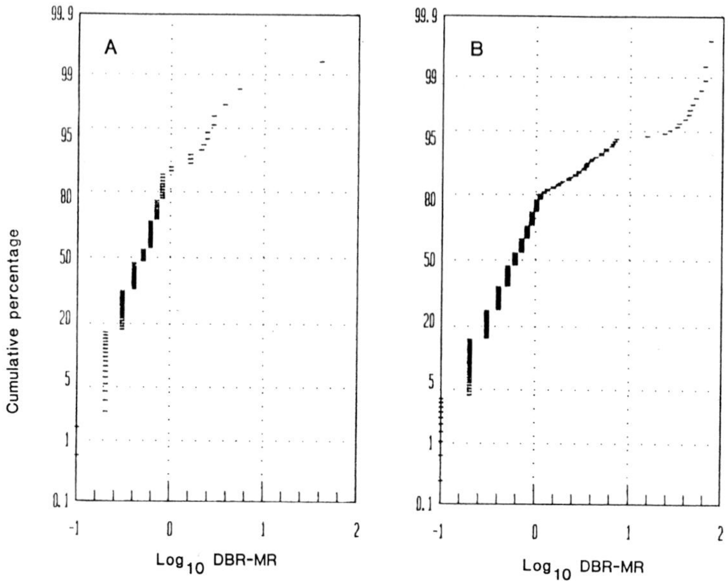

Figure 1. Normal probability distribution of debrisoquine metabolic ratios (DBR-MR) in patients (A) and controls (B) in villages endemic for Balkan nephropathy

they married (approximately 20 years of age), it was assumed that husbands and wives would have been exposed to similar dietary and other environmental factors for at least 40 years. If an individual's metabolic oxidative potential is an important determinant of susceptibility to BEN, then we would expect the spouse with the lower metabolic ratio to have a higher risk of developing the disease. We studied 11 families, in seven of which one spouse had BEN and in four of which one spouse had suspected BEN; the other spouse had no evidence of the disease. In all but one case, the spouse with the lower metabolic ratio developed BEN (Figure 2).

We also investigated siblings, one of whom had BEN. The link between BEN and metabolic ratio in this group might be expected to be weaker, as environmental influences may vary between siblings after they have left the parental home. In seven of eight pairs, however, an association was seen (Figure 3).

The population of BEN patients thus differed from healthy subjects with respect to their metabolism of debrisoquine. This group contained fewer subjects with impaired oxidative ability and more with exceptionally enhanced oxidative ability. The question arises of the biological significance of this difference in relation to the etiology of BEN and/or UUT. Although the case for the involvement of a fungal toxin as

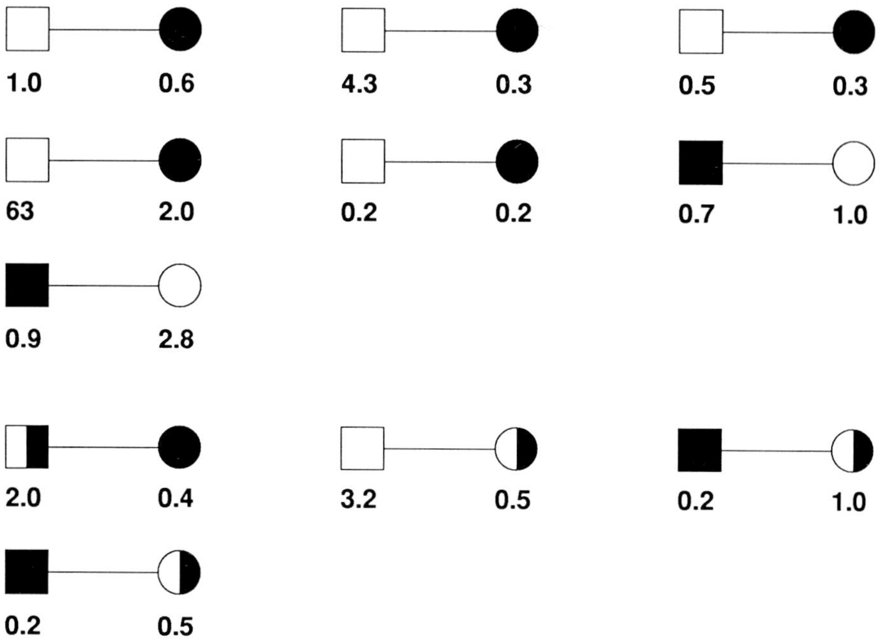

Figure 2. Occurrence of Balkan endemic nephropathy and metabolic ratios for debrisoquine (numbers below symbols) in married couples[a]

[a]Squares, husbands; circles, wives. Filled symbols, patients with Balkan endemic nephropathy; half-filled symbols, subjects suspected of having Balkan endemic nephropathy; empty symbols, subjects with no disease

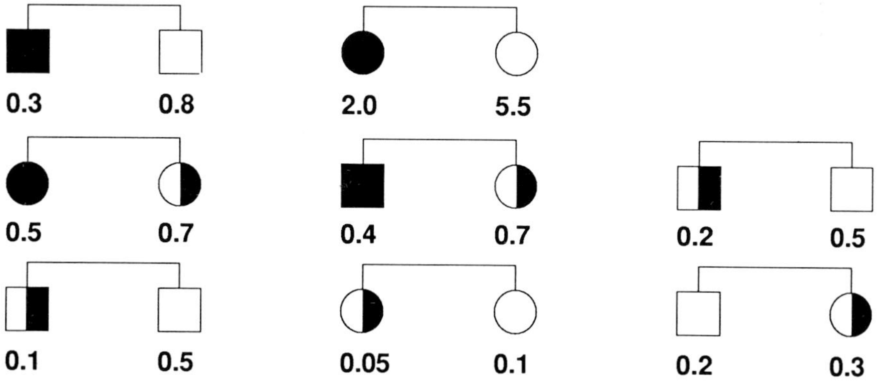

Figure 3. Occurrence of Balkan endemic nephropathy and metabolic ratios for debrisoquine (numbers below symbols) in siblings[a]

[a]Squares, brothers; circles, sisters. Filled symbols, patients with Balkan endemic nephropathy; half-filled symbols, subjects suspected of having Balkan endemic nephropathy; empty symbols, subjects with no disease

the causative agent in BEN remains to be proven, it is tempting to speculate that the metabolic oxidation of such an environmental toxicant might result in its activation to a highly nephrotoxic metabolite. Thus, in a population exposed to similar levels of the toxic agent over a period of many years, the individuals most at risk of developing BEN and/or UUT would be those with the greatest potential for forming the nephrotoxic and/or carcinogenic metabolite. We judged that one way of circumventing the problem of heterogeneous exposures (which could mask the effect of oxidating capacity) was to examine the incidence of BEN in subjects who had shared the same environment for a long time (for example, married couples) and to relate this to oxidative capacity.

The results are consistent with the hypothesis that some individuals, because of genetically determined status, are more likely to develop BEN and/or UUT due to their capacity to activate an agent present in their environment. This agent must now be identified before the precise role of variable metabolic activation in its nephrotoxicity can be fully understood. Ochratoxin A, a potent nephrotoxic carcinogen produced by a number of fungal species, has been implicated by many investigators as a possible etiological factor for BEN and/or UUT, but whether or not it requires metabolism before expressing its full nephrotoxicity and/or carcinogenicity is unknown.

The power of modern techniques of molecular biology should be used to obtain more precise information about genetically determined ability to metabolize xenobiotics. We are planning to genotype the same population groups, in and outside the area of BEN, using the simple DNA-based genetic assay described by Gough (1990) for debrisoquine hydroxylase.

Acknowledgements

This work was carried out in collaboration with I.S. Stoyanov, I.I. Stoichev and N.D. Kalcheva, National Oncological Centre, Medical Academy, Sofia, Bulgaria; J.C. Ritchie, M.J. Crothers and B. Cartmel, Pharmacology Department, St Mary's Hospital Medical School, London, United Kingdom; T.A. Connors and J.B. Greig, Toxicology Unit, Medical Research Council Laboratories, Carshalton, Surrey, United Kingdom; and M. Castegnaro and J. Michelon, International Agency for Research on Cancer, Lyon, France.

References

Gough, A.C., Miles, J.S., Spurr, N.K., Moss, J.E., Gaedigk, A., Eichelbaum, M. & Wolf, C.R. (1990) Identification of the primary gene defect at the cytochrome P450 CYP2D locus. *Nature*, **347**, 773–775

Idle, J.R., Mahgoub, A., Angelo, M.M., Doring, L.G., Lancaster, R. & Smith, R.L. (1979) The metabolism of (14C)-debrisoquine in man. *Br. J. Clin. Pharmacol.*, **7**, 257–266

Mahgoub, A., Idle, J.R., Dring, L.G., Lancaster, R. & Smith, R.L. (1977) Polymorphic hydroxylation of debrisoquine in man. *Lancet*, **ii**, 584–586

Nikolov, I.G. (1986) Concept of poor and extensive metabolizers (Abstract 3305). In: *Abstracts of Lectures, Symposium and Free Communication, 14th International Cancer Congress, Budapest, 21–27 August 1986*, Vol. 3, Basel, Karger, p. 853

Ritchie, J.C., Crothers, M.J., Idle, J.R., Greig, J.B., Connors, T.A., Nicolov, I.G. & Chernozemsky, I.N. (1982) Evidence for an inherited metabolic susceptibility to endemic (Balkan) nephropathy. In: Strahinjic, S. & Stefanovic, V., eds, *Current Research in Endemic (Balkan) Nephropathy*, Niš, University Press, pp. 23–27

Mycotoxins, Endemic Nephropathy and Urinary Tract Tumours
Ed. M. Castegnaro, R. Pleština, G. Dirheimer, I.N. Chernozemsky & H. Bartsch
Lyon, International Agency for Research on Cancer, pp. 297-304

CHARACTERIZATION OF THE CYTOCHROME P450 ISOZYME THAT METABOLIZES OCHRATOXIN A, USING METABOLIC INDUCERS, INHIBITORS AND ANTIBODIES

E. Hietanen[1], H. Bartsch[2], J.-C. Béréziat[2], M. Castegnaro[2] & J. Michelon[2]

[1]Department of Clinical Physiology, University of Turku, 20520 Turku, Finland; and [2]International Agency for Research on Cancer, Lyon, France

Summary

The phenotypic pattern of drug biotransformation is determined by both host and environmental factors. Debrisoquine is a good probe for phenotyping individuals, as its metabolism is not affected by age, gender, smoking habits or alcohol intake. People with Balkan endemic nephropathy or with urinary tract tumours in endemic areas are more frequently extensive metabolizers of debrisoquine than are healthy people. This finding has led to studies of the possible relationship between the metabolism of ochratoxin A and its toxicity and carcinogenicity on experimental models. Ochratoxin A is metabolized mainly in the liver into R- and S-isomers of 4-and 10-hydroxyochratoxin A, and the reaction is catalysed by cytochrome P450 haemoprotein. Animal species that are genetically different in their capacity to metabolize debrisoquine differ similarly in their capacity to metabolize ochratoxin A: female DA rats that are poor metabolizers of debrisoquine also poorly metabolize ochratoxin A, as assayed by urinary excretion of both the parent compound and of 4-hydroxyochratoxin A. Ochratoxin A hydroxylase activity is low in DA rat liver (and kidney) but is inducible by phenobarbital and 3-methylcholanthrene; debrisoquine hydroxylase is not known to be inducible by enzyme inducers. The reaction of ochratoxin A hydroxylase thus resembles those induced by 3-methylcholanthrene and catalysed by cytochrome P450IA. Ochratoxin A hydroxylase activity was further characterized in the livers of B_6 and D_2 mice that had been treated with typical enzyme inducers. Ochratoxin A hydroxylase was weakly inducible by phenobarbital, 3-methylcholanthrene and 2,4,7,8-tetrachlorodibenzodioxin. When the activity of various metabolic inhibitors was compared, benzoflavone was found to inhibit phenacetin *O*-deethylase activity in both control and 3-methylcholanthrene-treated mice; bufuranol-1-hydroxylase was strongly inhibited by benzoflavone in both B_6 and D_2 mice, and ochratoxin A hydroxylase was strongly inhibited by this compound in D_2 strain and moderately inhibited in controls of the B_6 strain. Metyrapone inhibited these enzymes relatively weakly. Three monoclonal antibodies (1-7-1 against cytochrome P450IA1 and 2, 2-66-3 against P450IIB and 1-91-3 against P450IIE) were used to characterize the cytochrome P450 isozyme responsible for the 4-hydroxylation of ochratoxin A in mice treated with various inducers. Only 1-7-1 could inhibit ochratoxin A 4-hydroxylation, and the inhibition was not dependent on pretreatment of the mice. The results suggest that the isozyme that catalyses ochratoxin A hydroxylation is not identical to $P450_{db}$ and may share common epitopes with cytochrome P450IA.

Introduction

The cytochrome P450-dependent monooxygenase system catalyses the oxidation of numerous drugs, chemicals and environmental and endogenous compounds. Although many cytochrome P450 isozymes may have overlapping catalytic activities or interfering immunological properties, their catalytic and immunological properties and their amino acid sequences can be used to identify different isozymes clearly. The number of genes that determines enzyme activity varies (Nebert & Weber, 1990). Drug biotransformation activity is determined by both host and environmental factors, yielding a phenotypic drug biotransformation pattern. Depending on the isozyme, one or more genes controls the P450-catalysed activity; some isozymes are controlled mainly genetically, and some are strongly dependent on environmental factors which induce or inhibit enzyme activity.

Genetic polymorphism is determined as a monogenic or Mendelian trait in a population as at least two phenotypes, each representing the actions of different alleles at a single gene locus (Vogel & Motulsky, 1979). Debrisoquine has limited use as an antihypertensive drug; its large dose range led to the observation of a genetic polymorphism in its 4-hydroxylation (Mahgoub *et al.*, 1977), as measured by the percentage ratio of excretion of debrisoquine and 4-hydroxydebrisoquine. This polymorphism was confirmed subsequently, and it was suggested that the hydroxylation of debrisoquine to its 4-hydroxy derivative is regulated in humans by two genes which act at a single gene locus (Price-Evans *et al.*, 1980) such that the poor metabolizer phenotype is autosomally recessive. Debrisoquine is a good probe drug in the sense that its metabolism is not affected by factors such as age, gender, smoking habits or alcohol intake (see Kallio, 1990, for a review).

The hydroxylation of debrisoquine has been shown experimentally to be catalysed by a cytochrome P450IID6 isozyme that is regulated by a gene located on chromosome 22 (Nebert *et al.*, 1989). The cDNA library has been constructed from the mRNA (Gonzales *et al.*, 1988), and cDNA analysis of the livers of poor metabolizers showed that they have at least three variant mRNAs, which deviate from the normal P450IID6 mRNA, leading to synthesis of unstable P450IID6 proteins that cannot be detected by anti-db1 antibody in humans. It has subsequently become evident that many drugs cosegregate with debrisoquine, sometimes representing very different oxidative processes (Kallio, 1990). Interestingly, people who are extensive metabolizers have a higher risk of developing cancers of the liver, gastrointestinal tract and lung than people of the poor metabolizer phenotype (for references, see Kallio, 1990); however, in a recent study, the association with lung cancer was not confirmed (Speirs *et al.*, 1990).

An association has been found between the presence of the environmental contaminant, ochratoxin A, in mouldy grain and other foods and the incidences of Balkan endemic nephropathy and of renal tumours. Furthermore, the concentrations of ochratoxin A in blood samples from subjects living in areas of high risk in the Balkans were higher and more frequent than in people living in low-risk areas (Castegnaro *et al.*, 1990). Patients with Balkan endemic nephropathy or renal tumours in these areas were also more frequently extensive metabolizers of debrisoquine than healthy persons. This finding led us to study further the possible relationship between the metabolism of ochratoxin and its toxicity and carcinogenicity in experimental models, and its relationship to the metabolism of debrisoquine.

Ochratoxin A is metabolized mainly in the liver to R- and S-isomers of 4- and 10-hydroxyochratoxin A; the reaction is catalysed by cytochrome P450 haemoprotein, possibly *via* intestinal hydrolysis into ochratoxin α. We have used metabolic inducers and inhibitors and monoclonal antibodies against different cytochrome P450 isozymes to characterize further the metabolism of ochratoxin A and to compare it to that of debrisoquine, in order to explore their close cosegregation in animal models and in humans.

Materials and Methods

Animals

Adult male B_6 and D_2 mice were used to characterize the regulation of ochratoxin A hydroxylase *in vitro*. Animals were treated with typical enzyme inducers, then microsomes were prepared by the conventional ultracentrifugation method for enzyme assays, as described elsewhere (Hietanen *et al.*, 1986).

Enzyme assays

The cytochrome P450 concentration of liver microsomes was measured as described by Omura and Sato (1964). Ochratoxin A hydroxylase was assayed as described by Störmer *et al.* (1983), bufuranol-1-hydroxylase according to Gut *et al.* (1984) and phenacetin *O*-deethylase by the method of Gillam and Reilly (1988).

Inhibition of ochratoxin A hydroxylase by various monoclonal antibodies (kindly donated by Dr H.V. Gelboin, Laboratory of Molecular Carcinogenesis, National Cancer Institute, National Institutes of Health, Bethesda, MD, USA) and by benzoflavone (a specific competitive inhibitor of cytochrome P450IA-catalysed oxidations) and metyrapone was compared with that of other monooxygenases. The optimal monoclonal antibody concentrations were used as described previously (Hietanen *et al.*, 1986a), and the concentrations of metabolic inhibitors were such as to produce maximal enzyme inhibition.

Results

We showed previously (Hietanen *et al.*, 1986b) that ochratoxin A hydroxylase is inducible in rat liver by 3-methylcholanthrene and phenobarbital. In the present study in mice, ochratoxin A hydroxylase was inducible by phenobarbital, 3-methylcholathrene and 2,4,7,8-tetrachlorodibenzodioxin over a fairly modest range (Tables 1 and 2).

Inhibition of three monooxygenases by the metabolic inhibitors benzoflavone and metyrapone was compared in control and in 3-methylcholanthrene-pretreated mouse liver microsomes (Table 2). Benzoflavone moderately inhibited phenacetin *O*-deethylase activity (catalysed by cytochrome P450IA) in both control and methylcholanthrene-pretreated mice and fairly strongly inhibited bufuranol-1-hydroxylase in both B_6 and D_2 mice. Ochratoxin A hydroxylase was strongly inhibited by benzoflavone in D_2 mice and moderately in B_6 controls. Metyrapone inhibited these enzymes relatively weakly.

Three monoclonal antibodies (1-7-1 against cytochrome P450IA1 and 2, 2-66-3 against P450IIB and 1-91-3 against P450IIE) were used to characterize the cytochrome P450 isozyme responsible for the 4-hydroxylation of ochratoxin A in mice treated with various inducers (Table 3).

Table 1. Inducibility of ochratoxin 4-hydroxylase in the livers of B_6 and D_2 mice by typical enzyme inducers

Strain	Inducer	pmol 4-hydroxyochratoxin A/min per nmol P450	Fold induction
B_6	Control	29.1	1.0
	Pyrazole	21.0	0.7
	Phenobarbital	14.2	0.5
	Methylcholanthrene	28.8	1.0
D_2	Control	20.4	1.0
	Pyrazole	14.0	0.7
	Phenobarbital	28.9	1.4
	Methylcholanthrene	51.1	2.5
	TCDD	79.8	3.9

Table 2. Inhibition of ochratoxin 4-hydroxylase, phenacetin *O*-deethylase and bufuranol-1-hydroxylase activities by cytochrome P450IA (benzoflavone, 40 μM) or nonspecific cytochrome P450 (metyrapone, 40 μM) monooxygenase inhibitors in liver of B_6 and D_2 mice[a]

Inhibitor	Ochratoxin A hydroxylase		Phenacetin *O*-deethylase		Bufuranol-1-hydroxylase	
	Control	MC	Control	MC	Control	MC
Strain B_6						
None	100 (1)	100(1.5)	100(1)	100(1)	100(1)	100(1.9)
Benzoflavone	46	304	46	55	48	26
Metyrapone	72	99	102	106	115	90
Strain D_2						
None	100(1.34)	100(0.69)	100(1.2)	100(1.4)	100(0.85)	100(1.23)
Benzoflavone	23	30	63	84	48	26
Metyrapone	50	95	75	81	115	90

[a]Remaining activity as compared to that without the metabolic inhibitor is shown in control and 3-methylcholanthrene (MC)-pretreated mice. Numbers in parentheses, activity relative to that in B_6 control livers

Only 1-7-1 was able to inhibit ochratoxin A 4-hydroxylation, and the inhibition was not dependent on pretreatment of the mice.

Discussion

The toxicity of ochratoxin A has not been elucidated completely. It induced DNA single-strand breaks in mice and rats but was not mutagenic in tests in *Salmonella typhimurium*; it induced sister chromatid exchange in CHO cells in the presence of an exogenous metabolic system but not without it (Boorman, 1988). In a rat liver microsomal system, lipid peroxidation was induced to a similar extent by ochratoxin A and ochratoxin C but not by other ochratoxin derivatives, nor did the latter intervene in calcium homeostasis (Rahimtula *et al.*,

Table 3. Inhibition of ochratoxin 4-hydroxylase activity by monoclonal antibodies against cytochrome P450IA(1-7-1), P450IIB(2-66-3) and P450IIE(1-91-3) in livers of B_6 and D_2 mice pretreated with various enzyme inducers. Activities are shown as % remaining activity without monoclonal antibody.

Strain	Inducer	Monoclonal antibody		
		1-7-1	2-66-3	1-91-3
B_6	Control	67	113	159
	Pyrazole	–	–	155
	Phenobarbital	–	143	–
	Methylcholanthrene	99	–	–
D_2	Control	64	133	125
	Pyrazole	66	117	82
	Phenobarbital	57	102	102
	TCDD	57	119	92
	Methylcholanthrene	59	–	–

TCDD, 2,4,7,9-tetrachlorodibenzodioxin

1988; Khan *et al.*, 1989). Whether lipid peroxidation is a primary event in the toxicity of ochratoxin A remains to be clarified, as does its relationship to the metabolic conversion of ochratoxin A.

In a two-year carcinogenicity study in Fischer 344/N rats, ochratoxin A caused a dose-dependent increase in the incidence of both benign and malignant kidney tumours, more frequently in males than in females (Boorman, 1988). When the data were adjusted to historical data on the debrisoquine metabolic rate in the same strain, males were found to be faster metabolizers than females (Castegnaro *et al.*, 1990). Similarly, as mice are much slower metabolizers of debrisoquine than rats, they were much more resistant to ochratoxin A-induced kidney tumours (Castegnaro *et al.*, 1990).

As reviewed above and as shown previously (Castegnaro *et al.*, 1990), animal species that are genetically different in their capacity to metabolize debrisoquine seem to differ in a similar way in terms of ochratoxin A metabolism: female DA rats, which are poor metabolizers of debrisoquine, are also poor metabolizers of ochratoxin A, as assayed by the urinary excretion of ochratoxin A and 4-hydroxyochratoxin A. In a study using organs from differently induced female DA strain rats (poor metabolizers) and female Lewis rats (extensive metabolizers of debrisoquine) (Hietanen *et al.*, 1986b), we found that ochratoxin A hydroxylase activity was low in DA rat livers (and kidneys) but was inducible by phenobarbital and 3-methylcholanthrene; debrisoquine hydroxylase is not known to be inducible by enzyme inducers. In this respect, the reactions of ochratoxin A hydroxylase resemble those induced by 3-methylcholanthrene and catalysed by cytochrome P450IA.

These results suggest that the isozyme that catalyses ochratoxin A hydroxylation is not identical to cytochrome P450IID and may share common epitopes with cytochrome P450IA. Its similarity to cytochrome P450IA is supported by the following results: (i) its inducibility by enzyme inducers (although rather weak), (ii) the correlation with P450IA-catalysed oxidation reactions, (iii) its partial inhibition by the monoclonal antibody 1-7-1 and (iv) its inhibition *in vitro* by benzoflavone. It also has properties in common with P450IID: (i) genetic cosegregation in rats and (ii) inhibition in human livers by an antibody against P450IID (Hietanen *et al.*, 1985); however, the cytochrome P450 isozyme that catalyses ochratoxin A oxidation differs from P450IA in the lack of genetic cosegregation in D2 and B6 mice, and it differs from P450IID in its inducibility.

References

Boorman, G., ed. (1988) *NTP Technical Report on the Toxicology and Carcinogenesis Studies of Ochratoxin A* (NTP Technical Report 358; NIH Publication No. 88-2813), Research Triangle Park, NC, National Toxicology Program

Castegnaro. M., Chernozemsky, I.N., Hietanen, E. & Bartsch, H. (1990) Are mycotoxins risk factors for endemic nephropathy and associated urothelial cancers? *Arch. Geschwulstforsch.*, **60**, 295–303

Gillam, E.M. & Reilly, P.E.B. (1988) Phenacetin O-deethylation by human liver microsomes: kinetics and propranolol inhibition. *Xenobiotica*, **18**, 95–104

Gonzales, F.J., Vilbois, F., Hardwick, J.P., McBride, W., Nebert, D.W., Gelboin, H.V. & Meyer, U.A. (1988) Human debrisoquine 4-hydroxylase (P450IID1): cDNA and deduced amino acid sequence and assignment of CYP2D locus to chromosome 22. *Genomics*, **2**, 174–179

Gut, J., Gasser, R., Dayer, P., Kronbach, T., Catin, T. & Meyer, U.A. (1984) Debrisoquine-type polymorphism of drug oxidation: purification from human liver of a cytochrome P450 isozyme with high activity for bufuranol hydroxylation. *FEBS Lett.*, **173**, 287–290

Hietanen, E., Bartsch, H., Castegnaro, M., Malaveille, C., Michelon, J. & Broussolle, L. (1985) Use of antibodies (Ab) against cytochrome P-450 isozymes to study genetic polymorphism in drug oxidations. *J. Pharm. Clin.*, **4**, 71–78

Hietanen, E., Malaveille, C., Friedman, F.K., Park, S.S., Béréziat, J.-C., Brun, G., Bartsch, H. & Gelboin, H.V. (1986a) Monoclonal antibody-directed analysis of cytochrome P-450-dependent monooxygenases and mutagen activation in the livers of DBA/2 and C57BL/6 mice. *Cancer Res.*, **46**, 524–531

Hietanen, E., Malaveille, C., Camus, A.-M., Béréziat, J.-C., Brun, G., Castegnaro, M., Michelon, J., Idle, J.R. & Bartsch, H. (1986b) Interstrain comparison of hepatic and renal microsomal carcinogen metabolism and liver S9-mediated mutagenicity in DA and Lewis rats phenotyped as poor and extensive metabolizers of debrisoquine. *Drug Metab. Disposition*, **14**, 118–126

Kallio, J. (1990) *Genetic Polymorphism of Debrisoquine Oxidation in Man* (University of Turku Publication 63), Academic Dissertation

Khan, S., Martin, M., Bartsch, H. & Rahimtula, A.D. (1989) Perturbation of liver microsomal calcium homeostasis by ochratoxin A. *Biochem. Pharmacol.*, **38**, 67–72

Mahgoub, A., Idle, J.R., Dring, L.G., Lancaster, R. & Smith, R.L. (1977) Polymorphic hydroxylation of debrisoquine in man. *Lancet*, **ii**, 584–58

Nebert, D.W. & Weber, W.W. (1990) Pharmacogenetics. In: Pratt, W.B. & Taylor, P., eds, *Principles of Drug Action*, 3rd ed., New York, Churchill Livingstone, pp. 469–531

Nebert, D.W., Nelson, D.R., Adesnik, M., Coon, M.J., Estabrook, R.W., Gonzalez, F.J., Guengerich, F.P., Gunsalus, I.C., Johnson, E.F., Kemper, B., Levin, W., Phillips, I.R., Sato, R. & Waterman, R.M. (1989) The P450 superfamily: updated listing of all genes and recommended nomenclature for the chromosomal loci. *DNA*, **8**, 1–13

Omura, T. & Sato, R. (1964) The carbon monoxide binding pigment of liver microsomes. *J. Biol. Chem.*, **239**, 2370–2378

Price-Evans, D.A., Mahgoub, A., Sloan, T.P., Idle, J.R. & Smith, R.L. (1989) A family and population study of the genetic polymorphism of debrisoquine oxidation in a white British population. *J. Med. Genet.*, **17**, 102–105

Rahimtula, A.D., Béréziat, J.-C., Bussachini-Griot, V. & Bartsch, H. (1988) Lipid peroxidation as a possible cause of ochratoxin A toxicity. *Biochem. Pharmacol.*, **37**, 4469–4477

Speirs, C.J., Murray, S., Davies, D.S., Biola Mabadeje, A.F. & Boobis, A.R. (1990) Debrisoquine oxidation phenotype and susceptibility to lung cancer. *Br. J. Clin. Pharmacol.*, **29**, 101–109

Strömer, F.C., Stören, O., Hansen, C.E., Pedersen, J.I. & Aasen, A.J. (1983) Formation of (4R) and (4S)-4-hydroxyochratoxin A and 10-hydroxyochratoxin A from ochratoxin A by rabbit liver microsomes. *Appl. Environ. Microbiol.*, **45**, 1183–1187

Vogel, F. & Motulsky, A.G. (1979) *Human Genetics. Problems and Approaches*, Berlin, Springer

RISK ASSESSMENT FOR HUMAN EXPOSURE TO OCHRATOXIN A

Mycotoxins, Endemic Nephropathy and Urinary Tract Tumours
Ed. M. Castegnaro, R. Pleština, G. Dirheimer, I.N. Chernozemsky & H. Bartsch
Lyon, International Agency for Research on Cancer, pp. 307-320

RISK ASSESSMENT OF OCHRATOXIN A RESIDUES IN FOOD

T. Kuiper-Goodman

Toxicological Evaluation Division, Bureau of Chemical Safety, Food Directorate, Health Protection Branch, Health and Welfare Canada, Tunney's Pasture, Ottawa, Canada

Summary

Ochratoxin A is a mycotoxin that has been found to occur in foods of plant origin, in edible animal tissues and in human sera and tissues. The ability of ochratoxin A to move up the food chain is associated with its long half-life in certain edible animal species. In this presentation, approaches for the evaluation of the health risks due to the presence of ochratoxin A in food products are described. The major target for ochratoxin A toxicity in all mammalian species tested is the kidney, and endemic nephropathies affecting livestock as well as humans have been attributed to ochratoxin A. Ochratoxin A is also teratogenic, and in the fetus the major target is the developing central nervous system. Recent studies have provided 'clear evidence' for the carcinogenicity of ochratoxin A in two rodent species. It was found to be non-mutagenic in various microbial and mammalian gene mutation assays, but weak genotoxic activity to mammalian cells was noted. In addition, ochratoxin A was found to suppress immune function. On the basis of a carcinogenicity study with ochratoxin A in rats, reported from the National Toxicology Program in the USA, the estimated tolerable daily intake of ochratoxin A in humans ranges from 1.5 to 5.7 ng/kg bw per day, depending on the method of extrapolation used. The worst-case estimate for daily exposure to ochratoxin A from the consumption of pork-based food products and cereal foods for young Canadian children, the highest consumption group on a body weight basis, is probably less than 1.5 ng/kg body weight per day (mean of eaters). In view of the toxic properties of ochratoxin A, it is recommended that exposure to this toxin be kept to a minimum.

Introduction

Ochratoxin A is a mycotoxin found in a variety of food crops in temperate climates. It is the causal agent of porcine nephropathy, and its presence in human foodstuffs in certain regions of the Balkans is associated with a human nephropathy endemic in those regions. Ochratoxin A is nephrotoxic in all monogastric mammalian species tested and is a renal carcinogen in the two rodent species tested. It also causes liver lesions and is teratogenic as well as immunotoxic. Widespread human exposure to low levels of this mycotoxin is evident from surveys of human sera conducted in several European countries and in Canada. It is therefore necessary to assess whether such exposure represents a potential health risk.

In assessing the risk represented by the presence of ochratoxin A residues in human foodstuffs, one examines the likelihood (probability) that harm might occur as a result of consumption of food or food products containing ochratoxin A. Such risk assessments have two major components (Figure 1)—exposure assessment, which may be

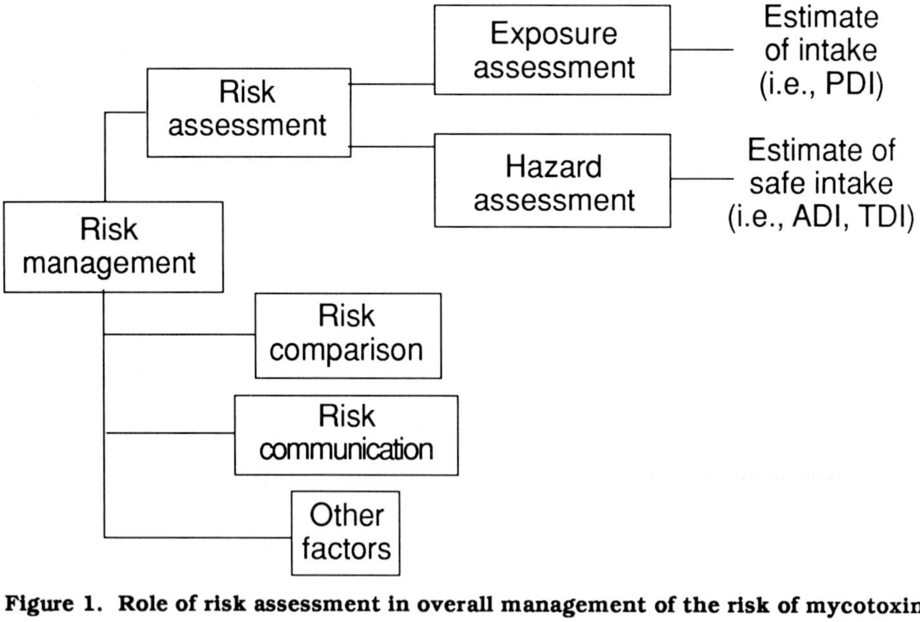

Figure 1. Role of risk assessment in overall management of the risk of mycotoxins

PDI, probable daily intake; TDI, tolerable daily intake; ADI, acceptable daily intake different for different countries or even groups of individuals within countries, and hazard assessment, which can also vary within subgroups of the population, owing, for instance, to differences in genetic make up or in exposure to viruses. The term *hazard* refers to an intrinsic property of a chemical and to the occurrence of toxic effects in a particular species with a certain degree of exposure. *Risk* implies some uncertainty, and in absolute terms it is the predicted frequency of occurrence of adverse effects resulting from a given exposure; in relative terms, as used here, it is the expression of exposure (probable daily intake, PDI) relative to the dose that is considered to be safe. *Safety* or *safe dose* (i.e., acceptable daily intake, ADI, or tolerable daily intake, TDI) reflects the probability that injury will not result under specific conditions of exposure. A detailed review and risk assessment of ochratoxin A, which includes a discussion of the chemistry, mycology and natural occurrence of ochratoxin A as well as its toxicity, metabolic disposition and role in porcine, avian and human nephropathies, was published recently (Kuiper-Goodman & Scott, 1989). The WHO Joint Expert Committee on Food Additives has also recently reviewed this mycotoxin, with the aim of arriving at a TDI (WHO, 1991a), and has prepared a monograph (WHO, 1991b). Reviews have also been published by the International Programme on Chemical Safety (WHO, 1990) and by the Deutsche Forschungsgemeinschaft (Frank *et al.*, 1990), but a TDI was not established .

Exposure Assessment

Human exposure to ochratoxin A is widespread, as evidenced from surveys of human serum samples conducted in several European countries. Recent studies in Manitoba, Canada, yet to be confirmed (see Frohlich, this symposium), have also indicated the presence of low levels of ochratoxin A in about 40% of persons sampled. The widespread occurrence in human serum is not surprising and can be attributed to the probable long half-life of ochratoxin A (Hagelberg *et al.*, 1989).

Estimates of human exposure based on levels in serum or tissue samples reflect both recent and chronic exposure at the level of the individual and are very useful in epidemiological studies. Comparisons help to identify populations at higher risk due to greater exposure to ochratoxin A, although these measurements do not tell us what the intake of ochratoxin A is. For this, a pharmacokinetic model applicable to humans is necessary. Such a model, which relates continuous intake of ochratoxin A to plasma concentration, plasma clearance and bioavailability of ochratoxin A, presumably at steady-state, based on the probable long half-life of ochratoxin A, was recently used by Breitholz *et al.* (1991), who found that the daily intake calculated in this way (expressed as nanograms per kilogram body weight per day) was about

40% greater than the human plasma concentration (expressed as nanograms per millilitre). Comparison of the results obtained from different laboratories for different populations has been difficult because of the use of methods with different detection limits and sensitivities and of different ways of calculating means. There does not appear to be a large difference between the ranges of values found in Germany (Bauer & Gareis, 1987), Sweden (Breitholz *et al.*, 1991), Denmark (Hald, 1989), Poland (Goliński & Grabarkiewicz-Szczęsna, 1985) and recent, as yet unconfirmed results obtained in Canada; but values found in some endemic areas of the Balkan have been higher (Petkova-Bocharova *et al.*, 1988).

To discover the possible source of ochratoxin A and to help in setting tolerances or guidelines for minimally allowable residue limits, the level of ochratoxin A in food commodities has been measured, and the information has been used to estimate human exposure. The estimates of occurrence, combined with food consumption patterns obtained by dietary recall for people of different ages and genders of the Canadian population (Health and Welfare Canada, 1976), have been used to estimate the probable daily intake (Kuiper-Goodman, 1990; Figure 1). The value obtained can be compared to the usually more reliable estimate of feed intake in animal feeding studies. Preliminary data indicate that exposure to ochratoxin A in Canada is, in general, probably less than 1.5 ng/kg body weight per day. Further monitoring of grains, pork serum and commercial animal- and grain-derived food products, with better detection limits, are in progress in order to refine the exposure asssessment.

Hazard Assessment

Absorption and metabolic disposition

Data on the absorption and metabolic disposition of ochratoxin A have been reviewed (Kuiper-Goodman & Scott, 1989; WHO, 1991b) and are discussed further by Galtier (this volume). The fraction of ochratoxin A bound to serum albumin and other serum macromolecules constitutes a mobile reserve of mycotoxin which is available for release to the tissues for a long time. Differences in serum half-life may be related in part to differences in absorption, differences in peak plasma values and species differences in degree of binding to serum macromolecules, including albumin. Both biliary excretion and glomerular filtration play important roles in the plasma clearance of ochratoxin A in rats. In different species, the relative contribution of each excretory route is influenced by the degree of serum macromolecular binding and differences in the degree of enterohepatic recycling of ochratoxin A (Hagelberg *et al.*, 1989).

Ochratoxin A is hydrolysed to the non-toxic ochratoxin α at various sites. In rats, the major excretory products are ochratoxin α (in both urine and faeces), ochratoxin A and the 4R-hydroxy-ochratoxin A epimer;

in the urine, these represented 25–27%, 6% and 1–1.5% of the administered dose, respectively (Støren *et al.*, 1982). Many researchers have considered that the toxicity of ochratoxin A is due to one of its metabolites. It appears, however, that in the rat ochratoxin A itself, rather than one of its metabolites, may be the active toxic agent, since the known metabolites are equally or less toxic than ochratoxin A itself.

Very few data are available on the metabolic disposition of ochratoxin A in humans. It may have a long serum half-life in humans in view of its strong binding to human serum macromolecules (Bauer & Gareis, 1987; Hagelberg *et al.*, 1989).

Effects on enzymes and other biochemical parameters

The primary effect of ochratoxin A is to inhibit protein synthesis. Secondarily, RNA and DNA synthesis may be inhibited. The inhibition of protein synthesis is specific and occurs at the post-transcription level, so that ochratoxin A has a direct effect on the translation step involving competitive inhibition of phenylalanine-$tRNA^{Phe}$ synthetase, so that amino acylation and peptide elongation are stopped (Röschenthaler *et al.*, 1984). This reaction is fundamental for all living organisms. In it, ochratoxin A may be regarded as an analogue of phenylalanine.

Toxicological studies

In the evaluation of toxicological studies, the highest experimental dose at which no effect is observed is the no-observed-effect level (NOEL); the lowest level at which effects are observed is the lowest-observed-effect level (LOEL). The NOEL is not really an intrinsic property of the chemical under study but is highly sensitive to the choice of dose level and the number of animals per dose group in a study. To overcome problems in establishing a NOEL, we have estimated, where necessary, a statistically derived no-effect-level (NEL) (Kuiper-Goodman, 1990).

Acute toxicity: In studies of acute toxicity, dogs and pigs were found to be more sensitive to the effects of ochratoxin A than mice or rats, as indicated by the LD_{50} values (Table 1).

Short-term studies: Ochratoxin A was nephrotoxic in all monogastric mammalian species tested, and the results of about 12 short-term studies, reviewed by Kuiper-Goodman and Scott (1989) and the WHO Joint Expert Comittee on Food Additives (WHO, 1991a), showed that the effects were dose-related. At high doses, changes in renal function involved increases in blood urea nitrogen, changes in renal volume, increased urinary protein and glucose; changes in renal pathology involved degenerative changes in the entire tubular system, but especially in the proximal convoluted tubules. Thickening of the tubular

Table 1. LD_{50} values for ochratoxin A in various species[a]

Species	LD_{50} (mg/kg body weight)		
	Oral route	Intraperitoneal route	Intravenous route
Mouse	46–58.3	22–40.1	25.7–33.8
Rat	20–30.3	12.6	12.7
Rat neonate	3.9		
Dog	0.2		
Pig	1		
Chicken	3.3		

[a] From Kuiper-Goodman & Scott (1989)

basement membrane was a prominent feature at higher dose levels in some studies in rats (Munro *et al.*, 1974). The most sensitive effects in the kidney were changes in urinary enzymes (LOEL = 0.008 mg/kg body weight for pigs) and morphological changes, such as karyomegaly and eosinophilia in proximal convoluted tubules in rats (LOEL = 0.015 mg/kg bw; Munro *et al.*, 1974) (Table 2). At higher dose levels, the latter effects persisted even after rats had been on a control diet for 16 weeks. Changes in other organ systems, such as the liver, were seen with higher doses of ochratoxin A (for review, see Kuiper-Goodman & Scott, 1989).

Teratology and reproduction: It is now well established that ochratoxin A is a potent teratogen in mice, rats, hamsters and chickens, but not in pigs (for review, see Kuiper-Goodman & Scott, 1989). Species differences in susceptibility to teratogenic effects can be attributed in part to species differences in placental transfer of ochratoxin A, and severity of effects depended on the route of exposure and the time of administration during gestation.

When given to mice by gavage, ochratoxin A was teratogenic only when administered on day 8 or 9 of gestation. Arora and Frölen (1981) found that mice were most susceptible when treated by gavage on day 9 of gestation, with a LOEL = 1 mg/kg body weight (Table 2); no NOEL was established.

In teratogenicity studies with ochratoxin A, the central nervous system was one of the most susceptible targets of ochratoxin A, and it was affected at the time of early organogenesis. In adults, however, the kidney is the major target. Since ochratoxin A is a potent kidney carcinogen in adult rats (see below), it could be a transplacental central nervous system carcinogen. Studies should be designed to explore this hypothesis.

Ochratoxin A is also a developmental neurotoxin, and more elaborate postnatal neurotoxicity testing is recommended. No adequate study of the effects of ochratoxin A on reproduction has yet been conducted.

Table 2. Summary of effects observed in studies with laboratory animals

Effect	Species	Duration of of study	LOEL[a] (mg/kg bw/day)	NOEL[b] (mg/kg bw/day)
Renal function	Pig	5–90 days	0.008	–
Karyomegaly of proximal convo-luted tubules	Rat	90 days	0.015	–
Progressive nephropathy	Pig	2 years	0.041	–
Overt craniofacial anomalies	Mouse	Day 9[c]	1	–
Kidney tumours	Mouse	2 years	4.4	0.13
Kidney tumours	Rat	2 years	0.07	0.02
Necrosis of lymphoid tissues of thymus and tonsils	Dog	14 days	0.1	–
Decreased antibody response	Mouse	50 days		0.5[d]

[a]Lowest-observed-effect level
[b]No-observed-effect level
[c]Single dose on day 9 of gestation
[d]Only dose tested

Genotoxicity: The genotoxicity of ochratoxin A was reviewed recently (Kuiper-Goodman & Scott, 1989). It was not mutagenic in various microbial and mammalian gene mutation assays, either with or without exogenous metabolic activation. It gave negative or weakly positive results in other assays for genotoxicity, conducted *in vitro* or *in vivo*.

Carcinogenicity: Most of the early carcinogenicity studies, reviewed by the IARC (1983), were considered to be inadequate for evaluation. Subsequent studies in mice and rats have been evaluated critically (Kuiper-Goodman & Scott, 1989; WHO, 1991b). In these studies, in addition to the NOEL and/or the LOEL, the TD_{50} has been estimated as an indicator of carcinogenic potency. The TD_{50} has been defined as the dose rate at which 50% of the animals would be expected to have developed a particular tumour or combination of tumours if the compound being tested were given for a 'standard lifespan' (24 months in rodents), corrected for intercurrent mortality and background incidence of tumours (Peto *et al.*, 1984). TD_{50} s have been determined in about 3300 studies on 884 chemicals. The relationship between the NOEL and the TD_{50} gives an indication of the slope of the dose-response curve. Like the NOEL, the TD_{50} does not reflect biological aspects of tumours, such as whether they are benign or malignant, the number of target sites or the species affected, and it is not necessarily experimentally achievable.

In a study in which ochratoxin A was fed to male $B6C3F_1$ mice (Bendele *et al.*, 1985), only two dose levels were used—1 and 40 mg/kg of diet, equivalent to approximately 150 and 6000 µg/kg body weight per day, respectively. A high incidence of renal adenomas and carcinomas (combined; 31/49; 62%) was found at the highest dose level; no renal tumour was found at the lower dose level nor in controls, and the incidence in historical controls was $< 0.1\%$. Although 150 µg/kg body weight per day is thus clearly the level at which no tumorigenic effect was observed and could be considered the NOEL on both biological and statistical grounds, it is likely that the true NOEL would be considerably higher; indeed the NEL was found to be 760 µg/kg body weight per day.

In a recently completed carcinogenicity study of the National Toxicology Program (Boorman, 1989), groups of 80 male and female Fischer 344/N rats were administered ochratoxin A in corn oil by gavage at 0, 21, 70 or 210 µg/kg body weight per day on five days per week for 9, 15 and 24 months. Since renal tubular adenomas are generally considered to progress naturally to carcinomas, the two lesions may be considered together in estimating the potency of a carcinogenic effect (Bannasch & Zerban, 1986) (Table 3, Figure 2). The LOEL for the combined incidence in male rats was 70 µg/kg body weight, and the NOEL was 21 µg/kg body weight. The incidence of renal tubular cell adenomas and carcinomas observed in the male rats was the highest seen in any study of the National Cancer Institute or National Toxicology Program to date. At the highest dose level, many of the renal tumours were multiple or bilateral.

The significance of the ochratoxin A-induced rat renal carcinoma is increased by the high frequency of metastases, the majority of which could be attributed to metastases from renal-cell carcinomas, whereas metastasis of this type of tumor is normally a rare event (Bannasch & Zerban, 1986). Metastases were seen mainly in the lungs and lymph nodes. In female rats given the high dose, there was also an increased incidence in the multiplicity of fibroadenomas in the mammary gland (14/50 compared to 4–5/50 in controls and in animals given lower doses)

Data provided in the report indicate that the maximum tolerated dose was reached but not exceeded. No compound-related clinical sign was noted, and the results of haematological and serum chemical analyses showed no effects of biological significance. Urinalysis indicated a mild to moderate change in the ability to concentrate urine but no other change in kidney function.

Table 3. Incidence of renal adenomas and combined incidence of renal adenomas and carcinomas induced by dietary ochratoxin A in Fischer rats[a]

Sex	Dose (µg/kg bw)	Response	
		Adenomas	Adenomas and carcinomas
Males	Historical controls	3/248	
	0	1/50	1/50
	21	1/51	1/51
	70	6/51	20/51
	210	10/50	36/50
Females	Historical controls	0/248	0/248
	0	0/50	0/50
	21	0/51	0/51
	70	1/50	2/50
	210	5/50	8/50

[a] From Boorman (1989)

Non-neoplastic lesions involved mainly the kidney. Chronic diffuse nephropathy, common in old rats, was seen at about the same incidence in all groups of males and females. In addition, karyomegaly was seen in males and females, and it was the most consistent finding at the two highest dose levels at the time of the interim 9- and 15-month sacrifices, as well as in a 13-week preliminary study. Karyomegaly was also seen in 14- and 90-day studies conducted by the Health Protection Branch (Canada) in all groups of rats treated with ochratoxin A, given at dose levels of 0.2–5 mg/kg of diet, equivalent to 10–250 µg/kg body weight for the 90-day study (Munro *et al.*, 1974). Karyomegaly is not considered to occur spontaneously in this species, except in one strain of Wistar rats in which familial adenomas develop (Richardson & Woodward, 1986). Karyomegaly has been induced in rats by several unusual amino acids; ochratoxin A is similarly an amino acid derivative (of phenylalanine). It has been postulated that the configuration of the carbon in the α-position is responsible for the toxic effects of these unusual amino acids (Jones *et al.*, 1987). It should be noted that many compounds that induce renal karyomegaly are also renal carcinogens.

It is not possible at this time to assess clearly the non-neoplastic (and preneoplastic) effects of ochratoxin A on the kidney. This information will be important for understanding the sequence of steps in the development of kidney neoplasia. The doses of ochratoxin A used in the carcinogenicity study described above were much lower than those used in the studies in mice. Although they were sufficient to cause histopathological changes, there was no serious functional damage to the rat kidney. It is therefore unlikely that, in the rat, kidney damage *per se* was responsible for the observed neoplasia.

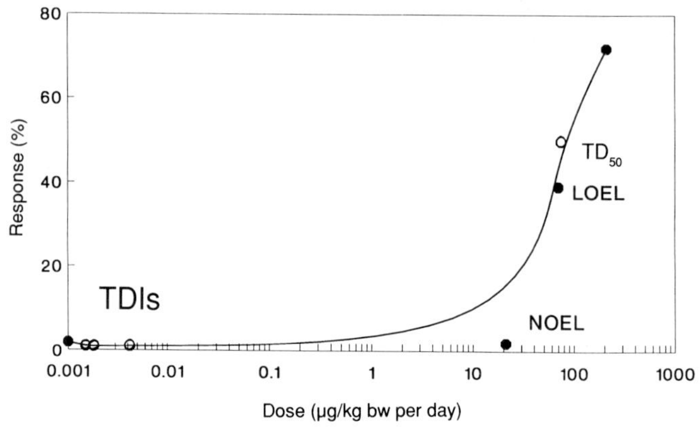

Figure 2. Dose-response relationship of ochratoxin A-induced renal tumours (adenomas and carcinomas combined) in male rats[a]

[a]From Boorman (1989). Experimentally observed values, closed circles; risk assessment parameters, open circles. TDI, tolerable daily intake; NOEL, no-observed-effect level; LOEL, lowest-observed-effect level; TD_{50}, dose at which 50% of animals are expected to develop tumours if the test compound is given for a 'standard lifespan' and corrected for intercurrent mortality and background incidence of tumours

Immunosuppression is unlikely to be the primary factor in the induction of renal tumours by ochratoxin A but could be of importance in the development and metastasis of these tumours. Suppression of natural killer cell activity may also play a role. More work is required before the basic mechanisms underlying the carcinogenicity of ochratoxin A are fully understood.

Human data

Epidemiological data implicating ochratoxin A in the causation of Balkan nephropathy were insufficient for consideration in the health hazard assessment at the time of writing.

Extrapolation

The endpoint of hazard assessment is estimation of a safe dose, such as an ADI or TDI (Figure 1). Because only a relatively small group of animals can be used in a bioassay, dose levels are exaggerated so as to increase the chance of detecting any toxic effect of the test chemical. Thus, in order to use the results of animal experiments to determine the potential for human disease, the following extrapolations must be made: (i) extrapolation from high doses, within the dose ranges used in animal experiments, to low doses, usually outside the experimental range, to which humans might be exposed; (ii) extrapolation from test species to humans; and (iii) extrapolation to the most sensitive subgroup of humans. Of these, the biological aspects of extrapolation from test species to humans are often the most problematic, and such extrapolations may be subject to uncertainties of several orders of magnitude (Kuiper-Goodman, 1990). We use the numerical aspects of extrapolation as a first step, through two major approaches: a 'NOEL/ safety factor' approach and a 'quantitative' or 'mathematical risk assessment approach', using mathematical modelling to estimate a virtually safe dose (VSD) (Kuiper-Goodman, 1990).

NOEL/safety factor approach: In the first approach, the NOEL (or the NEL, if more appropriate) is determined, and a safety factor is chosen to take into account empirical values for the three aspects of extrapolation mentioned. The approach may also include an 'uncertainty factor' when the experimental data are incomplete or when the lowest dose had an effect, so that no NOEL was established, yet a decision must be made. For non-carcinogenic effects, safety factors generally range from 100 to 1000. It may also be necessary to add a factor for severity or irreversibility of disease. Thus, for carcinogenic effects, larger safety factors, of 1000–5000, are generally established on a case-by-case basis, by taking account of the weight of evidence of all the biological aspects of the induced tumours. The estimated TDIs based on the NOEL in carcinogenicity studies divided by 5000 are 4.2 (male rat) and 26 ng/kg (male mouse); and the TDIs based on the NEL/5000 are 5.7 and 152 ng/kg, respectively (Table 4). Although there are problems associated with determination of the NOEL, which can be overcome in part by determining the NEL, and the choice of safety factor appears to some people as somewhat subjective, the safety factor approach has a long history of use and continues to be useful.

Table 4. Parameters for risk assessment of ochratoxin A

Parameter (unit/kg body weight)	Rat (male)	Mouse (male)
TD_{50} (µg)	74	4050
NOEL (µg)	21	> 150[a]
NEL (µg)	28.5	760
NOEL/5000 (ng)	4.2	26
NEL/5000 (ng)	5.7	152
TD_{50}/50 000 (ng)	1.5	81
VSD (lower confidence limit; risk level = 10^{-5})(ng)	1.8	16
Human exposure (ng)[b]	< 1.5	
Species differences	Human/pig more sensitive	
Genotoxicity	Weak	
Human effects	Probably	

[a]Uncertain
[b]Worst case estimate

VSD approach: This approach was developed to estimate the risk of very low doses (in the nonobservable region of the dose-response curve) of those carcinogenic substances for which it was considered that there was no threshold dose below which no tumour formation would occur, and to deal with the insufficient statistical power of animal experiments when studying rare diseases of high severity and irreversibility. Each approach gives a more or less conservative, similar estimate of a safe exposure level, such as an ADI or TDI, depending on the size of the safety factor in the former approach or the type of model and the level of allowable risk (generally ranging from 10^{-5} to 10^{-8}) in the latter. 'Level of risk' in this case does not represent true risk but may be chosen to correspond to concern about the effects noted.

For ochratoxin A, non-model-dependent robust linear extrapolation has been used on the responses in the NOEL region of the dose-response curve to obtain the lower confidence limit of the VSD (Krewski *et al.*, 1986). Like the NOEL/safety factor approach, the robust linear approach ignores data in the upper part of the dose-response curve. Thus, for the carcinogenicity study with ochratoxin A in male Fischer rats (Boorman, 1989), the same VSD was reached even though there were differences in the TD_{50} (Table 3). The incidence of renal adenomas alone gave a TD_{50} of 647 µg/kg body weight per day, whereas for the combined incidence of adenomas and carcinomas the TD_{50} was 74 µg/kg body weight per day; the lower 95% confidence limit for the VSD in both cases, however, was 0.18 ng/kg body weight per day (risk level, 10^{-6}) (Kuiper-Goodman & Scott, 1989) or 1.8 ng/kg body weight per day (risk level, 10^{-5}). In the carcinogenicity study in mice, the VSD was 16 ng/kg body weight per day, or 10 times lower if a risk level of $1:10^{-6}$ was chosen (Table 4).

TD_{50}/safety factor as an additional approach: Since the ADI or TDI is an intrinsic property of a chemical substance, it should be possible to determine it more precisely by taking into account all the biological and numerical data. Although carcinogenic potency was considered previously in the weight of evidence, it should be possible also to use the TD_{50} in quantitative aspects of the risk assessment. On the basis of the relationship between the TD_{50} and the NOEL or NEL, we proposed previously that the TD_{50} divided by 50 000 could be used as another means of estimating the TDI (Kuiper-Goodman, 1990). The estimated TDIs for ochratoxin A using the TD_{50}/50 000 are 1.5 (male rat) and 81 ng/kg body weight per day (male mouse) (Table 4).

The above estimates for the TDI of ochratoxin A are based on the results of carcinogenicity studies. The TDI can also be determined on the basis of other lesions. Thus, the WHO Joint Expert Committee on Food Additives (WHO, 1990) recently derived a TDI for ochratoxin A based on the low LOEL for kidney changes observed in pigs, to which a safety factor of 500 was applied. Instead of using the term TDI, the Committee preferred the term 'provisional tolerable weekly intake', to reflect uneven consumption patterns. This value was determined to be 112 ng/kg body weight per week.

The various estimates for the TDI of ochratoxin A (based on data from studies of rats and of pigs) are within one order of magnitude if one uses the VSD at a risk level of $1:10^{-5}$. Depending on the confidence we have in the individual estimates, we either use the range of estimates or choose a single estimate. A weight-of-evidence approach, using biological factors derived from the above and other animal and human studies, and information on the shape of the dose-response curve are used to assess whether the numerical estimates so derived should be moved upward or downward.

Risk Assessment

The experimental data presented above show that ochratoxin A is a toxic substance, with nephrotoxic, immunosuppressive, teratogenic and carcinogenic effects in many species. The overall estimated TDI for ochratoxin A is 1.5–5.7 ng/kg body weight per day (Table 4), on the basis of the National Toxicology Program carcinogenicity study. In Canada, exposure to ochratoxin A is probably less than 1.5 ng/kg body weight per day, but further studies with lower detection limits, involving more food commodities, are in progress to better define the overall residue profile of ochratoxin A in cereal grains, animal feeds, animal food products and human blood. Such data are required to better assess dietary exposure and to ascertain the need for regulatory measures and other control mechanisms.

References

Arora, R.G.& Frölén, H. (1981) Interference of mycotoxins with prenatal development of the mouse. 2. Ochratoxin A induced teratogenic effects in relation to the dose and stage of gestation. *Acta Vet. Scand.*, **22**, 535–552

Bannasch, P & Zerban, H. (1986) Renal cell adenoma and carcinoma, rat. In: Jones, T.C., Mohr, U. & Hunt, R.D., eds, *Urinary System*, New York, Springer-Verlag, pp. 112–139

Bauer, J. & Gareis, M. (1987) Ochratoxin A in the food chain. *Z. Veterinärmed. B*, **34**, 613–627 (in German)

Bendele, A.M., Carlton, W.W., Krogh, P. & Lillehøj, E.B. (1985a) Ochratoxin A carcinogenesis in the (C57BL/6J xC3H)F_1 mouse. *J. Natl Cancer Inst.*, **75**, 733–742

Boorman, G., ed. (1989) *NTP Technical Report on the Toxicology and Carcinogenesis Studies of Ochratoxin A (CAS NO. 303-47-9) in F344/N Rats (Gavage Studies)* (NIH Publication No. 88-2813), Research Triangle Park, NC, National Toxicology Program, US Department of Health and Human Services, National Institutes of Health

Breitholz, A., Olsen, M., Dahlbäck, A. & Hult, K. (1991) Plasma ochratoxin A levels in three Swedish populations surveyed using an ion-pair HPLC technique. *Food Addit. Contam.*, **8**, 183–192

Frank, H.K., Dirheimer, G., Grunow, W., Netter, K.J., Osswald, H. & Schlatter, J. (1990) *Ochratoxin A: Occurrence and Toxicological Evaluation*, Deutsche Forshungsgemeinschaft, Weinheim, VCH

Goliński, P & Grabarkiewicz-Szczęsna, J. (1985) The first Polish cases of the detection of ochratoxin A residues in human blood. *Rocz. Panstw. Zakl. Hig.*, **36**, 378–381 (in Polish)

Hagelberg, S., Hult, K. & Fuchs, R. (1989) Toxicokinetics of ochratoxin A in several species and its plasma-binding properties. *J. Appl. Toxicol.*, **9**, 91–96

Hald, B. (1989) Human exposure to ochratoxin A. *Bioactive Mol.*, **10**, 57–67

Health and Welfare Canada (1976) *Nutrition Canada Food Consumption Patterns Report*, Department of Supply and Services, Ottawa

IARC (1983) *IARC Monographs on the Evaluation of the Carcinogenic Risk of Chemicals to Humans*, Vol. 31, *Some Food Additives, Feed Additives and Naturally Occurring Substances*, Lyon, pp. 191–206

Jones, G.P., Hooper, P.T., Rivett, D.E., Tucker, D.J., Lambert, G. & Billett, A. (1987) Nephrotoxic activity in rats fed diets containing DL-3-(N-phenylethylamino)-alanine. *Aust. J. Biol. Sci.*, **40**, 115–123

Krewski, D., Murdoch, D. & Dewanji, A. (1986) Statistical modelling and extrapolation of carcinogenesis data. In: Moolgavkar, S.H. & Prentice, R.L., eds, *Modern Statistical Methods in Chronic Disease Epidemiology*, New York, Wiley Interscience, pp. 259–282

Kuiper-Goodman, T. (1990) Uncertainties in the risk assessment of three mycotoxins: aflatoxin, ochratoxin, and zearalenone. *Can. J. Physiol. Pharmacol.*, **68**, 1017–1024

Kuiper-Goodman, T. & Scott, P M. (1989) Risk assessment of the mycotoxin ochratoxin A. *Biomed. Environ. Sci.*, **2**, 179–248

Munro, I.C., Moodie, C.A., Kuiper-Goodman, T., Scott, P.M. & Grice, H.C. (1974) Toxicologic changes in rats fed graded dietary levels of ochratoxin A. *Toxicol. Appl. Pharmacol.*, **28**, 180–188

Petkova-Bocharova, T., Chernozemsky, I.N. & Castegnaro, M. (1988) Ochratoxin A in human blood in relation to Balkan endemic nephropathy and urinary system tumours in Bulgaria. *Food Addit. Contam.*, **5**, 299–301

Peto, R., Pike, M.C., Bernstein, L., Gold, L.S. & Ames, B.W. (1984) The TD50: a proposed general convention for the numerical description of the carcinogenic potency of chemicals in chronic-exposure animal experiments. *Environ. Health Perspect.*, **58**, 1–8

Richardson, J.A. & Woodward, J.C. (1986) Renal tubular karyocytomegaly, rat. In: Jones, T.C., Mohr, U. & Hunt, R.D., eds, *Urinary System*, New York, Springer-Verlag, pp. 189–192

Röschenthaler, R., Creppy, E.E. & Dirheimer, G. (1984) Ochratoxin A: on the mode of action of a ubiquitous mycotoxin. *J. Toxicol.*, **3**, 53–86

Støren, O., Holm, H. & Størmer, F.C. (1982) Metabolism of ochratoxin A by rats. *Appl. Environ. Microbiol.*, **44**, 785–789

WHO (1990) *Selected Mycotoxins: Ochratoxins, Trichothecenes, Ergot* (Environmental Health Criteria 105), Geneva, pp. 27–69

WHO (1991a) *Evaluation of Certain Food Additives and Contaminants. Thirty-seventh Report of the Joint FAO/WHO Expert Comitte on Food Additives* (WHO Technical Report Series 806), Geneva, pp. 29–31

WHO (1991b) *Ochratoxin A* (WHO Food Additive Series No. 28), Geneva (in press)

Mycotoxins, Endemic Nephropathy and Urinary Tract Tumours
Ed. M. Castegnaro, R. Pleština, G. Dirheimer, I.N. Chernozemsky & H. Bartsch
Lyon, International Agency for Research on Cancer, pp. 321-325

RISK ESTIMATION FOR OCHRATOXIN A IN EUROPEAN COUNTRIES

H.K. Frank

Formerly: Federal Research Centre for Nutrition, Ettlingen, Germany

Summary

Ochratoxin A is a mycotoxin produced by fungi occurring frequently on cereals. Their growth, and the associated toxin production, are closely correlated to the degree of moisture to which they are exposed, which itself is dependent on weather conditions at harvest and techniques for drying and storage. Eighteen species of fungi found in foods can produce ochratoxin A; all of these also produce other mycotoxins. The distribution of these mycotoxins in foods, which probably differs according to climate, has not yet been studied. The lowest doses of ochratoxin A that are toxic in experimental animals are four to five times higher than those found even in hyperendemic regions. This finding casts doubt on the hypothesis that ochratoxin A is the sole cause of Balkan endemic nephropathy and renal tumours. It is therefore not possible to estimate the risk from this mycotoxin for Europe or for individual European countries. Data on the occurrence of both ochratoxin A and other toxins and on the quantities consumed should be collected systematically; these can then be used to assess post-harvest techniques.

Introduction

The presence of ochratoxin A in food and feed depends exclusively on the growth of ochratoxin A-producing fungi in or on produce after harvest. Such fungi occur frequently in European climates. The risk from ochratoxin A is thus 'home-made' and not 'imported' as for aflatoxins. In the absence of fungal growth, there will be no mycotoxins and no risk!

Conditions of Ochratoxin A Production

Ten species of *Penicillium* and eight of *Aspergillus* that can produce ochratoxin A occur more or less frequently in soils, from which they can contaminate crops. These fungi also synthesize other mycotoxins (Table 1), resulting in combination effects, which can be predicted only with difficulty, if at all (Deutsche Forschungsgemeinschaft, 1990). Several mycotoxins are always present in mouldy produce, even if only ochratoxin A is demonstrated.

The growth conditions for some ochratoxin A-producing fungi are listed in Table 2. Growth can continue during the winter. Ochratoxin A production does not usually begin at temperatures below 10 °C, but above that the output is high, because by that time the mycelium is sufficiently developed and has entered the idiophase.

The relative air humidity during long-term (one year or more) storage should not exceed 0.65 over the entire period. In practice, therefore, the water content of wheat and maize should be no higher than 13–13.5%, that of oats, 12.5–13% and that of barley, 13.5–14%. A water content of 15% or more, frequently found under commercial conditions, precludes long storage. Thus, the risk that ochratoxin A and concomitant toxins will be produced is highest in places where the relative air humidity is over 65% for a long time, during which unprotected produce will absorb water from the atmosphere. A relative air humidity level of 0.70 is reached quickly, allowing the xerophilic *Aspergillus glaucus* group to grow and to create a climate in which the less xerotolerant mycotoxin producers can grow as well. The presence of insects promotes this process. Regional differences in the frequency of toxicogenic and nontoxicogenic isolates of fungal species are also important (Cvetnić & Pepeljnjak, 1990).

Ochratoxin Uptake and Disease

The average daily uptake of ochratoxin A by the population of central Europe is about 1 ng/kg body weight (Frank, this volume). The values are probably somewhat lower in southern Europe, higher in northern Europe and twice or three times higher in the Danube lowlands.

These values are substantially different from the doses that provoke definable disease in experimental animals. In rats, the lowest nephrotoxic dose is 10 000 times higher, and the lowest dose that leads to increased tumour frequency is 70 000 times higher. The difference from the initial oral dose that induces immunosuppression is difficult to estimate.

Unfortunately, the quantities of ochratoxin A taken up in regions hyperendemic for Balkan endemic nephropathy and renal tumours are unknown, even among patients with these conditions, although detailed

Table 1. Mycotoxins produced by ochratoxin A-synthesizing *Penicillia* and *Aspergilli*[a]

Fungus	Mycotoxin	Action
P. chrysogenum	Patulin	Haemorrhagic
	Penciillic acid	Nephrotoxic
	Penicillin	Antibiotic
P. commune	Penicillic acid	Nephrotoxic
P. crustosum	Viomellein	Nephrotoxic
	Xanthomegnin	Nephrotoxic
P. cyclopium	Cyclopiazonic acid	Neurotoxic
cf. P. cyclopium	Patulin	Haemorrhagic
aurantiogriseum	Citrinin	Nephrotoxic
	Penicillic acid	Nephrotoxic
	Viomellein	Nephrotoxic
	Xanthomegnin	Nephrotoxic
	Emodin	Laxative
	Tremortin A, B	Tremorgenic
	Rugulosin	Hepatotoxic, carcinogenic
P. palitans	Citrinin	Nephrotoxic
	Penicillic acid	Nephrotoxic
	Penitrem A	Tremorgenic
P. purpurrescens	Citrinin	Nephrotoxic
	Penitrem A	Tremorgenic
	Kojic acid	Mutagenic
P. purpurogenum	Rubratoxin A, B	Nephrotoxic, carcinogenic
	Kojic acid	Mutagenic
P. varaibila	Patulin	Haemorrhagic
	Rugulosin	Hepatotoxic, carcinogenic
	Skyrin	Antibiotic
P. verrucosum	Penicillic acid	Nephrotoxic
	Viomellein	Nephrotoxic
	Xanthomegnin	Nephrotoxic
P. viridicatum	Penicillic acid	Nephrotoxic
	Citrinin	Nephrotoxic
	Viomellein	Nephrotoxic
	Xanthomegnin	Nephrotoxic
	Cyclopiazonic acid	Neurotoxic
	Penitrem A	Tremorgenic
	Griseofulvin	Antimycotic
A. alliaceus	Kojic acid	Mutagenic
	Penicillic acid	Nephrotoxic
A. glaucus	Kojic acid	Mutagenic
A. meeleus	Penicillic acid	Nephrotoxic
A. ochraceus	Penicillic acid	Nephrotoxic
	Viomellein	Nephrotoxic
	Xanthomegnin	Nephrotoxic
	Kojic acid	Mutagenic
A. ostianus	Penicillic acid	Nephrotoxic
A. petrakii	Penicillic acid	Nephrotoxic
A. sclerotiorum	Penicillic acid	Nephrotoxic
A. sulphureus	Penicillic acid	Nephrotoxic

[a]Adapted from Deutsche Forschungsgemeinschaft (1990)

Table 2. Growth conditions for some ochratoxin A-producing moulds[a]

Fungus	Temperature (°C)		Minimal relative
	Optimal	Minimal	humidity
A. glaucus	30	–8	0.65
A. ochraceus	28–32	6	0.80
P. chrysogenum	25–28	–4	0.85
P. cyclopium	25–30	–2	0.84
P. viridicatum	21–23	–2	0.81

[a]From Reiss (1986)

studies of annual changes in the contamination of foods are available (Krogh *et al.*, 1977; Pavlović *et al.*, 1979; Petkova-Bocharova & Castegnaro, 1987). The levels are not in the range of the lowest doses that produce symptoms in experimental animals; however, the chronic influence of subacute doses over a long period and possible effects of combinations of ochratoxin A with the other nephrotoxins listed in Table 1 have not been taken into consideration. Little is also known about general and individual eating habits in these regions. Furthermore, the fact that the distribution of mycotoxins in bulk grain in small stores is extremely inhomogeneous should be taken into account.

Localization of Risks in Europe

If mycotoxins are the cause of Balkan endemic nephropathy and renal tumours, some regions of Europe have a particularly high risk; these are climatic regions with long periods of greater than 65% humidity, fog and rain, especially at harvest time. Denmark, Sweden, the British Isles, Germany's coastal areas and the Danube lowlands in Yugoslavia, Romania and Bulgaria have such conditions. Tholstrup and Rasmussen (1990) demonstrated clearly the association between climatic conditions in Denmark during and after grain harvesting and the ochratoxin A content of grain in 1986–89. Endemic nephropathy has, however, been observed only in the Balkan region of Europe (Deutsche Forschungsgemeinschaft, 1990).

Inadequate drying and unfavourable storage conditions for grains and leguminous seeds, i.e., conditions that lead to a high water content in the stored produce, are thus responsible for fungal growth and toxin production in endemic regions. Many reports have indicated an association between high water content, fungal growth and toxin release in pig and poultry feed, which is usually stored less carefully than grain for human consumption, and ochratoxin A concentration in the serum of slaughtered pigs (Hult *et al.*, 1980; Marquardt *et al.*, 1988). This finding corresponds closely to the high incidences of Balkan endemic nephropathy and renal tumours among populations in endemic regions who share agricultural produce that they have stored for their own use.

The risk from ochratoxin A is hence minimized when harvest produce is stored under dry conditions and subsequent exposure to humidity from the air or condensation is prevented. The post-harvest techniques necessary for achieving such conditions are known: they involve expense, however, and watchfulness. Our observations would indicate that they are used in most parts of Europe, at least in regard to food produce.

References

Cvetnić, Z. & Pepeljnjak, S. (1990) Ochratoxinogenicity of Aspergillus ochraceus strains from nephropathic and non-nephropathic areas in Yugoslavia. *Mycopathologia*, **110**, 93–100

Deutsche Forschungsgemeinschaft (1990) *Ochratoxin A*, Weinheim, VCH Verlag (in German)

Hult, K., Hökby, E., Gaterbeck, S. & Rutquist, L. (1980) Ochratoxin A in blood from slaughter pigs in Sweden. *Appl. Environ. Microbiol.*, **39**, 828–830

Krogh, P., Hald, B., Pleština, R. & Čeović, S. (1977) Balkan (endemic) nephropathy and foodborn ochratoxin A: preliminary results of a survey of foodstuffs. *Acta Pathol. Microbiol. Scand. Sect. B*, **85**, 238–240

Marquardt, R.R., Frohlich, A.A., Sreemannarayana, O., Abramson, D. & Bernatsky, A. (1988) Ochratoxin A in blood from slaughter pigs in western Canada. *Can. J. Vet. Res.*, **52**, 186–190

Pavlović, M.R., Pleština, R. & Krogh, P. (1979) Ochratoxin A contamination of foodstuffs in an area with Balkan (endemic) nephropathy. *Acta Pathol. Microbiol. Scand. Sect. B*, **87**, 243–246

Petkova-Bocharova, T. & Castegnaro, M. (1987) Ochratoxin A in human blood in relation to Balkan endemic nephropathy and urinary tumours in Bulgaria. *Eur. J. Cancer Clin. Oncol.*, **23**, 1788–1790

Reiss, J. (1986) *Moulds*, Berlin, Springer Verlag (in German)

Tholstrup, B. & Rasmussen, G. (1990) Ochratoxin A in cereals 1986–1989. *Levnedsmiddelstyrelsen*, **199** (in Danish)

Mycotoxins, Endemic Nephropathy and Urinary Tract Tumours
Ed. M. Castegnaro, R. Pleština, G. Dirheimer, I.N. Chernozemsky & H. Bartsch
Lyon, International Agency for Research on Cancer, pp. 327-329

RISK EVALUATION OF OCHRATOXIN A BY THE JOINT FAO/WHO EXPERT COMMITTEE ON FOOD ADDITIVES

J.L. Herrman

International Programme on Chemical Safety, World Health Organization, Geneva, Switzerland

The Joint FAO/WHO Expert Committee on Food Additives (JECFA) has been meeting regularly since 1956. During that time, 38 meetings have been held. Originally, the Committee evaluated only food additives, but since the early 1970s food contaminants have been evaluated as well. Recently, meetings of JECFA have been convened to evaluate residues of veterinary drugs in food.

The WHO Secretariat for these meetings has been the responsibility of the International Programme on Chemical Safety, which has been a joint venture of the World Health Organization, the International Labour Organisation and the United Nations Environment Programme since the time of its inception in 1980. The Central Unit of the Programme is located at WHO Headquarters in Geneva.

JECFA advises FAO, WHO and their Member States on the safety of food additives, contaminants and veterinary drug residues. JECFA also advises the Codex Alimentarius Commission through two commodity committees, the Codex Committee on Food Additives and Contaminants (CCFAC) and the Codex Committee on Residues of Veterinary Drugs in Foods. Contaminants are placed on the agenda of JECFA either because they were placed on the priority list by CCFAC or because their evaluation was requested by a FAO or WHO Member State. The CCFAC uses the evaluations of contaminants by JECFA as the basis for establishing guideline levels in food commodities that move in international trade.

The main endpoint of assessment by JECFA of food contaminants is the 'provisional tolerable weekly intake'. The term 'provisional' is used to indicate the tentative nature of the evaluation, in view of the paucity of reliable data on the consequences of human exposure. Tolerable intakes are expressed on a weekly basis because the contaminants may accumulate within the body over a period of time. On any particular day, consumption of food containing above-average levels of the contaminant may exceed the proportionate share of its weekly tolerable intake; however, such daily variations are of less concern than prolonged intake of contaminants because of possible accumulation.

Ochratoxin A was placed on the list of priorities for evaluation by JECFA at the Twenty-first Session of CCFAC in 1989. Accordingly, ochratoxin A was placed on the agenda of the thirty-seventh meeting of JECFA, which was held in June 1990 and was evaluated at that time. The report of the meeting has been published as No. 806 in the *WHO Technical Report Series*. A monograph which summarizes the data on which the Committee relied for its evaluation will be published in the *WHO Food Additives Series*.

Extensive studies on the metabolic disposition and toxicology of ochratoxin A, and limited information on the association between exposure to ochratoxin A and chronic human nephropathy, were reviewed by the Committee. Metabolic studies showed that the serum half-time varies widely among species, from 4 to more than 500 h. In blood, ochratoxin A is predominantly bound to serum albumin and other unidentified molecules. Tissue concentrations of ochratoxin A residues decrease in the order kidney > liver > muscle > fat. Studies of acute toxicity show that pigs and dogs are the most sensitive species. Short-term studies in rats, dogs and pigs showed that the dominant pathological effects occurred in the kidneys. Kidney tumours were observed in mice and rats.

The effects of ochratoxin A that were considered to be the most significant by the Committee were (i) deterioration of renal function in pigs, for which the lowest-observed-effect level (LOEL) was 0.008 mg/kg body weight per day (no no-observed-effect level (NOEL) was obtained in this study); (ii) karyomegaly of the proximal tubular cells in rats, for which the LOEL was 0.015 mg/kg body weight per day (no NOEL obtained); (iii) progressive nephropathy in pigs, for which the NOEL was 0.008 mg/kg body weight per day; (iv) kidney tumours in mice (NOEL = 0.13 mg/kg body weight per day) and rats (NOEL = 0.02 mg/kg body weight per day); (v) necrosis of lymphoid tissues of the thymus and tonsils in dogs, for which the LOEL was 0.1 mg/kg body weight per day (no NOEL obtained); and (vi) decreased antibody response in mice, for which the NOEL was 0.5 mg/kg body weight per day. On the basis of these findings, the Committee concluded that pigs were the most sensitive species.

The Committee reviewed the available data on chronic human nephropathy, which is endemic in the Balkans and has been linked to exposure to ochratoxin A, as indicated by the presence of ochratoxin A residues in local foodstuffs as well as in the blood of inhabitants. The finding that some individuals and village populations have detectable ochratoxin A residues in the blood but have no evidence of nephropathy, however, suggests that either the effects of ochratoxin A are delayed or the disease is caused by more than one factor. About one-third of people who die with Balkan endemic nephropathy have papillomas and/or carcinomas of the renal pelvis, ureter or bladder. Since no quantitative estimate of ochratoxin A intake in the diet was available for these patients, the available studies in humans could not be used as the basis for establishing the tolerable intake of ochratoxin A.

After reviewing all of the data, JECFA concluded that, in assessing the tolerable intake of ochratoxin A, a 500-fold margin of safety should be applied to the LOEL of 0.008 mg/kg of body weight per day. On this basis, a provisional tolerable weekly intake of 112 ng/kg of body weight was established.

At its Twenty-third Session in March of this year, CCFAC decided to gather more information from governments on the occurrence of ochratoxin A in various food commodities before deciding whether to establish Codex guideline levels.

References

WHO (1991) *Evaluation of Certain Food Additives and Contaminants (Thirty-seventh Report of the Joint FAO/WHO Expert Committee on Food Additives)* (WHO Technical Report Series No. 806), Geneva, pp 29–31

WHO (1991) *Toxicological Evaluation of Certain Food Additives and Contaminants* (WHO Food Additives Series No. 28), Geneva

Mycotoxins, Endemic Nephropathy and Urinary Tract Tumours
Ed. M. Castegnaro, R. Pleština, G. Dirheimer, I.N. Chernozemsky & H. Bartsch
Lyon, International Agency for Research on Cancer, pp. 331-336

WORLDWIDE REGULATIONS FOR OCHRATOXIN A

H.P. van Egmond

National Institute of Public Health and Environmental Protection, Laboratory for Residue Analysis, Bilthoven,The Netherlands

Summary

Currently, some 60 countries have enacted or proposed regulations for levels of mycotoxins in food and animal feed. Various factors may influence the establishment of tolerances for certain mycotoxins, such as the availability of data on dietary exposure and on toxicology, the distribution of the mycotoxins over commodities, the availability of analytical methodology, legislation in countries with which trading occurs and a sufficient food supply. Most of the existing regulations for mycotoxins concern aflatoxins, but the number of countries that regulate other mycotoxins as well is growing. Of the nephrotoxic mycotoxins, ochratoxin A was the only one for which regulations existed in 1990. At least 11 countries have proposed or official limits for ochratoxin A. The acceptable levels range from 1 to 50 µg/kg for food and from 100 to 1000 µg/kg for animal feed. The scientific basis for the established regulations appears to be weak, and a rationalization of tolerance levels for ochratoxin A would be highly desirable.

Introduction

Food legislation serves a very useful purpose: to safeguard the health and economic interests of the ultimate consumer. At present, food laws often include specific legislation that imposes limits or tolerances on the concentrations of specific contaminants in foods, such as mycotoxins. In this paper, the factors that may influence the establishment of regulations for mycotoxins are discussed and an overview is given of current regulations for ochratoxin A.

Factors that Influence the Regulation of Mycotoxins

The following factors may play a role in establishing limits and regulations for mycotoxins:

Survey data

Data on the occurrence of mycotoxins and the possible carry-over of mycotoxins from feeds into animal products may provide a basis for exposure assessment. Such data also make it possible to estimate the effects of enforcement on the availability of food, including animal products, and of feeds.

Toxicological data

Without toxicological information there can be no assessment of whether the substance in question is indeed hazardous. Measures are most logically taken on the basis of specific toxicological effects, except when cancer is the basis for concern, as is the case with aflatoxins and ochratoxins. Assuming that there is no threshold level for these mycotoxins below which some effect cannot occur, any small dose will entail a proportionally small probability of inducing some effect. A zero tolerance would then be appropriate; but the problem is that mycotoxins are natural contaminants that cannot be eliminated completely without banning the susceptible food or feed. This makes regulatory judgements particularly difficult.

Methods of analysis

To make inspection of commodities possible, accurate methods of analysis must be available. Without reliable methods of analysis, the fulfilment of established tolerances is not feasible. It should also be borne in mind that a tolerance cannot in fact be lower than the actual limit of detection of the method of analysis used.

Distribution throughout commodities

The distribution of mycotoxins in products may pose difficult problems for the establishment of regulatory criteria. If the distribution is nonhomogeneous, as is the case with many mycotoxins, there is a good chance that the mycotoxin concentration in the batch to be inspected will be incorrectly estimated, due to difficulties of representative sampling. The risk to both consumer and producer must be considered when establishing sampling and analytical criteria for products in which mycotoxins are nonhomogeneously distributed.

Trade contacts

The regulations in force in other countries with which trade contacts exist must be considered. Unnecessarily strict regulatory actions may make it difficult for importing countries to obtain supplies of essential commodities, such as food grains and animal feeds. Exporting countries may have difficulties in finding markets for their products.

Food supply

The regulatory philosophy should not jeopardize the availability of basic commodities at reasonable prices. Especially in developing countries, where food supplies are already limited, drastic legal measures could lead to shortages of food and higher prices.

Weighing the various factors that play a role in the decision-making process for establishing tolerances for mycotoxins is not easy. Despite the dilemmas, regulations have been established in the past decades in many countries.

Regulations on Ochratoxin A

Currently, about 60 countries have enacted or proposed regulations for levels of mycotoxins in food and feed. Most of the existing regulations refer to aflatoxins, but the number of other mycotoxins for which regulations are being developed is growing. The number of countries that regulate ochratoxin A is also increasing rapidly (Figure 1), whereas there are no regulations for other nephrotoxic mycotoxins.

Figure 1. Geographical situation of countries with (proposed) regulations or guidelines for ochratoxin A (black areas)

In Table 1, the current worldwide regulations for ochratoxin A in foods, animal feeds and animal organs are summarized. The information originated from publications by van Egmond (1989), Stoloff *et al.* (1991) and Gilbert (1991) and from personal communications. The author apologizes for any errors in the characterization of the current situation of ochratoxin A regulations and would appreciate receiving criticisms and comments from readers.

Table 1. Maximum tolerated levels for ochratoxin A in food, animal feed and animal organs (1990)

Country	Commodity	Tolerance (μg/kg)	Remarks	Reference
Brazil	Rice, barley, beans, maize	50		van Egmond (1989)
Czecho-slovakia	All foods except the following:	20		Stoloff *et al.* (1991)
	infant foods (calculated on product ready for consumption)	1		
	children's foods	5		
Denmark	Pork kidney	25	Whole carcass condemned if ochratoxin A content > 25 μg/kg	Buchmann & Hald (1985)
	Pork kidney	10	Kidney, liver and other visceral organs condemned if ochratoxin A content 10–25 μg/kg	
	Pork kidney	< 10 (detectable)	Only kidneys condemned	
France	Cereals	30	Proposal	Conseil Superieur (1990)
Hungary	All foods	20		Stoloff *et al.* (1991)
Israel	Grain for feed	300		Stoloff *et al.* (1991)

(cont.)

Table 1 *(continued)*

Netherlands	Cereals	3	Proposal	State Supervisory Agency of Public Health (1990)
Romania	All foods, all feeds	5		van Egmond (1989)
Sweden	Complete feed for swine	100		National Board of Agriculture (1987)
	Complete feed for poultry	1000		
United Kingdom	Cereals	10	Industrial guideline	S. Patel (1990) (personal communication)
Greece	Coffee	20		Anon. (1990)

Discussion

The tolerance levels established for ochratoxin A in various countries appear to vary widely and depend on country and commodity. Little information on how these levels were set is made public. The study of Stoloff *et al.* (1991) on the rationales for establishing limits and regulations yielded only a Canadian risk assessment (Kuiper-Goodman & Scott, 1989), although Canada had not established any guideline levels or limits for ochratoxin in foods. The authors recommended that better data should be produced on the exposure of humans and animals to ochratoxin A; they estimated tolerable daily intakes in humans that ranged from 0.2 to 4.2 ng/kg body weight on the basis of a recent carcinogenicity study with ochratoxin A in rats.

The FAO/WHO Joint Expert Committee on Food Additives recently established a provisional tolerable weekly intake level of 112 ng/kg body weight. This was based on the lowest effect level found for kidney damage in the pig, the most sensitive species (WHO, 1991). This advice was used in the Netherlands as the basis for a proposed tolerance level for ochratoxin A in cereals.

So far, epidemiological data on human kidney disease and human blood levels have not been taken into account in published studies of risk assessment.

Conclusion

Although at least 11 countries had (proposed) regulations for ochratoxin A in 1990, the scientific basis for these (varying) regulations seems to be weak and the available scientific knowledge may not have been fully used. A rationalization of tolerance levels is highly desirable.

References

Anon. (1990) Determination of residues of mycotoxins in agricultural products. Ministerial degree no. 90434. *Official Gazette 727/B/1990* (in Greek)

Buchmann, N.B. & Hald, B. (1985) Analysis, occurrence and control of ochratoxin A residues in Danish pig kidneys. *Food Addit. Contam.*, **2**, 193–199

Conseil Supérieur d'Hygiene Publique de France. Section de l'Alimentation (1990) *The Mycotoxins* (Meetings of 8 March 1988 and 13 February 1990, Ministère de la Solidarité, de la Santé et de la Protection Sociale, Direction Générale de la Santé, Sous-Direction de la Prévention Générale et de l'Environnement), Paris, pp.15–20 (in French)

van Egmond, H.P. (1989) Current situation on regulations for mycotoxins. Overview of tolerances and status of standard methods of sampling and analysis. *Food Addit. Contam.*, **6**, 139–188

Gilbert, F. (1991) Regulatory aspects of mycotoxins in the EC and the USA. In: *Proceedings International Conference on Fungi and Mycotoxins in Stored Products, Bangkok, Thailand, 23–26 April 1991* (in press)

Kuiper-Goodman, T. & Scott, P.M. (1989) Risk assessment of the mycotoxin ochratoxin A. *Biomed. Environ. Sci.*, **2**, 179–248

National Board of Agriculture (1987) *Code of Statutes of the National Board of Agriculture* (Veterinary Regulations 42 Vb 19), Stockholm

Stoloff, L., van Egmond, H.P. & Park D.L. (1991) Rationales for the establishment of limits and regulations for mycotoxins. *Food Addit. Contam.*, **8**, pp. 213–222

State Supervisory Agency of Public Health (1990) Advisory letter of 12 November 1990, The Hague (in Dutch)

WHO (1991) *Evaluation of Certain Food Additives and Contaminants. Thirty-seventh Report of the Joint FAO/WHO Expert Committee on Food Additives* (WHO Technical Report Series 806), Geneva, pp. 28–31

Author index

Subject index

PUBLICATIONS OF THE INTERNATIONAL AGENCY FOR RESEARCH ON CANCER

Scientific Publications Series

(Available from Oxford University Press through local bookshops)

No. 1 **Liver Cancer**
1971; 176 pages (*out of print*)

No. 2 **Oncogenesis and Herpesviruses**
Edited by P.M. Biggs, G. de-Thé and L.N. Payne
1972; 515 pages (*out of print*)

No. 3 ***N*-Nitroso Compounds: Analysis and Formation**
Edited by P. Bogovski, R. Preussman and E.A. Walker
1972; 140 pages (*out of print*)

No. 4 **Transplacental Carcinogenesis**
Edited by L. Tomatis and U. Mohr
1973; 181 pages (*out of print*)

No. 5/6 **Pathology of Tumours in Laboratory Animals, Volume 1, Tumours of the Rat**
Edited by V.S. Turusov
1973/1976; 533 pages; £50.00

No. 7 **Host Environment Interactions in the Etiology of Cancer in Man**
Edited by R. Doll and I. Vodopija
1973; 464 pages; £32.50

No. 8 **Biological Effects of Asbestos**
Edited by P. Bogovski, J.C. Gilson, V. Timbrell and J.C. Wagner
1973; 346 pages (*out of print*)

No. 9 ***N*-Nitroso Compounds in the Environment**
Edited by P. Bogovski and E.A. Walker
1974; 243 pages; £21.00

No. 10 **Chemical Carcinogenesis Essays**
Edited by R. Montesano and L. Tomatis
1974; 230 pages (*out of print*)

No. 11 **Oncogenesis and Herpesviruses II**
Edited by G. de-Thé, M.A. Epstein and H. zur Hausen
1975; Part I: 511 pages
Part II: 403 pages; £65.00

No. 12 **Screening Tests in Chemical Carcinogenesis**
Edited by R. Montesano, H. Bartsch and L. Tomatis
1976; 666 pages; £45.00

No. 13 **Environmental Pollution and Carcinogenic Risks**
Edited by C. Rosenfeld and W. Davis
1975; 441 pages (*out of print*)

No. 14 **Environmental *N*-Nitroso Compounds. Analysis and Formation**
Edited by E.A. Walker, P. Bogovski and L. Griciute
1976; 512 pages; £37.50

No. 15 **Cancer Incidence in Five Continents, Volume III**
Edited by J.A.H. Waterhouse, C. Muir, P. Correa and J. Powell
1976; 584 pages; (*out of print*)

No. 16 **Air Pollution and Cancer in Man**
Edited by U. Mohr, D. Schmähl and L. Tomatis
1977; 328 pages (*out of print*)

No. 17 **Directory of On-going Research in Cancer Epidemiology 1977**
Edited by C.S. Muir and G. Wagner
1977; 599 pages (*out of print*)

No. 18 **Environmental Carcinogens. Selected Methods of Analysis. Volume 1: Analysis of Volatile Nitrosamines in Food**
Editor-in-Chief: H. Egan
1978; 212 pages (*out of print*)

No. 19 **Environmental Aspects of *N*-Nitroso Compounds**
Edited by E.A. Walker, M. Castegnaro, L. Griciute and R.E. Lyle
1978; 561 pages (*out of print*)

No. 20 **Nasopharyngeal Carcinoma: Etiology and Control**
Edited by G. de-Thé and Y. Ito
1978; 606 pages (*out of print*)

No. 21 **Cancer Registration and its Techniques**
Edited by R. MacLennan, C. Muir, R. Steinitz and A. Winkler
1978; 235 pages; £35.00

No. 22 **Environmental Carcinogens. Selected Methods of Analysis. Volume 2: Methods for the Measurement of Vinyl Chloride in Poly(vinyl chloride), Air, Water and Foodstuffs**
Editor-in-Chief: H. Egan
1978; 142 pages (*out of print*)

No. 23 **Pathology of Tumours in Laboratory Animals. Volume II: Tumours of the Mouse**
Editor-in-Chief: V.S. Turusov
1979; 669 pages (*out of print*)

No. 24 **Oncogenesis and Herpesviruses III**
Edited by G. de-Thé, W. Henle and F. Rapp
1978; Part I: 580 pages, Part II: 512 pages (*out of print*)

Prices, valid for September 1991, are subject to change without notice

No. 25 **Carcinogenic Risk. Strategies for Intervention**
Edited by W. Davis and C. Rosenfeld
1979; 280 pages (*out of print*)

No. 26 **Directory of On-going Research in Cancer Epidemiology 1978**
Edited by C.S. Muir and G. Wagner
1978; 550 pages (*out of print*)

No. 27 **Molecular and Cellular Aspects of Carcinogen Screening Tests**
Edited by R. Montesano, H. Bartsch and L. Tomatis
1980; 372 pages; £29.00

No. 28 **Directory of On-going Research in Cancer Epidemiology 1979**
Edited by C.S. Muir and G. Wagner
1979; 672 pages (*out of print*)

No. 29 **Environmental Carcinogens. Selected Methods of Analysis. Volume 3: Analysis of Polycyclic Aromatic Hydrocarbons in Environmental Samples**
Editor-in-Chief: H. Egan
1979; 240 pages (*out of print*)

No. 30 **Biological Effects of Mineral Fibres**
Editor-in-Chief: J.C. Wagner
1980; **Volume 1:** 494 pages; **Volume 2:** 513 pages; £65.00

No. 31 ***N*-Nitroso Compounds: Analysis, Formation and Occurrence**
Edited by E.A. Walker, L. Griciute, M. Castegnaro and M. Börzsönyi
1980; 835 pages (*out of print*)

No. 32 **Statistical Methods in Cancer Research. Volume 1. The Analysis of Case-control Studies**
By N.E. Breslow and N.E. Day
1980; 338 pages; £20.00

No. 33 **Handling Chemical Carcinogens in the Laboratory**
Edited by R. Montesano *et al.*
1979; 32 pages (*out of print*)

No. 34 **Pathology of Tumours in Laboratory Animals. Volume III. Tumours of the Hamster**
Editor-in-Chief: V.S. Turusov
1982; 461 pages; £39.00

No. 35 **Directory of On-going Research in Cancer Epidemiology 1980**
Edited by C.S. Muir and G. Wagner
1980; 660 pages (*out of print*)

No. 36 **Cancer Mortality by Occupation and Social Class 1851-1971**
Edited by W.P.D. Logan
1982; 253 pages; £22.50

No. 37 **Laboratory Decontamination and Destruction of Aflatoxins B_1, B_2, G_1, G_2 in Laboratory Wastes**
Edited by M. Castegnaro *et al.*
1980; 56 pages; £6.50

No. 38 **Directory of On-going Research in Cancer Epidemiology 1981**
Edited by C.S. Muir and G. Wagner
1981; 696 pages (*out of print*)

No. 39 **Host Factors in Human Carcinogenesis**
Edited by H. Bartsch and B. Armstrong
1982; 583 pages; £46.00

No. 40 **Environmental Carcinogens. Selected Methods of Analysis. Volume 4: Some Aromatic Amines and Azo Dyes in the General and Industrial Environment**
Edited by L. Fishbein, M. Castegnaro, I.K. O'Neill and H. Bartsch
1981; 347 pages; £29.00

No. 41 ***N*-Nitroso Compounds: Occurrence and Biological Effects**
Edited by H. Bartsch, I.K. O'Neill, M. Castegnaro and M. Okada
1982; 755 pages; £48.00

No. 42 **Cancer Incidence in Five Continents, Volume IV**
Edited by J. Waterhouse, C. Muir, K. Shanmugaratnam and J. Powell
1982; 811 pages (*out of print*)

No. 43 **Laboratory Decontamination and Destruction of Carcinogens in Laboratory Wastes: Some *N*-Nitrosamines**
Edited by M. Castegnaro *et al.*
1982; 73 pages; £7.50

No. 44 **Environmental Carcinogens. Selected Methods of Analysis. Volume 5: Some Mycotoxins**
Edited by L. Stoloff, M. Castegnaro, P. Scott, I.K. O'Neill and H. Bartsch
1983; 455 pages; £29.00

No. 45 **Environmental Carcinogens. Selected Methods of Analysis. Volume 6: *N*-Nitroso Compounds**
Edited by R. Preussmann, I.K. O'Neill, G. Eisenbrand, B. Spiegelhalder and H. Bartsch
1983; 508 pages; £29.00

No. 46 **Directory of On-going Research in Cancer Epidemiology 1982**
Edited by C.S. Muir and G. Wagner
1982; 722 pages (*out of print*)

No. 47 **Cancer Incidence in Singapore 1968–1977**
Edited by K. Shanmugaratnam, H.P. Lee and N.E. Day
1983; 171 pages (*out of print*)

No. 48 **Cancer Incidence in the USSR (2nd Revised Edition)**
Edited by N.P. Napalkov, G.F. Tserkovny, V.M. Merabishvili, D.M. Parkin, M. Smans and C.S. Muir
1983; 75 pages; £12.00

No. 49 **Laboratory Decontamination and Destruction of Carcinogens in Laboratory Wastes: Some Polycyclic Aromatic Hydrocarbons**
Edited by M. Castegnaro *et al.*
1983; 87 pages; £9.00

No. 50 **Directory of On-going Research in Cancer Epidemiology 1983**
Edited by C.S. Muir and G. Wagner
1983; 731 pages (*out of print*)

No. 51 **Modulators of Experimental Carcinogenesis**
Edited by V. Turusov and R. Montesano
1983; 307 pages; £22.50

No. 52 **Second Cancers in Relation to Radiation Treatment for Cervical Cancer: Results of a Cancer Registry Collaboration**
Edited by N.E. Day and J.C. Boice, Jr
1984; 207 pages; £20.00

No. 53 **Nickel in the Human Environment**
Editor-in-Chief: F.W. Sunderman, Jr
1984; 529 pages; £41.00

No. 54 **Laboratory Decontamination and Destruction of Carcinogens in Laboratory Wastes: Some Hydrazines**
Edited by M. Castegnaro *et al.*
1983; 87 pages; £9.00

No. 55 **Laboratory Decontamination and Destruction of Carcinogens in Laboratory Wastes: Some *N*-Nitrosamides**
Edited by M. Castegnaro *et al.*
1984; 66 pages; £7.50

No. 56 **Models, Mechanisms and Etiology of Tumour Promotion**
Edited by M. Börzsönyi, N.E. Day, K. Lapis and H. Yamasaki
1984; 532 pages; £42.00

No. 57 ***N*-Nitroso Compounds: Occurrence, Biological Effects and Relevance to Human Cancer**
Edited by I.K. O'Neill, R.C. von Borstel, C.T. Miller, J. Long and H. Bartsch
1984; 1013 pages; £80.00

No. 58 **Age-related Factors in Carcinogenesis**
Edited by A. Likhachev, V. Anisimov and R. Montesano
1985; 288 pages; £20.00

No. 59 **Monitoring Human Exposure to Carcinogenic and Mutagenic Agents**
Edited by A. Berlin, M. Draper, K. Hemminki and H. Vainio
1984; 457 pages; £27.50

No. 60 **Burkitt's Lymphoma: A Human Cancer Model**
Edited by G. Lenoir, G. O'Conor and C.L.M. Olweny
1985; 484 pages; £29.00

No. 61 **Laboratory Decontamination and Destruction of Carcinogens in Laboratory Wastes: Some Haloethers**
Edited by M. Castegnaro *et al.*
1985; 55 pages; £7.50

No. 62 **Directory of On-going Research in Cancer Epidemiology 1984**
Edited by C.S. Muir and G. Wagner
1984; 717 pages (*out of print*)

No. 63 **Virus-associated Cancers in Africa**
Edited by A.O. Williams, G.T. O'Conor, G.B. de-Thé and C.A. Johnson
1984; 773 pages; £22.00

No. 64 **Laboratory Decontamination and Destruction of Carcinogens in Laboratory Wastes: Some Aromatic Amines and 4-Nitrobiphenyl**
Edited by M. Castegnaro *et al.*
1985; 84 pages; £6.95

No. 65 **Interpretation of Negative Epidemiological Evidence for Carcinogenicity**
Edited by N.J. Wald and R. Doll
1985; 232 pages; £20.00

No. 66 **The Role of the Registry in Cancer Control**
Edited by D.M. Parkin, G. Wagner and C.S. Muir
1985; 152 pages; £10.00

No. 67 **Transformation Assay of Established Cell Lines: Mechanisms and Application**
Edited by T. Kakunaga and H. Yamasaki
1985; 225 pages; £20.00

No. 68 **Environmental Carcinogens. Selected Methods of Analysis. Volume 7. Some Volatile Halogenated Hydrocarbons**
Edited by L. Fishbein and I.K. O'Neill
1985; 479 pages; £42.00

No. 69 **Directory of On-going Research in Cancer Epidemiology 1985**
Edited by C.S. Muir and G. Wagner
1985; 745 pages; £22.00

No. 70 **The Role of Cyclic Nucleic Acid Adducts in Carcinogenesis and Mutagenesis**
Edited by B. Singer and H. Bartsch
1986; 467 pages; £40.00

No. 71 **Environmental Carcinogens. Selected Methods of Analysis. Volume 8: Some Metals: As, Be, Cd, Cr, Ni, Pb, Se Zn**
Edited by I.K. O'Neill, P. Schuller and L. Fishbein
1986; 485 pages; £42.00

No. 72 **Atlas of Cancer in Scotland, 1975–1980. Incidence and Epidemiological Perspective**
Edited by I. Kemp, P. Boyle, M. Smans and C.S. Muir
1985; 285 pages; £35.00

No. 73 **Laboratory Decontamination and Destruction of Carcinogens in Laboratory Wastes: Some Antineoplastic Agents**
Edited by M. Castegnaro *et al.*
1985; 163 pages; £10.00

No. 74 **Tobacco: A Major International Health Hazard**
Edited by D. Zaridze and R. Peto
1986; 324 pages; £20.00

No. 75 **Cancer Occurrence in Developing Countries**
Edited by D.M. Parkin
1986; 339 pages; £20.00

No. 76 **Screening for Cancer of the Uterine Cervix**
Edited by M. Hakama, A.B. Miller and N.E. Day
1986; 315 pages; £25.00

No. 77 **Hexachlorobenzene: Proceedings of an International Symposium**
Edited by C.R. Morris and J.R.P. Cabral
1986; 668 pages; £50.00

No. 78 **Carcinogenicity of Alkylating Cytostatic Drugs**
Edited by D. Schmähl and J.M. Kaldor
1986; 337 pages; £25.00

No. 79 **Statistical Methods in Cancer Research. Volume III: The Design and Analysis of Long-term Animal Experiments**
By J.J. Gart, D. Krewski, P.N. Lee, R.E. Tarone and J. Wahrendorf
1986; 213 pages; £20.00

No. 80 **Directory of On-going Research in Cancer Epidemiology 1986**
Edited by C.S. Muir and G. Wagner
1986; 805 pages; £22.00

No. 81 **Environmental Carcinogens: Methods of Analysis and Exposure Measurement. Volume 9: Passive Smoking**
Edited by I.K. O'Neill, K.D. Brunnemann, B. Dodet and D. Hoffmann
1987; 383 pages; £35.00

No. 82 **Statistical Methods in Cancer Research. Volume II: The Design and Analysis of Cohort Studies**
By N.E. Breslow and N.E. Day
1987; 404 pages; £30.00

No. 83 **Long-term and Short-term Assays for Carcinogens: A Critical Appraisal**
Edited by R. Montesano, H. Bartsch, H. Vainio, J. Wilbourn and H. Yamasaki
1986; 575 pages; £48.00

No. 84 **The Relevance of *N*-Nitroso Compounds to Human Cancer: Exposure and Mechanisms**
Edited by H. Bartsch, I.K. O'Neill and R. Schulte- Hermann
1987; 671 pages; £50.00

No. 85 **Environmental Carcinogens: Methods of Analysis and Exposure Measurement. Volume 10: Benzene and Alkylated Benzenes**
Edited by L. Fishbein and I.K. O'Neill
1988; 327 pages; £35.00

No. 86 **Directory of On-going Research in Cancer Epidemiology 1987**
Edited by D.M. Parkin and J. Wahrendorf
1987; 676 pages; £22.00

No. 87 **International Incidence of Childhood Cancer**
Edited by D.M. Parkin, C.A. Stiller, C.A. Bieber, G.J. Draper, B. Terracini and J.L. Young
1988; 401 pages; £35.00

No. 88 **Cancer Incidence in Five Continents Volume V**
Edited by C. Muir, J. Waterhouse, T. Mack, J. Powell and S. Whelan
1987; 1004 pages; £50.00

No. 89 **Method for Detecting DNA Damaging Agents in Humans: Applications in Cancer Epidemiology and Prevention**
Edited by H. Bartsch, K. Hemminki and I.K. O'Neill
1988; 518 pages; £45.00

No. 90 **Non-occupational Exposure to Mineral Fibres**
Edited by J. Bignon, J. Peto and R. Saracci
1989; 500 pages; £45.00

No. 91 **Trends in Cancer Incidence in Singapore 1968–1982**
Edited by H.P. Lee , N.E. Day and K. Shanmugaratnam
1988; 160 pages; £25.00

No. 92 **Cell Differentiation, Genes and Cancer**
Edited by T. Kakunaga, T. Sugimura, L. Tomatis and H. Yamasaki
1988; 204 pages; £25.00

No. 93 **Directory of On-going Research in Cancer Epidemiology 1988**
Edited by M. Coleman and J. Wahrendorf
1988; 662 pages (*out of print*)

No. 94 **Human Papillomavirus and Cervical Cancer**
Edited by N. Muñoz, F.X. Bosch and O.M. Jensen
1989; 154 pages; £19.00

No. 95 **Cancer Registration: Principles and Methods**
Edited by O.M. Jensen, D.M. Parkin, R. MacLennan, C.S. Muir and R. Skeet
1991; 288 pages; £28.00

No. 96 **Perinatal and Multigeneration Carcinogenesis**
Edited by N.P. Napalkov, J.M. Rice, L. Tomatis and H. Yamasaki
1989; 436 pages; £48.00

No. 97 **Occupational Exposure to Silica and Cancer Risk**
Edited by L. Simonato, A.C. Fletcher, R. Saracci and T. Thomas
1990; 124 pages; £19.00

No. 98 **Cancer Incidence in Jewish Migrants to Israel, 1961–1981**
Edited by R. Steinitz, D.M. Parkin, J.L. Young, C.A. Bieber and L. Katz
1989; 320 pages; £30.00

No. 99 **Pathology of Tumours in Laboratory Animals, Second Edition, Volume 1, Tumours of the Rat**
Edited by V.S. Turusov and U. Mohr
740 pages; £85.00

No. 100 **Cancer: Causes, Occurrence and Control**
Editor-in-Chief L. Tomatis
1990; 352 pages; £24.00

No. 101 **Directory of On-going Research in Cancer Epidemiology 1989/90**
Edited by M. Coleman and J. Wahrendorf
1989; 818 pages; £36.00

No. 102 **Patterns of Cancer in Five Continents**
Edited by S.L. Whelan and D.M. Parkin
1990; 162 pages; £25.00

No. 103 **Evaluating Effectiveness of Primary Prevention of Cancer**
Edited by M. Hakama, V. Beral, J.W. Cullen and D.M. Parkin
1990; 250 pages; £32.00

No. 104 **Complex Mixtures and Cancer Risk**
Edited by H. Vainio, M. Sorsa and A.J. McMichael
1990; 442 pages; £38.00

No. 105 **Relevance to Human Cancer of *N*-Nitroso Compounds, Tobacco Smoke and Mycotoxins**
Edited by I.K. O'Neill, J. Chen and H. Bartsch
1991; 614 pages; £70.00

No. 106 **Atlas of Cancer Incidence in the German Democratic Republic**
Edited by W.H. Mehnert, M. Smans and C.S. Muir
Publ. due 1992; c.328 pages; £42.00

No. 107 **Atlas of Cancer Mortality in the European Economic Community**
Edited by M. Smans, C.S. Muir and P. Boyle
Publ. due 1991; approx. 230 pages; £35.00

No. 108 **Environmental Carcinogens: Methods of Analysis and Exposure Measurement. Volume 11: Polychlorinated Dioxins and Dibenzofurans**
Edited by C. Rappe, H.R. Buser, B. Dodet and I.K. O'Neill
1991; 426 pages; £45.00

No. 109 **Environmental Carcinogens: Methods of Analysis and Exposure Measurement. Volume 12: Indoor Air Contaminants**
Edited by B. Seifert, B. Dodet and I.K. O'Neill
Publ. due 1992; approx. 400 pages

No. 110 **Directory of On-going Research in Cancer Epidemiology 1991**
Edited by M. Coleman and J. Wahrendorf
1991; 753 pages; £38.00

No. 111 **Pathology of Tumours in Laboratory Animals, Second Edition, Volume 2, Tumours of the Mouse**
Edited by V.S. Turusov and U. Mohr
Publ. due 1992; approx. 500 pages

No. 112 **Autopsy in Epidemiology and Medical Research**
Edited by E. Riboli and M. Delendi
1991; 288 pages; £25.00

No. 113 **Laboratory Decontamination and Destruction of Carcinogens in Laboratory Wastes: Some Mycotoxins**
Edited by M. Castegnaro, J. Barek, J.-M. Frémy, M. Lafontaine, M. Miraglia, E.B. Sansone and G.M. Telling
1991; 64 pages; £11.00

No. 114 **Laboratory Decontamination and Destruction of Carcinogens in Laboratory Wastes: Some Polycyclic Heterocyclic Hydrocarbons**
Edited by M. Castegnaro, J. Barek, J. Jacob, U. Kirso, M. Lafontaine, E.B. Sansone, G.M. Telling and T. Vu Duc
1991; 50 pages; £8.00

No. 115 **Mycotoxins, Endemic Nephropathy and Urinary Tract Tumours**
Edited by M. Castegnaro, R. Pleština, G. Dirheimer, I.N. Chernozemsky and H Bartsch
1991; 340 pages; £45.00

IARC MONOGRAPHS ON THE EVALUATION OF CARCINOGENIC RISKS TO HUMANS

(Available from booksellers through the network of WHO Sales Agents)

Volume 1 **Some Inorganic Substances, Chlorinated Hydrocarbons, Aromatic Amines, *N*-Nitroso Compounds, and Natural Products**
1972; 184 pages (*out of print*)

Volume 2 **Some Inorganic and Organometallic Compounds**
1973; 181 pages (out of print)

Volume 3 **Certain Polycyclic Aromatic Hydrocarbons and Heterocyclic Compounds**
1973; 271 pages (*out of print*)

Volume 4 **Some Aromatic Amines, Hydrazine and Related Substances, *N*-Nitroso Compounds and Miscellaneous Alkylating Agents**
1974; 286 pages;
Sw. fr. 18.-/US $14.40

Volume 5 **Some Organochlorine Pesticides**
1974; 241 pages (*out of print*)

Volume 6 **Sex Hormones**
1974; 243 pages (*out of print*)

Volume 7 **Some Anti-Thyroid and Related Substances, Nitrofurans and Industrial Chemicals**
1974; 326 pages (*out of print*)

Volume 8 **Some Aromatic Azo Compounds**
1975; 375 pages;
Sw. fr. 36.-/US $28.80

Volume 9 **Some Aziridines, *N*-, *S*- and *O*-Mustards and Selenium**
1975; 268 pages;
Sw.fr. 27.-/US $21.60

Volume 10 **Some Naturally Occurring Substances**
1976; 353 pages (*out of print*)

Volume 11 **Cadmium, Nickel, Some Epoxides, Miscellaneous Industrial Chemicals and General Considerations on Volatile Anaesthetics**
1976; 306 pages (*out of print*)

Volume 12 **Some Carbamates, Thiocarbamates and Carbazides**
1976; 282 pages;
Sw. fr. 34.-/US $27.20

Volume 13 **Some Miscellaneous Pharmaceutical Substances**
1977; 255 pages;
Sw. fr. 30.-/US$ 24.00

Volume 14 **Asbestos**
1977; 106 pages (*out of print*)

Volume 15 **Some Fumigants, The Herbicides 2,4-D and 2,4,5-T, Chlorinated Dibenzodioxins and Miscellaneous Industrial Chemicals**
1977; 354 pages;
Sw. fr. 50.-/US $40.00

Volume 16 **Some Aromatic Amines and Related Nitro Compounds - Hair Dyes, Colouring Agents and Miscellaneous Industrial Chemicals**
1978; 400 pages;
Sw. fr. 50.-/US $40.00

Volume 17 **Some *N*-Nitroso Compounds**
1987; 365 pages;
Sw. fr. 50.-/US $40.00

Volume 18 **Polychlorinated Biphenyls and Polybrominated Biphenyls**
1978; 140 pages;
Sw. fr. 20.-/US $16.00

Volume 19 **Some Monomers, Plastics and Synthetic Elastomers, and Acrolein**
1979; 513 pages;
Sw. fr. 60.-/US $48.00

Volume 20 **Some Halogenated Hydrocarbons**
1979; 609 pages (*out of print*)

Volume 21 **Sex Hormones (II)**
1979; 583 pages;
Sw. fr. 60.-/US $48.00

Volume 22 **Some Non-Nutritive Sweetening Agents**
1980; 208 pages;
Sw. fr. 25.-/US $20.00

Volume 23 **Some Metals and Metallic Compounds**
1980; 438 pages (*out of print*)

Volume 24 **Some Pharmaceutical Drugs**
1980; 337 pages;
Sw. fr. 40.-/US $32.00

Volume 25 **Wood, Leather and Some Associated Industries**
1981; 412 pages;
Sw. fr. 60-/US $48.00

Volume 26 **Some Antineoplastic and Immunosuppressive Agents**
1981; 411 pages;
Sw. fr. 62.-/US $49.60

Volume 27 **Some Aromatic Amines, Anthraquinones and Nitroso Compounds, and Inorganic Fluorides Used in Drinking Water and Dental Preparations**
1982; 341 pages;
Sw. fr. 40.-/US $32.00

Volume 28 **The Rubber Industry**
1982; 486 pages;
Sw. fr. 70.-/US $56.00

Volume 29 **Some Industrial Chemicals and Dyestuffs**
1982; 416 pages;
Sw. fr. 60.-/US $48.00

Volume 30 **Miscellaneous Pesticides**
1983; 424 pages;
Sw. fr. 60.-/US $48.00

Volume 31 **Some Food Additives, Feed Additives and Naturally Occurring Substances**
1983; 314 pages;
Sw. fr. 60-/US $48.00

Volume 32 **Polynuclear Aromatic Compounds, Part 1: Chemical, Environmental and Experimental Data**
1984; 477 pages;
Sw. fr. 60.-/US $48.00

Volume 33 **Polynuclear Aromatic Compounds, Part 2: Carbon Blacks, Mineral Oils and Some Nitroarenes**
1984; 245 pages;
Sw. fr. 50.-/US $40.00

Volume 34 **Polynuclear Aromatic Compounds, Part 3: Industrial Exposures in Aluminium Production, Coal Gasification, Coke Production, and Iron and Steel Founding**
1984; 219 pages;
Sw. fr. 48.-/US $38.40

Volume 35 **Polynuclear Aromatic Compounds, Part 4: Bitumens, Coal-tars and Derived Products, Shale-oils and Soots**
1985; 271 pages;
Sw. fr. 70.-/US $56.00

Volume 36 **Allyl Compounds, Aldehydes, Epoxides and Peroxides**
1985; 369 pages;
Sw. fr. 70.-/US $70.00

Volume 37 **Tobacco Habits Other than Smoking: Betel-quid and Areca-nut Chewing; and some Related Nitrosamines**
1985; 291 pages;
Sw. fr. 70.-/US $56.00

Volume 38 **Tobacco Smoking**
1986; 421 pages;
Sw. fr. 75.-/US $60.00

Volume 39 **Some Chemicals Used in Plastics and Elastomers**
1986; 403 pages;
Sw. fr. 60.-/US $48.00

Volume 40 **Some Naturally Occurring and Synthetic Food Components, Furocoumarins and Ultraviolet Radiation**
1986; 444 pages;
Sw. fr. 65.-/US $52.00

Volume 41 **Some Halogenated Hydrocarbons and Pesticide Exposures**
1986; 434 pages;
Sw. fr. 65.-/US $52.00

Volume 42 **Silica and Some Silicates**
1987; 289 pages;
Sw. fr. 65.-/US $52.00

Volume 43 **Man-Made Mineral Fibres and Radon**
1988; 300 pages;
Sw. fr. 65.-/US $52.00

Volume 44 **Alcohol Drinking**
1988; 416 pages;
Sw. fr. 65.-/US $52.00

Volume 45 **Occupational Exposures in Petroleum Refining; Crude Oil and Major Petroleum Fuels**
1989; 322 pages;
Sw. fr. 65.-/US $52.00

Volume 46 **Diesel and Gasoline Engine Exhausts and Some Nitroarenes**
1989; 458 pages;
Sw. fr. 65.-/US $52.00

Volume 47 **Some Organic Solvents, Resin Monomers and Related Compounds, Pigments and Occupational Exposures in Paint Manufacture and Painting**
1990; 536 pages;
Sw. fr. 85.-/US $68.00

Volume 48 **Some Flame Retardants and Textile Chemicals, and Exposures in the Textile Manufacturing Industry**
1990; 345 pages;
Sw. fr. 65.-/US $52.00

Volume 49 **Chromium, Nickel and Welding**
1990; 677 pages;
Sw. fr. 95.–/US$76.00

Volume 50 **Pharmaceutical Drugs**
1990; 415 pages;
Sw. fr. 65.–/US$52.00

Volume 51 **Coffee, Tea, Mate, Methylxanthines and Methylglyoxal**
1991; 513 pages;
Sw. fr. 80.–/US$64.00

Volume 52 **Chlorinated Drinking-water; Chlorination By-products; Some Other Halogenated Compounds; Cobalt and Cobalt Compounds**
1991; 544 pages;
Sw. fr. 80.–/US$64.00

Supplement No. 1
Chemicals and Industrial Processes Associated with Cancer in Humans (IARC Monographs, Volumes 1 to 20)
1979; 71 pages; (*out of print*)

Supplement No. 2
Long-term and Short-term Screening Assays for Carcinogens: A Critical Appraisal
1980; 426 pages;
Sw. fr. 40.-/US $32.00

Supplement No. 3
Cross Index of Synonyms and Trade Names in Volumes 1 to 26
1982; 199 pages (*out of print*)

Supplement No. 4
Chemicals, Industrial Processes and Industries Associated with Cancer in Humans (IARC Monographs, Volumes 1 to 29)
1982; 292 pages (*out of print*)

Supplement No. 5
Cross Index of Synonyms and Trade Names in Volumes 1 to 36
1985; 259 pages;
Sw. fr. 46.-/US $36.80

Supplement No. 6
Genetic and Related Effects: An Updating of Selected IARC Monographs from Volumes 1 to 42
1987; 729 pages;
Sw. fr. 80.-/US $64.00

Supplement No. 7
Overall Evaluations of Carcinogenicity: An Updating of IARC Monographs Volumes 1-42
1987; 434 pages;
Sw. fr. 65.-/US $52.00

Supplement No. 8
Cross Index of Synonyms and Trade Names in Volumes 1 to 46 of the IARC Monographs
1990; 260 pages;
Sw. fr. 60.-/US $48.00

IARC TECHNICAL REPORTS*

No. 1 **Cancer in Costa Rica**
Edited by R. Sierra, R. Barrantes, G. Muñoz Leiva, D.M. Parkin, C.A. Bieber and N. Muñoz Calero
1988; 124 pages;
Sw. fr. 30.-/US $24.00

No. 2 **SEARCH: A Computer Package to Assist the Statistical Analysis of Case-control Studies**
Edited by G.J. Macfarlane, P. Boyle and P. Maisonneuve (in press)

No. 3 **Cancer Registration in the European Economic Community**
Edited by M.P. Coleman and E. Démaret
1988; 188 pages;
Sw. fr. 30.-/US $24.00

No. 4 **Diet, Hormones and Cancer: Methodological Issues for Prospective Studies**
Edited by E. Riboli and R. Saracci
1988; 156 pages;
Sw. fr. 30.-/US $24.00

No. 5 **Cancer in the Philippines**
Edited by A.V. Laudico, D. Esteban and D.M. Parkin
1989; 186 pages;
Sw. fr. 30.-/US $24.00

No. 6 **La genèse du Centre International de Recherche sur le Cancer**
Par R. Sohier et A.G.B. Sutherland
1990; 104 pages
Sw. fr. 30.-/US $24.00

No. 7 **Epidémiologie du cancer dans les pays de langue latine**
1990; 310 pages
Sw. fr. 30.-/US $24.00

No. 8 **Comparative Study of Anti-smoking Legislation in Countries of the European Economic Community**
Edited by A. Sasco
1990; c. 80 pages
Sw. fr. 30.-/US $24.00
(English and French editions available) (in press)

DIRECTORY OF AGENTS BEING TESTED FOR CARCINOGENICITY (Until Vol. 13 Information Bulletin on the Survey of Chemicals Being Tested for Carcinogenicity)*

No. 8 Edited by M.-J. Ghess, H. Bartsch and L. Tomatis
1979; 604 pages; Sw. fr. 40.-

No. 9 Edited by M.-J. Ghess, J.D. Wilbourn, H. Bartsch and L. Tomatis
1981; 294 pages; Sw. fr. 41.-

No. 10 Edited by M.-J. Ghess, J.D. Wilbourn and H. Bartsch
1982; 362 pages; Sw. fr. 42.-

No. 11 Edited by M.-J. Ghess, J.D. Wilbourn, H. Vainio and H. Bartsch
1984; 362 pages; Sw. fr. 50.-

No. 12 Edited by M.-J. Ghess, J.D. Wilbourn, A. Tossavainen and H. Vainio
1986; 385 pages; Sw. fr. 50.-

No. 13 Edited by M.-J. Ghess, J.D. Wilbourn and A. Aitio 1988; 404 pages; Sw. fr. 43.-

No. 14 Edited by M.-J. Ghess, J.D. Wilbourn and H. Vainio
1990; 370 pages; Sw. fr. 45.-

NON-SERIAL PUBLICATIONS †

Alcool et Cancer
By A. Tuyns (in French only)
1978; 42 pages; Fr. fr. 35.-

Cancer Morbidity and Causes of Death Among Danish Brewery Workers
By O.M. Jensen
1980; 143 pages; Fr. fr. 75.-

Directory of Computer Systems Used in Cancer Registries
By H.R. Menck and D.M. Parkin
1986; 236 pages; Fr. fr. 50.-

* Available from booksellers through the network of WHO sales agents.

†Available directly from IARC